GEOSTATISTICS FOR ENGINEERS AND EARTH SCIENTISTS

GEOSTATISTICS FOR ENGINEERS AND EARTH SCIENTISTS

by

Ricardo A. Olea

Kansas Geological Survey
The University of Kansas
Lawrence, Kansas USA

Kluwer Academic Publishers
Boston • Dordrecht • London

Distributors for North, Central and South America:
Kluwer Academic Publishers
101 Philip Drive
Assinippi Park
Norwell, Massachusetts 02061 USA
Telephone (781) 871-6600
Fax (781) 871-6528
E-Mail <kluwer@wkap.com>

Distributors for all other countries:
Kluwer Academic Publishers Group
Distribution Centre
Post Office Box 322
3300 AH Dordrecht, THE NETHERLANDS
Telephone 31 78 6392 392
Fax 31 78 6546 474
E-Mail <services@wkap.nl>

Electronic Services <http://www.wkap.nl>

Library of Congress Cataloging-in-Publication Data

Olea, R. A. (Ricardo A.)
Geostatistiscs for engineers and earth scientists / by Ricardo A. Olea.
p. cm.
Includes bibliographical references and index.
ISBN 0-7923-8523-3 (alk. paper)
1. Geology--Statistical methods. 2. Kriging. I. Title.
QE33.2.S82054 1999
550'.72--dc21 99-24689
CIP

Printed on acid-free paper.

Printed in the United States of America

To Lucila

CONTENTS

List of Mathematical Definitions

List of Theorems

List of Corollaries

List of Lemmas

PREFACE

This textbook is an outgrowth of notes that I prepared for engineering students attending my graduate-level course taught at The University of Kansas in the United States. Geostatistics builds upon other disciplines, a command of which is basic for a ready understanding of its concepts. Prerequisite to enrollment in my class is the successful completion of the equivalent of (American) college-level courses in matrix algebra, probability, and statistics. Rather than including a lengthy introductory review of such subjects here, I have decided to require from the readers the same level of understanding expected from my students. Tutorial deficits or memory lapses should be remedied by referring to the abundant literature on these topics.

I am indebted to Yun-Shen Yu for inviting me to lecture at his School of Engineering. For their comments, I thank my students, who have encouraged me to publish the class notes. I am most grateful for remarks by Frederik P. Agterberg and George Christakos which contributed to the improvement of an early draft of this text. Special thanks are due to André Journel, who introduced me to geostatistics in my native Santiago de Chile almost 30 years ago and who was an insightful and demanding reviewer. Many of the suggestions of my colleagues were incorporated into the final text, but I reserved the right, and assume the blame, for disagreeing in a few instances.

Preparation of this textbook could not have been possible without support from the Kansas Geological Survey, in particular Jo Anne DeGraffenreid for the careful editing of the final manuscript and preparation of the camera-ready copy; Patricia Acker for her assistance in the preparation of all final illustrations; Geoffrey Bohling for his meticulous review of the camera-ready copy; and Dana Adkins-Heljeson for the preparation of a site on the Internet containing all color illustrations and data used in the exercises.

Ricardo A. Olea
Lawrence, Kansas

Chapter 1
INTRODUCTION

Geostatistics can be regarded as a collection of numerical techniques that deal with the characterization of spatial attributes, employing primarily random models in a manner similar to the way in which time series analysis characterizes temporal data. The French engineer Georges Matheron—at the time with the Bureau de Recherches Géologiques et Minières—coined the word *géostatistique*, inspired by the clear meaning and success of the older terms geochemistry and geophysics in which the prefix *geo-* was added to the name of some classical body of knowledge to denote an application of such knowledge to the modeling and understanding of processes of interest in earth sciences and technology (Matheron, 1962, p. 22; Journel and Huijbregts, 1978, p. *vi*). Both geostatistics and time series analysis are conceptually and historically related and primarily address the situation in which inferences must be drawn from autocorrelated data that are insufficient for obtaining precise results. Unlike time series analysis, however, geostatistics was not originally developed by mainstream statisticians, which accounts for its terminology and a notation that at first sight may be unfamiliar to scientists with training in classical statistics (Christensen, 1991, p. 262).

As in time series analysis, most geostatistical techniques rely on random variables to model the uncertainty associated with such assessments. The idea of using random variables to characterize uncertain deterministic events may be counterintuitive, but it is by no means unique to geostatistics and time series analysis. Yet feeling comfortable describing the real world through random variables has been a first stumbling block for many trying to understand geostatistics.

To compare the familiar deterministic realm and the stochastic world of geostatistics, let us consider the following thought experiment. Suppose that a commercial jet airplane flies at a constant altitude, direction, and cruising speed between two locations 1,000 km apart. Physics tells us that the time t to travel this distance is given by

$$t = \frac{1000}{v}$$

where v is the speed. If the speed is known, the simple model is completely determined and one can deduce the travel time whatever the speed, *e.g.*, 1 hr if the airplane happens to be flying at 1,000 km/hr. This is the deterministic realm, in which there is a unique and precise answer for every question.

From this experiment it follows that three conditions must be satisfied simultaneously in order to make predictions in the deterministic realm: flawless models are required to characterize the event of interest, assumptions must be honored, and all model parameters must be known. It is impossible to make an accurate deterministic assessment if these conditions are not satisfied. For example, if our airplane fails to honor the assumption of a straight flight path and follows a meandering route, the distance traveled will be greater and the predicted flight time will be optimistic.

If the value representing the parameter speed is unknown, our model is entirely useless for deterministic prediction. In such a case, the alternative to doing nothing (hence remaining in total ignorance) would be to use a stochastic model to make the best possible educated guess and arrive at an estimate—not as good, certainly, as a precise answer, yet better than complete uncertainty, especially when a decision must be made.

How do we proceed in the absence of a deterministic model or its parameters? Continuing with the thought experiment, we can turn the travel time for the airplane into a random variable or "variate," a collection of values with an associated probability describing how often each of those values may occur. There are several ways to assign such probabilities. In the case of airplane travel time, for example, based on known cruising speeds for commercial jets, one could decide that any velocity between 600 and 800 km/hr is feasible and equally probable, leading to the conclusion that the travel time would be between 1:15 and 1:40 hours, with a most likely time of 1 hr and 26 min.

Geostatistics deals with the characterization of spatial attributes—also known as regionalized variables—for which deterministic models are not possible because of the complexity of the natural processes and, often, the impossibility is compounded by high measurement costs. Typically the attributes are partially known at a limited number of sampling sites. Surveys always include a geographical location for each measurement of the attribute. Just as one can borrow from a table of airplane cruising speeds to reduce uncertainty about travel time, one can use geostatistical techniques to borrow values from an attribute's individual measurements and spatial fluctuation rates to minimize prediction uncertainty at unsampled sites, uncertainty that is not a property of the attribute itself, but a result of our imperfect knowledge.

Typically geostatistics uses models comprising one variate per attribute per site to model incomplete knowledge. This is a radical departure from classical multivariate statistics, in which one random variable per attribute for all sites is considered. Because geographic location varies continuously, the number of variates dealt with in geostatistics is infinite even for areas of interest of finite size. Such a collection of random variates is called a random function or a random field.

Definition 1.1

Let D be the spatial domain in an n-dimensional Euclidean space and let Ω be the permissible domain of variation—the sample space. A random function is the collection of random variables $\{Z(\boldsymbol{x}, \omega) : \boldsymbol{x} \in D, \omega \in \Omega\}$. □

For a fixed location $\boldsymbol{x}_i$ of $\boldsymbol{x}$, $Z(\boldsymbol{x}_i, \omega)$ is a function of ω, that is, $Z(\boldsymbol{x}_i, \omega)$ is a variate. In contrast, for a specific ω_j in Ω, $Z(\boldsymbol{x}, \omega_j)$ is a deterministic function of $\boldsymbol{x}$ that represents a possible observation of the random field and is called a realization. Finally $Z(\boldsymbol{x}_i, \omega_j)$ is merely a number. To simplify notation, it is customary to shorten the specifications. $\{Z(x, \omega) : \boldsymbol{x} \in D, \omega \in \Omega\}$ is referred to simply as $Z(\boldsymbol{x})$ and $\{Z(\boldsymbol{x}_i, \omega_j) : \boldsymbol{x}_i \in D, \omega_j \in \Omega\}$ as $z(\boldsymbol{x}_i)$.

A model is only a representation used to describe reality, and its merits can only be assessed by the quality of its results. Dealing with multiple values per site to predict another unique but unknown value is a notion that may take time to accept. But the fact is that stochastic models are able to synthesize a single, reasonable value out of a random variable, no matter how strange the model may appear at first glance. Not only that, but stochastic models can also assign measures of confidence to the inferences. Going back to our thought experiment, despite dealing with a multitude of speeds and traveling times, if required, the stochastic model is perfectly capable of predicting that the probability that the flight will take less than 1.5 hr is 66.6%—a single number plus a measure of confidence.

Any presentation of geostatistics must include kriging. Kriging is a collection of generalized linear regression techniques; it can be regarded as a normal or orthogonal projection of an unknown variate onto some linear manifold of the Hilbert space generated by the data. The generalization from classical linear regression comprises:

(a) Elimination of the assumption that the variates are independent and identically distributed.

Kriging makes no distributional assumptions and the variates are statistically correlated.

(b) Replacement of the random sampling assumption of classical statistics by non-random sampling.

Given an attribute, the classical assumption is to consider a set of observations as multiple realizations of one random variable, whereas geostatistics regards data as one partial realization of a random function comprising as many variates as there are observations.

(c) Direct assessment of attributes related to volumes orders of magnitude larger than the volumes of the observations that comprise the samples.

Historically this generalization accounts for the initial success of geostatistics in the 1960s when it was used to calculate ore reserves for which most data comprise assays of cylindrical cores up to a few inches in diameter and interest lies in the calculation of the amount of ore in mining units several cubic yards in size.

By analogy to the terminology of time series analysis, Cressie (1991, p. 106) calls kriging a *predictor*. The prevailing practice of geostatisticians is to refer to kriging as an *estimator*, a term that statisticians prefer to reserve for the inference of fixed but unknown parameters (Cressie, 1991, p. 26). Considering that kriging can be regarded as a generalized linear regression, perhaps the most sensible label would be *regressor*, which I do not use here, however, for this has not been the practice. Today there are many forms of kriging, and the number of geostatistical techniques continues to grow.

In the modeling of random functions, any sample is regarded as one possible partial realization of a model. Stochastic simulation generalizes the concept by allowing the generation of as many equally likely realizations per random function as necessary. Stochastic simulation is developing rapidly and has become a basic component of geostatistics. This book presents several simulation techniques, emphasizing their similarities and differences and their relationship to kriging.

The initial motivation for the preparation of my class notes was my desire to expand upon the limited scope of old textbooks by covering methods that I consider to be parsimonious and non-redundant, and that have proved through time to work satisfactorily for a variety of attributes and sampling schemes. What I have taught to my students—a reasonable number of techniques that may be mastered during a one-semester course—I now pass along to the readers. The sequence of topics is intended to whet the appetite by demonstrating results that may be put to immediate, practical use.

A desire to fulfill the need for a detailed explanation of geostatistical formulations, concentrating on the essentials and avoiding special situations of limited general interest, was a second motivation for preparing my notes. Derivations in mathematical books are unnecessarily general, resulting in a complexity that tends to intimidate the uninitiated, while explanations in geostatistical books are cryptically short—to the point that most readers end up taking the derivations for granted. Some readers of this book may consider it as a reference source and prefer not to read it straight through from cover to cover. Thus the presentation follows a modular approach. Each chapter is as self-contained as possible so that it may be read individually; excessive cross-references to previous results have been avoided, providing the opportunity to review similar derivations under slightly different circumstances. Guidelines and rules are offered whenever possible so that novice users may choose among alternative methods and select parameters without making subjective calls based on an experience that is not yet there.

Nothing is standard about the notation used by geostatisticians. Conventions used by various practitioners may differ from each other and from those used by classical statisticians. I have tried to be consistent and avoid even more variations in geostatistical notation. All terms have been defined, sometimes more than once, and adhere as much as possible to usage in the *Geostatistical Glossary and Multilingual Dictionary*, published under auspices of the International Association for Mathematical Geology (Olea, 1991). This volume is the closest thing to a standard reference in the field of geostatistics.

The ultimate test of understanding a procedure is being able to employ it. Problem exercises for trying out most techniques introduced here are included, as are the answers, which makes this textbook suitable for self-tutorial instruction. For ease in presentation of graphical displays, sampling domains are two-dimensional. To reduce printing costs all maps are gray-scale. Unfortunately, the poor resolution of such maps defeats to a large extent the very purpose of their use. Color versions of these maps that clearly illustrate the points may be viewed on the Internet. Readers are urged to access the site at http://www.kgs.ukans.edu/Mathgeo/Books/Geostat/index.html which also contains downloadable versions of the data sets in the Appendices.

The serious application of geostatistics requires the use of digital computers. Although rudimentary implementation from scratch is fairly straightforward for most geostatistical techniques, the coding of programs is only recommended as part of the practice required to master the principles of geostatistics. In the long run, for the pursuit of professional work, the user should employ software packages that have been thoroughly tested to handle any sampling scheme, which run as efficiently as possible, and which offer graphic capabilities for the analysis and display of results. Today some software of this nature is quite affordable or even free. For the cost of a regular hard-bound volume, one can get the Fortran source code of GSLIB (Deutsch and Journel, 1992) and Geo-EAS (Englund and Sparks, 1988) is available without charge from the United States Environmental Protection Agency.

Chapter 2

SIMPLE KRIGING

There are several forms of kriging, all of them initially formulated for the estimation of a continuous, spatial attribute at an unsampled site, preferably inside the convex hull defined by the location of the data. Figure 2.1 illustrates the case for a two-dimensional point sampling. Although extrapolations outside the convex hull are possible, they are unreliable. This points out a significant difference between kriging and time series analysis wherein the interest lies in predictions beyond the time span containing the data.

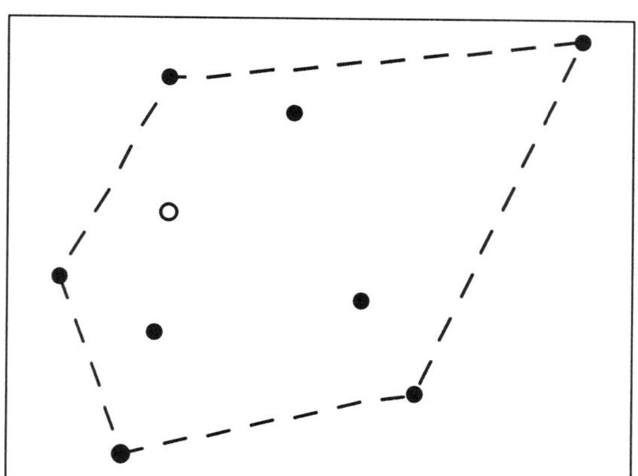

Figure 2.1. Dashed line showing the convex hull for an example of two-dimensional sampling. The solid dots denote sampling sites and the open circle the estimation location.

Kriging is a form of generalized linear regression for the formulation of an optimal estimator in a minimum mean square error sense. Upon defining the estimator and making some minimal assumptions about the underlying autocorrelated random function model, the first step in the formulation of simple kriging involves calculation of the estimation error variance, which turns out to be a function of the estimator weights. Minimization of such estimation variance reduces to finding weights that minimize the mean square error, which is accomplished by equating to zero the first derivative of the error with respect to each of the unknown weights. The final product in this formulation is a system of equations whose solution provides the simple kriging weights required by the estimator at a single location. Kriging requires the solution of as many systems of equations as locations in need of an estimation.

The roots of the method go back to the pioneering endeavors of Wold (1938), Kolmogorov (1939, 1941), and Wiener (1949) in research not directly related to regression and to the approach to gold mining valuation problems by Krige (1951), later popularized in geostatistics through work done by Matheron (1962, 1963). Luenberger (1969), focused on optimization problems and working independently from geostatisticians, published a complete theory of a vectorial approach to optimization, of which the kriging system of equations is a special application. As for the term "kriging," it was originally coined as *krigeage*, after Danie G. Krige, by Pierre Carlier at the French Commissariat à la Energie Atomique (Matheron 1962; Cressie, 1990).

Simple kriging is the most basic form of kriging in the sense that the model is the simplest in its mathematical formulation. The weakness of this simplicity is the same as that of any simple mathematical model—limited applicability and suboptimal results if an attempt is made to force usage beyond the embedded assumptions.

Properties of Linear Combinations of Variates

Let us begin by proving some fundamental properties of linear combinations of random variables.

Lemma 2.1

Let $Z(\boldsymbol{x})$ be a random variable of a continuous random function in Definition 1.1. Then for any coefficient λ

$$\mathrm{E}[\lambda Z(\boldsymbol{x})] = \lambda \mathrm{E}[Z(\boldsymbol{x})].$$

Proof

By the distributive property of the mathematical expectation operator $\mathrm{E}[\cdot]$,

$$\mathrm{E}[\lambda Z(\boldsymbol{x})] = \int_{-\infty}^{\infty} \lambda Z(\boldsymbol{x}) f(z)\, dz,$$

where $f(z)$ is the probability density function. Considering that the coefficient λ is a constant

$$\mathrm{E}[\lambda Z(\boldsymbol{x})] = \lambda \int_{-\infty}^{\infty} Z(\boldsymbol{x}) f(z)\, dz$$

and the proof follows because the integral is the expected value of $Z(\boldsymbol{x})$. □

Lemma 2.2

Let $Z(\boldsymbol{x}_i)$ be a random variable of a continuous random function in Definition 1.1. Then for any coefficient λ_i

$$\mathrm{E}\left[\sum_{i=1}^{k} \lambda_i Z(\boldsymbol{x}_i)\right] = \sum_{i=1}^{k} \lambda_i \mathrm{E}\left[Z(\boldsymbol{x}_i)\right].$$

Proof

By the distributive property of expectation

$$\mathrm{E}\left[\sum_{i=1}^{k} \lambda_i Z(\boldsymbol{x}_i)\right] = \mathrm{E}\left[\lambda_1 Z(\boldsymbol{x}_1)\right] + \mathrm{E}\left[\lambda_2 Z(\boldsymbol{x}_2)\right] + \ldots + \mathrm{E}\left[\lambda_k Z(\boldsymbol{x}_k)\right],$$

and by Lemma 2.1

$$\mathrm{E}\left[\sum_{i=1}^{k} \lambda_i Z(\boldsymbol{x}_i)\right] = \lambda_1 \mathrm{E}\left[Z(\boldsymbol{x}_1)\right] + \lambda_2 \mathrm{E}\left[Z(\boldsymbol{x}_2)\right] + \ldots + \lambda_k \mathrm{E}\left[Z(\boldsymbol{x}_k)\right].$$

The proof follows because the right-hand side of the equality is equal to

$$\sum_{i=1}^{k} \lambda_i \mathrm{E}\left[Z(\boldsymbol{x}_i)\right]. \quad \square$$

Lemma 2.3

Let $Z(\boldsymbol{x}_i)$ be a random variable of a continuous random function. Then for any coefficient λ_i

$$\mathrm{E}\left[\left\{\sum_{i=1}^{k} \lambda_i Z(\boldsymbol{x}_i)\right\}^2\right] = \sum_{i=1}^{k}\sum_{j=1}^{k} \lambda_i \lambda_j \mathrm{E}\left[Z(\boldsymbol{x}_i)\, Z(\boldsymbol{x}_j)\right].$$

Proof

Expanding the square

$$\mathrm{E}\left[\left\{\sum_{i=1}^{k} \lambda_i Z(\boldsymbol{x}_i)\right\}^2\right] = E\left[\sum_{i=1}^{k} \lambda_1 \lambda_i Z(\boldsymbol{x}_1)\, Z(\boldsymbol{x}_i) + \sum_{i=1}^{k} \lambda_2 \lambda_i Z(\boldsymbol{x}_2)\, Z(\boldsymbol{x}_i) \right.$$
$$\left. + \ldots + \sum_{i=1}^{k} \lambda_k \lambda_i Z(\boldsymbol{x}_k)\, Z(\boldsymbol{x}_i)\right].$$

Considering that one can regard $\lambda_i\lambda_j$ as a new constant and $Z(\boldsymbol{x}_i)\,Z(\boldsymbol{x}_j)$ as a new variable, by the distributive property of the expectation and by Lemma 2.2

$$\mathrm{E}\left[\left\{\sum_{i=1}^{k}\lambda_i Z(\boldsymbol{x}_i)\right\}^2\right] = \lambda_1\lambda_i\sum_{i=1}^{k}\mathrm{E}\left[Z(\boldsymbol{x}_1)\,Z(\boldsymbol{x}_i)\right] + \lambda_2\lambda_i\sum_{i=1}^{k}\mathrm{E}\left[Z(\boldsymbol{x}_2)\,Z(\boldsymbol{x}_i)\right] + \ldots + \lambda_k\lambda_i\sum_{i=1}^{k}\mathrm{E}\left[Z(\boldsymbol{x}_k)\,Z(\boldsymbol{x}_i)\right]$$

and the proof follows by condensing the expression using a second summation index. □

Lemma 2.4

Let $Z(\boldsymbol{x}_i)$ be a random variable of a continuous random function. Then for any coefficient λ_i

$$\mathrm{Var}\left[\sum_{i=1}^{k}\lambda_i Z(\boldsymbol{x}_i)\right] = \sum_{i=1}^{k}\sum_{j=1}^{k}\lambda_i\lambda_j \mathrm{Cov}\big(Z(\boldsymbol{x}_i)\,,Z(\boldsymbol{x}_j)\big).$$

Proof

By the definition of variance

$$\mathrm{Var}\left[\sum_{i=1}^{k}\lambda_i Z(\boldsymbol{x}_i)\right] = \mathrm{E}\left[\left\{\sum_{i=1}^{k}\lambda_i Z(\boldsymbol{x}_i)\right\}^2\right] - \left\{\mathrm{E}\left[\sum_{i=1}^{k}\lambda_i Z(\boldsymbol{x}_i)\right]\right\}^2,$$

and by Lemmas 2.2 and 2.3

$$\mathrm{Var}\left[\sum_{i=1}^{k}\lambda_i Z(\boldsymbol{x}_i)\right] = \sum_{i=1}^{k}\sum_{j=1}^{k}\lambda_i\lambda_j\mathrm{E}\left[Z(\boldsymbol{x}_i)\,Z(\boldsymbol{x}_j)\right] - \sum_{i=1}^{k}\sum_{j=1}^{k}\lambda_i\lambda_j\mathrm{E}\left[Z(\boldsymbol{x}_i)\right]\mathrm{E}\left[Z(\boldsymbol{x}_j)\right].$$

Factoring

$$\mathrm{Var}\left[\sum_{i=1}^{k}\lambda_i Z(\boldsymbol{x}_i)\right] = \sum_{i=1}^{k}\sum_{j=1}^{k}\lambda_i\lambda_j\left\{\mathrm{E}\left[Z(\boldsymbol{x}_i)\,,Z(\boldsymbol{x}_j)\right] - \mathrm{E}\left[Z(\boldsymbol{x}_i)\right]\mathrm{E}\left[Z(\boldsymbol{x}_j)\right]\right\},$$

and the fact that the difference is equal to the covariance of $Z(\boldsymbol{x}_i)$ and $Z(\boldsymbol{x}_j)$ proves the lemma. □

Notice that this result is completely general, requiring no assumptions about the random function.

Assumptions and Definitions

Formulation and application of simple kriging based on a random function model requires some minimum set of assumptions, without which it is not possible to find the optimal weights.

Assumption 2.1

The sampling is a partial realization of a random function $Z(\boldsymbol{x})$ where $\boldsymbol{x}$ denotes spatial location. □

Assumption 2.2

The random function is second order stationary, which implies that moments involving up to two variates are insensitive to any joint spatial translation, depending only on Euclidean distance.

$$\begin{aligned} &\mathrm{E}[Z(\boldsymbol{x})] = m \\ &\mathrm{E}[(Z(\boldsymbol{x}) - m)(Z(\boldsymbol{x}+\boldsymbol{h}) - m)] = \mathrm{E}[Z(\boldsymbol{x})\,Z(\boldsymbol{x}+\boldsymbol{h})] - m^2 \\ &\qquad = \mathrm{Cov}(\boldsymbol{x}, \boldsymbol{x}+\boldsymbol{h}) = \mathrm{Cov}(\boldsymbol{h}), \end{aligned}$$

where $\mathrm{E}[\cdot]$ denotes expected value, m is a scalar constant, and $\boldsymbol{h}$ is a vectorial distance in the sampling space. The constant m is the mean and $\mathrm{Cov}[\cdot]$ is the covariance of the random function. □

Assumption 2.3

Unique to simple kriging is the assumption that the mean is known. □

Definition 2.1

Let Z be a second order stationary random function with mean m. The estimator $\hat{Z}(\boldsymbol{x}_0)$ at site $\boldsymbol{x}_0$ is given by the following linear combination of random variables at sites $\boldsymbol{x}_i$ considered in the sampling

$$\hat{Z}(\boldsymbol{x}_0) = m + \sum_{i=1}^{k} \lambda_i (Z(\boldsymbol{x}_i) - m). \quad \square$$

The Estimation Variance

Calculation of optimal weights requires the prior study of some properties of second order stationary random functions. Rather than interrupting the main derivation to make definitions and prove basic results, I will perform that chore "up front." This scheme occasionally may give the reader the idea that we are digressing onto irrelevant topics, but this is not the case; this approach will add coherency to the main derivations.

Definition 2.2

The residual $Y(\boldsymbol{x})$ of a random function $Z(\boldsymbol{x})$ is the difference between the random function and its expected value

$$Y(\boldsymbol{x}) = Z(\boldsymbol{x}) - \mathrm{E}[Z(\boldsymbol{x})]. \quad \square$$

The residual is also a random function. The expected value $\mathrm{E}[Z(\boldsymbol{x})]$ of a random function is called the *drift*. By Assumption 2.2, simple kriging assumes that the drift is the same regardless of location.

Lemma 2.5

Let $Y(\boldsymbol{x})$ be the residual of random variable $Z(\boldsymbol{x})$. Then

$$\mathrm{E}[Y(\boldsymbol{x})] = 0.$$

Proof

By Definition 2.2,

$$Y(\boldsymbol{x}) = Z(\boldsymbol{x}) - \mathrm{E}[Z(\boldsymbol{x})],$$

and taking the expectation of both sides,

$$\mathrm{E}[Y(\boldsymbol{x})] = \mathrm{E}\big[Z(\boldsymbol{x}) - \mathrm{E}[Z(\boldsymbol{x})]\big].$$

Expanding the difference, because $\mathrm{E}[Z(\boldsymbol{x})]$ is constant,

$$\mathrm{E}[Y(\boldsymbol{x})] = \mathrm{E}[Z(\boldsymbol{x})] - \mathrm{E}[Z(\boldsymbol{x})]$$

$$\mathrm{E}[Y(\boldsymbol{x})] = 0. \quad \square$$

Lemma 2.6

Let $\mathrm{Cov}(\boldsymbol{x}_i, \boldsymbol{x}_j)$ be the covariance of a second order stationary random function $Z(\boldsymbol{x})$ and let $\mathrm{Cov}_Y(\boldsymbol{x}_i, \boldsymbol{x}_i)$ be the covariance of its residuals. Then

$$\mathrm{Cov}(\boldsymbol{x}_i, \boldsymbol{x}_i) = \mathrm{Cov}_Y(\boldsymbol{x}_i, \boldsymbol{x}_i).$$

Proof

By Assumption 2.2,

$$\mathrm{Cov}(\boldsymbol{x}_i, \boldsymbol{x}_j) = \mathrm{E}[Z(\boldsymbol{x}_i)\, Z(\boldsymbol{x}_j)] - m^2.$$

From Definition 2.2,

$$\mathrm{Cov}\left(\boldsymbol{x}_i, \boldsymbol{x}_j\right) = \mathrm{E}\big[\left(m + Y\left(\boldsymbol{x}_i\right)\right)\left(m + Y\left(\boldsymbol{x}_j\right)\right)\big] - m^2 \,,$$

but expanding the expectation, because according to Assumption 2.2 the mean m is a constant,

$$\mathrm{Cov}\left(\boldsymbol{x}_i, \boldsymbol{x}_j\right) = m^2 + m\mathrm{E}\left[Y\left(\boldsymbol{x}_i\right)\right] + m\mathrm{E}\left[Y\left(\boldsymbol{x}_j\right)\right] + \mathrm{E}\left[Y\left(\boldsymbol{x}_i\right) Y\left(\boldsymbol{x}_j\right)\right] - m^2 \,.$$

By Lemma 2.5 and cancellation of the m^2 terms

$$\mathrm{Cov}\left(\boldsymbol{x}_i, \boldsymbol{x}_j\right) = \mathrm{E}\left[Y\left(\boldsymbol{x}_i\right) Y\left(\boldsymbol{x}_j\right)\right]$$

and the proof follows because, according to Lemma 2.5, the mean of the residuals is zero. □

Now we are ready to derive a general expression for the variance for the linear estimator in Definition 2.1.

Theorem 2.1

Let $\hat{Z}\left(\boldsymbol{x}_0\right)$ be the linear estimator in Definition 2.1 and let $\mathrm{Cov}\left(\boldsymbol{h}\right)$ be the covariance for the random function. Then the estimation variance $\sigma^2\left(\boldsymbol{x}_0\right)$ at site $\boldsymbol{x}_0$ is equal to

$$\sigma^2\left(\boldsymbol{x}_0\right) = \mathrm{Cov}\left(\boldsymbol{x}_0, \boldsymbol{x}_0\right) - 2\sum_{i=1}^{k} \lambda_i \mathrm{Cov}\left(\boldsymbol{x}_i, \boldsymbol{x}_0\right) + \sum_{i=1}^{k}\sum_{j=1}^{k} \lambda_i \lambda_j \mathrm{Cov}\left(\boldsymbol{x}_i, \boldsymbol{x}_j\right).$$

Proof

The estimation variance is equal to

$$\sigma^2\left(\boldsymbol{x}_0\right) = \mathrm{Var}\left[\hat{Z}\left(\boldsymbol{x}_0\right) - Z\left(\boldsymbol{x}_0\right)\right]$$

and by Definitions 2.1 and 2.2

$$\sigma^2\left(\boldsymbol{x}_0\right) = \mathrm{Var}\left[\sum_{i=1}^{k} \lambda_i Y\left(\boldsymbol{x}_i\right) - Y\left(\boldsymbol{x}_0\right)\right],$$

which is equivalent to

$$\sigma^2\left(\boldsymbol{x}_0\right) = \mathrm{Var}\left[\sum_{i=0}^{k} \lambda_i Y\left(\boldsymbol{x}_i\right)\right]$$

when $\lambda_0 = -1$. Now, by Lemma 2.4

$$\sigma^2\left(\boldsymbol{x}_0\right) = \sum_{i=0}^{k}\sum_{j=0}^{k} \lambda_i \lambda_j \mathrm{Cov}\big(Y\left(\boldsymbol{x}_i\right), Y\left(\boldsymbol{x}_j\right)\big)$$

but by Lemma 2.6, $\text{Cov}_Y(\cdot) = \text{Cov}(\cdot)$. Finally, separating the terms in $i = 0$ and expanding,

$$\sigma^2(\boldsymbol{x}_0) = \lambda_0^2 \text{Cov}(\boldsymbol{x}_0, \boldsymbol{x}_0) + 2\lambda_0 \sum_{i=1}^{k} \lambda_i \text{Cov}(\boldsymbol{x}_i, \boldsymbol{x}_0) + \sum_{i=1}^{k} \sum_{j=1}^{k} \lambda_i \lambda_j \text{Cov}(\boldsymbol{x}_i, \boldsymbol{x}_j)$$

and the proof follows because $\lambda_0 = -1$. □

Normal Equations

The purpose of simple kriging is to find a set of weights for the estimator in Definition 2.1 that yields the minimum mean square error given by Theorem 2.1. In the context of vectorial projections, the solution to such a system of equations is known as a set of normal equations (Luenberger, 1969, p. 56).

Definition 2.3

Let m be a positive integer, let $\{\alpha_1, \alpha_2, \ldots, \alpha_m\}$ be a set of real or complex numbers, and let $\{\boldsymbol{x}_1, \boldsymbol{x}_2, \ldots, \boldsymbol{x}_m\}$ be a set of points in an n-dimensional Euclidean space. Then the continuous function $\phi(\boldsymbol{x}_i, \boldsymbol{x}_j)$ is said to be positive definite or of positive type (Itô, 1987, p. 720) or nonnegative definite if

$$\sum_{i=1}^{m} \sum_{j=1}^{m} \alpha_i \alpha_j \phi(\boldsymbol{x}_i, \boldsymbol{x}_j) \geq 0. \quad \square$$

Theorem 2.2

Let λ_i be the weights for the simple kriging estimator in Definition 2.1 and let $\text{Cov}(\cdot)$ be the covariance for the random function of interest. If the covariance is positive definite, then (a) the weights that produce the minimum estimation variance are the solution to:

$$\left.\begin{aligned} \sum_{i=1}^{k} \lambda_i \text{Cov}(\boldsymbol{x}_i, \boldsymbol{x}_1) &= \text{Cov}(\boldsymbol{x}_0, \boldsymbol{x}_1) \\ \sum_{i=1}^{k} \lambda_i \text{Cov}(\boldsymbol{x}_i, \boldsymbol{x}_2) &= \text{Cov}(\boldsymbol{x}_0, \boldsymbol{x}_2) \\ \cdots\cdots\cdots&\cdots\cdots\cdots \\ \sum_{i=1}^{k} \lambda_i \text{Cov}(\boldsymbol{x}_i, \boldsymbol{x}_k) &= \text{Cov}(\boldsymbol{x}_0, \boldsymbol{x}_k) \end{aligned}\right\}$$

and (b) the kriging variance is never negative.

Proof

(a) From Theorem 2.1, the objective function $\sigma^2(\boldsymbol{x}_0)$ is a quadratic expression in the weights. In such a case, the necessary and sufficient condition to have a unique global minimum is that the term $\sum_{i=1}^{k}\sum_{j=1}^{k}\lambda_i\lambda_j\text{Cov}\left(\boldsymbol{x}_i,\boldsymbol{x}_j\right)$ must be larger than or equal to zero, which by Definition 2.3 is mathematically equivalent to saying that the covariance must be positive definite. The minimum mean square error is given by those weights that result in zero first derivatives of the kriging variance.

$$\frac{\partial\sigma^2\left(\boldsymbol{x}_0\right)}{\partial\lambda_i}=-2\text{Cov}\left(\boldsymbol{x}_i,\boldsymbol{x}_0\right)+2\sum_{j=1}^{k}\lambda_j\text{Cov}\left(\boldsymbol{x}_i,\boldsymbol{x}_j\right),\ \text{for } i=1,\ 2,\ \ldots,\ k.$$

The proof follows after setting each first derivative equal to zero, dividing both sides by 2, and adding $\text{Cov}\left(\boldsymbol{x}_i,\boldsymbol{x}_0\right)$ to equation i, for $i=1,\ 2,\ \ldots,\ k$.

(b) Theorem 2.1 can be rewritten as

$$\sigma^2\left(\boldsymbol{x}_0\right)=\sum_{i=0}^{k}\sum_{j=0}^{k}\lambda_i\lambda_j\text{Cov}\left(\boldsymbol{x}_i,\boldsymbol{x}_j\right)$$

with $\lambda_0=-1$. The proof then follows directly from the positive condition. □

The system has k equations and the same number of unknowns λ_i.

Minimum Mean Square Error

The expression for the estimation variance given by Theorem 2.1 is valid for any weights λ_i. The next lemma and theorem provide the value of the estimation variance for the special case of the optimal parameters.

Lemma 2.7

Let λ_i be the optimal weights for the simple kriging estimator in Definition 2.1. If $\text{Cov}(\cdot)$ is the covariance for the random function of interest, then,

$$\sum_{i=1}^{k}\sum_{j=1}^{k}\lambda_i\lambda_j\text{Cov}\left(\boldsymbol{x}_i,\boldsymbol{x}_j\right)=\sum_{i=1}^{k}\lambda_i\text{Cov}\left(\boldsymbol{x}_i,\boldsymbol{x}_0\right).$$

Proof

From Theorem 2.2,

$$\sum_{j=1}^{k}\lambda_j\text{Cov}\left(\boldsymbol{x}_i,\boldsymbol{x}_j\right)=\text{Cov}\left(\boldsymbol{x}_i,\boldsymbol{x}_0\right)\ ,\ \text{for } i=1,\ 2,\ \ldots,\ k.$$

Each of the k expressions is still true if one multiplies both sides of the ith equation by λ_i. Summing them up all together,

$$\sum_{i=1}^{k}\lambda_i\sum_{j=1}^{k}\lambda_j\text{Cov}\left(\boldsymbol{x}_i,\boldsymbol{x}_j\right)=\sum_{i=1}^{k}\lambda_i\text{Cov}\left(\boldsymbol{x}_i,\boldsymbol{x}_0\right).$$

Introducing λ_i under the second summation completes the proof. □

Theorem 2.3

Let λ_i be the optimal weights for the simple kriging estimator in Definition 2.1. If $\operatorname{Cov}(\cdot)$ is the covariance for the random function of interest, then the minimum mean square error for simple kriging $\sigma^2_{SK}(\boldsymbol{x}_0)$ is

$$\sigma^2_{SK}(\boldsymbol{x}_0) = \operatorname{Cov}(0) - \sum_{i=i}^{k} \lambda_i \operatorname{Cov}(\boldsymbol{x}_0, \boldsymbol{x}_i).$$

Proof

From Theorem 2.1,

$$\sigma^2(\boldsymbol{x}_0) = \operatorname{Cov}(\boldsymbol{x}_0, \boldsymbol{x}_0) - 2\sum_{i=1}^{k} \lambda_i \operatorname{Cov}(\boldsymbol{x}_i, \boldsymbol{x}_0) + \sum_{i=1}^{k}\sum_{j=1}^{k} \lambda_i \lambda_j \operatorname{Cov}(\boldsymbol{x}_i, \boldsymbol{x}_j);$$

but by Lemma 2.7,

$$\sigma^2_{SK}(\boldsymbol{x}_0) = \operatorname{Cov}(\boldsymbol{x}_0, \boldsymbol{x}_0) - 2\sum_{i=1}^{k} \lambda_i \operatorname{Cov}(\boldsymbol{x}_i, \boldsymbol{x}_0) + \sum_{i=1}^{k} \lambda_i \operatorname{Cov}(\boldsymbol{x}_i, \boldsymbol{x}_0).$$

The distance from a site to itself is zero. Addition of the last two terms proves the theorem. □

Algorithm

For a more compact display of the results let us define some matrices.

Definition 2.4

Let $\boldsymbol{x}_i$'s be the sampling sites of a sample subset of size k, $i = 1, 2, \ldots, k$ and let $\operatorname{Cov}(\boldsymbol{x}_i, \boldsymbol{x}_j)$'s be covariances. Then $\mathbf{C}$ is the matrix

$$\mathbf{C} = \begin{bmatrix} \operatorname{Cov}(\boldsymbol{x}_1, \boldsymbol{x}_1) & \operatorname{Cov}(\boldsymbol{x}_2, \boldsymbol{x}_1) & \cdots & \operatorname{Cov}(\boldsymbol{x}_k, \boldsymbol{x}_1) \\ \operatorname{Cov}(\boldsymbol{x}_1, \boldsymbol{x}_2) & \operatorname{Cov}(\boldsymbol{x}_2, \boldsymbol{x}_2) & \cdots & \operatorname{Cov}(\boldsymbol{x}_k, \boldsymbol{x}_2) \\ \cdots & \cdots & \cdots & \cdots \\ \operatorname{Cov}(\boldsymbol{x}_1, \boldsymbol{x}_k) & \operatorname{Cov}(\boldsymbol{x}_2, \boldsymbol{x}_k) & \cdots & \operatorname{Cov}(\boldsymbol{x}_k, \boldsymbol{x}_k) \end{bmatrix}. \quad \square$$

Definition 2.5

Let λ_i's be the optimal weights for the estimator in Definition 2.1 and let the prime stand for the transpose of the matrix. Then $\mathbf{X}$ is the matrix

$$\mathbf{X} = [\lambda_1 \quad \lambda_2 \quad \cdots \quad \lambda_k]'. \quad \square$$

Definition 2.6

Let Cov $(\cdot)$ be the covariance of the random function, $\boldsymbol{x}_0$ be the estimation location, and the $\boldsymbol{x}_i$'s be sampling sites of a sample subset of size k, $i = 1, 2, \ldots, k$. Then $\mathbf{c}$ is the matrix

$$\mathbf{c} = [\,\mathrm{Cov}\,(\boldsymbol{x}_0, \boldsymbol{x}_1) \quad \mathrm{Cov}\,(\boldsymbol{x}_0, \boldsymbol{x}_2) \quad \cdots \quad \mathrm{Cov}\,(\boldsymbol{x}_0, \boldsymbol{x}_k)\,]' . \quad \square$$

Definition 2.7

Let $Z(\boldsymbol{x}_i)$ be random variables of a random function with mean m and let $\boldsymbol{x}_i$ be sampling sites, $i = 1, 2, \ldots, k$. Then the matrix $\mathbf{Y}$ is

$$\mathbf{Y} = \begin{bmatrix} Z(\boldsymbol{x}_1) - m \\ Z(\boldsymbol{x}_2) - m \\ \cdots \\ Z(\boldsymbol{x}_k) - m \end{bmatrix} . \quad \square$$

In summary, simple kriging involves the steps in the following algorithm.

Algorithm 2.1

This is an algorithm to perform simple kriging under Assumptions 2.1, 2.2, and 2.3.

1. Calculate each term in matrix $\mathbf{C}$ in Definition 2.4.
2. Calculate each term in matrix $\mathbf{c}$ in Definition 2.6.
3. Solve the system of equations

$$\mathbf{C}\mathbf{X} = \mathbf{c} ,$$

where $\mathbf{X}$ is the matrix in Definition 2.5.

4. Compute the estimate

$$\hat{Z}_{SK}(\boldsymbol{x}_0) = m + \mathbf{Y}'\mathbf{X} = m + \mathbf{Y}'\mathbf{C}^{-1}\mathbf{c} .$$

5. Calculate the simple kriging estimation variance

$$\sigma^2_{SK}(\boldsymbol{x}_0) = \mathrm{Cov}\,(0) - \mathbf{c}'\mathbf{X} = \mathrm{Cov}\,(0) - \mathbf{c}'\mathbf{C}^{-1}\mathbf{c} . \quad \square$$

The solution to the kriging system of equations is unique if and only if matrix $\mathbf{C}$ in Definition 2.4 is not singular (Luenberger, 1969, p. 56), which is equivalent to requiring that the covariance must be positive definite. Considering that the formulation is independent from the physical nature of the spatial attribute, the algorithm is completely general and applies to the characterization of any spatial attribute honoring the assumptions.

EXERCISE 2.1

Given the sampling in Figure 2.2,

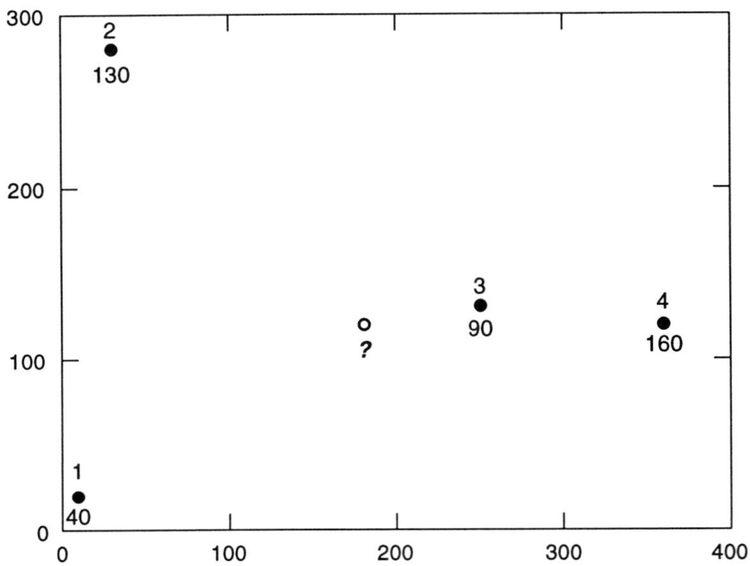

Figure 2.2. Sampling and estimation site for Exercise 2.1. The number above the symbol denotes the sampling number and the one below, the measurement.

Index	Easting	Northing	Measurement
1	10	20	40
2	30	280	130
3	250	130	90
4	360	120	160

for an attribute with a mean of 110 and covariance $\mathrm{Cov}\,(\boldsymbol{h}) = 2000\mathrm{e}^{-\frac{h}{250}}$, if the estimation location is $\boldsymbol{x}_0 = (180, 120)$, do the following by simple kriging:

1. Calculate the estimation weights.
2. What comment can you make about the estimation weights?
3. Find the estimate.
4. Compute the kriging estimation variance.

Solution

1. The matrix of distances $\mathbf{D}$ among the observations is

$$\mathbf{D} = \begin{bmatrix} 0 & & & \\ 260.8 & 0 & & \\ 264.0 & 266.3 & 0 & \\ 364.0 & 366.7 & 110.4 & 0 \end{bmatrix},$$

thus the covariance matrix is

$$\mathbf{C} = \begin{bmatrix} 2000 & & & \\ 704.8 & 2000 & & \\ 695.6 & 689.4 & 2000 & \\ 466.4 & 461.2 & 1285.8 & 2000 \end{bmatrix}.$$

The distance vector from the sampling sites to the estimation location is

$$\mathbf{d} = [\,197.2 \quad 219.3 \quad 70.7 \quad 180.0\,]',$$

hence,

$$\mathbf{c} = [\,908.7 \quad 831.8 \quad 1507.2 \quad 973.6\,]'.$$

Therefore, the weights are

$$\mathbf{X} = \begin{bmatrix} 2000 & & & \\ 704.8 & 2000 & & \\ 695.6 & 689.4 & 2000 & \\ 466.4 & 461.2 & 1285.8 & 2000 \end{bmatrix}^{-1} \begin{bmatrix} 908.6 \\ 831.8 \\ 1507.2 \\ 973.6 \end{bmatrix}$$
$$= [\,0.185 \quad 0.128 \quad 0.646 \quad -0.001\,]'.$$

2. (a) Although, in general, the farther away the estimate is from the sampling site the smaller its weight is, the weight for observation 4 breaks the rule. As shown in Figure 2.3, the absolute value of weight 4 is two orders of magnitude smaller than weights 1 and 2 despite observation 4 being closer to the estimation location than observations 1 and 2. This low value of weight 4 is the result of taking into account not only the distances between the estimate and the observations, but distances between observations as well.

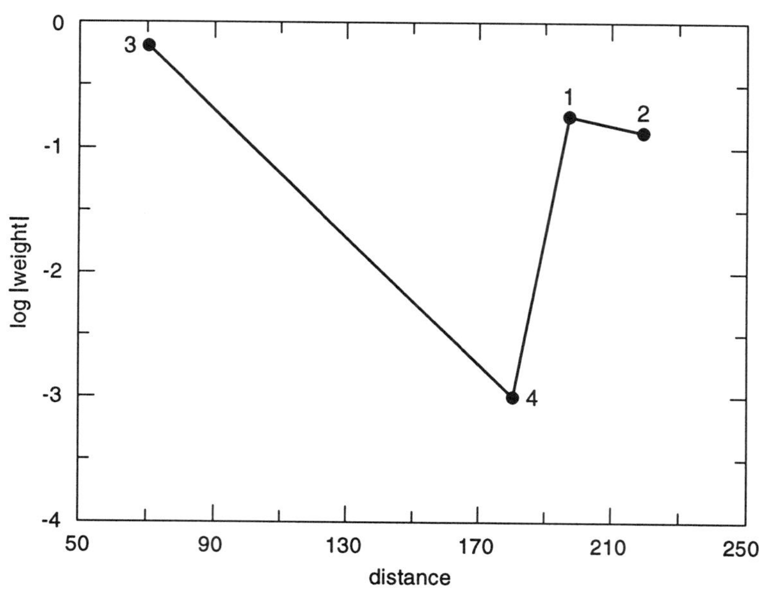

Figure 2.3. The weights in Exercise 2.1.

The situation is typical of kriging and has received the name *screen effect* because of its connection to collinear observations. Each time several observations are exactly or approximately in line and on the same side of the estimation site, such as $z(\boldsymbol{x}_3)$ and $z(\boldsymbol{x}_4)$ in Figure 2.2, the closest collinear observation takes the normal amount of weight and all other observations behind it tend to be ignored, receiving weights close to zero, as is the case of λ_4 in Figure 2.2.

(b) The smallest weight is negative. Simple kriging has no boundary constraints on the value of the weights. A linear combination such as the simple kriging estimator is said to be a non-convex combination if some of the coefficients are negative (Luenberger, 1969, p. 43).

(c) The possibility of having negative weights implies that the estimate is not confined to the data interval; depending on the application this can be an advantage or a drawback. For example, the estimate for an attribute that is defined only for values zero or larger may be negative.

3. From Step 4 in Algorithm 2.1,

$$\begin{aligned}\hat{Z}_{SK}(\boldsymbol{x}_0) &= 110 + [-70 \quad 20 \quad -20 \quad 50]' \, [0.185 \quad 0.128 \quad 0.646 \quad -0.001] \\ &= 86.7 \,.\end{aligned}$$

4. From Step 5 in Algorithm 2.1,

$$\begin{aligned}\sigma^2_{SK}(\boldsymbol{x}_0) &= 2000 - [908.6 \quad 831.8 \quad 1507.2 \quad 973.6]' \, [0.185 \quad 0.128 \quad 0.646 \quad -0.001] \\ &= 752.9 \,. \quad \square\end{aligned}$$

Properties

Besides the non-convexity and screen effect illustrated in Exercise 2.1 and the built-in optimality in a minimum mean square error sense, there are several properties of simple kriging that follow from the assumptions and the form of the estimator.

Theorem 2.4

The simple kriging estimator $\hat{Z}(\boldsymbol{x}_0)$ is unbiased.
Proof

From Definitions 2.1 and 2.2,

$$\mathrm{E}\left[\hat{Z}(\boldsymbol{x}_0) - Z(\boldsymbol{x}_0)\right] = \mathrm{E}\left[\sum_{i=1}^{k} \lambda_i Y(\boldsymbol{x}_i) - Y(\boldsymbol{x}_0)\right],$$

which by Lemma 2.2 is equivalent to

$$\mathrm{E}\left[\hat{Z}(\boldsymbol{x}_0) - Z(\boldsymbol{x}_0)\right] = \lambda_i \sum_{i=0}^{k} \mathrm{E}\left[Y(\boldsymbol{x}_i)\right],$$

if $\lambda_0 = -1$. Lemma 2.5 proves the theorem because all the expected values are zero. □

Kriging unbiasedness plus its optimality in a minimum mean square error sense have made it "BLUE"—a Best Linear Unbiased Estimator.

Theorem 2.5

Simple kriging is an exact interpolator.

Proof

Exact interpolation is the replication of an observation by an estimation at the site of the observation. Without losing any generality, let us assume that this location is $\boldsymbol{x}_1$. Because $\boldsymbol{x}_0 = \boldsymbol{x}_1$, from Step 3 in Algorithm 2.1,

$$\begin{bmatrix} \text{Cov}(0) & \text{Cov}(\boldsymbol{x}_2, \boldsymbol{x}_1) & \ldots & \text{Cov}(\boldsymbol{x}_k, \boldsymbol{x}_1) \\ \text{Cov}(\boldsymbol{x}_1, \boldsymbol{x}_2) & \text{Cov}(0) & \ldots & \text{Cov}(\boldsymbol{x}_k, \boldsymbol{x}_2) \\ \ldots & \ldots & \ldots & \ldots \\ \text{Cov}(\boldsymbol{x}_1, \boldsymbol{x}_k) & \text{Cov}(\boldsymbol{x}_2, \boldsymbol{x}_k) & \ldots & \text{Cov}(0) \end{bmatrix} \mathbf{X} = \begin{bmatrix} \text{Cov}(0) \\ \text{Cov}(\boldsymbol{x}_1, \boldsymbol{x}_2) \\ \ldots \\ \text{Cov}(\boldsymbol{x}_1, \boldsymbol{x}_k) \end{bmatrix}.$$

Notice that the first column in the square matrix is equal to the covariance vector. Thus, by simple inspection, it follows that the unique solution is:

$$\mathbf{X} = [1 \quad 0 \quad \cdots \quad 0]'.$$

Therefore by Definition 2.1,

$$\hat{Z}(\boldsymbol{x}_0) = m + Z(\boldsymbol{x}_1) - m$$

$$\hat{Z}(\boldsymbol{x}_0) = Z(\boldsymbol{x}_1). \quad \square$$

As shown in Figure 2.4, use of simple kriging in automatic contouring replaces the sampling—commonly irregular—by estimates at the nodes of a regular grid. Users sometimes complain about the inability of the grid to honor the data and blame kriging for the problem.

The source of the problem is not the inability of kriging to produce exact interpolations. Given that a node falls exactly at a sampling site, the value at the node will be that of the observation at such place. This situation, however, is rare and most likely the nodes will not coincide with the sampling locations, thus taking values other than those of the sampling. In the final stage, in a graphical display using contour lines, algorithms other than kriging used for interpolating from the nodes rarely do a perfect job of duplicating the data.

If honoring the data is important, the best thing to do when employing grids is to reduce their increment to the smallest feasible magnitude in order to minimize discrepancies or render them undetectable. A less recommendable practice is to move the observations to the closest node, which among other problems has the potential of resulting in more than one observation per node.

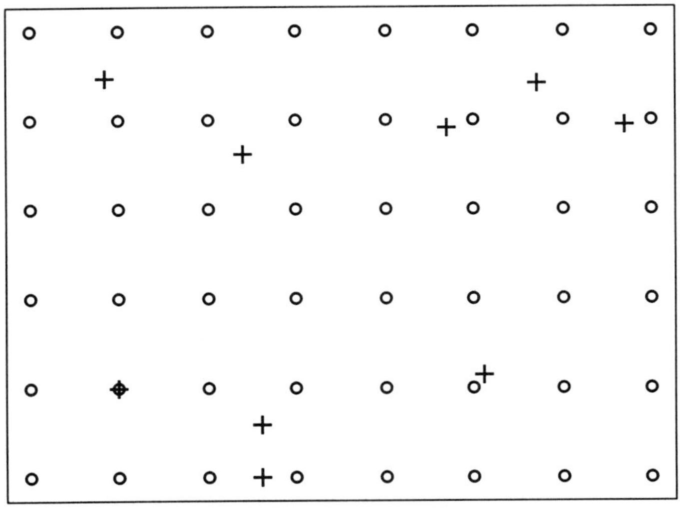

Figure 2.4. Example of square grid, denoted by open circles, associated with a sampling, denoted by plus signs.

Theorem 2.6

The kriging estimate is orthogonal to its error.

Proof

To say that an estimate $\hat{Z}(\boldsymbol{x})$ of $Z(\boldsymbol{x})$ is orthogonal to its estimation error is mathematically equivalent to saying that

$$\mathrm{E}\left[\hat{Z}(\boldsymbol{x}_0+\boldsymbol{h})\left\{\hat{Z}(\boldsymbol{x}_0)-Z(\boldsymbol{x}_0)\right\}\right]=0\,.$$

From Definition 2.1,

$$\begin{aligned}&\mathrm{E}\left[\hat{Z}(\boldsymbol{x}_0+\boldsymbol{h})\left\{\hat{Z}(\boldsymbol{x}_0)-Z(\boldsymbol{x}_0)\right\}\right]\\&\quad=\mathrm{E}\left[\left\{m+\sum_{i=1}^{k}\lambda_i\left(Z(\boldsymbol{x}_i)-m\right)\right\}\left\{\hat{Z}(\boldsymbol{x}_0)-Z(\boldsymbol{x}_0)\right\}\right]\,.\end{aligned}$$

Expanding,

$$\begin{aligned}\mathrm{E}\left[\hat{Z}(\boldsymbol{x}_0+\boldsymbol{h})\left\{\hat{Z}(\boldsymbol{x}_0)-Z(\boldsymbol{x}_0)\right\}\right]&=m\mathrm{E}\left[\hat{Z}(\boldsymbol{x}_0)-Z(\boldsymbol{x}_0)\right]\\&\quad+\mathrm{E}\left[\left\{\sum_{i=1}^{k}\lambda_i\left(Z(\boldsymbol{x}_i)-m\right)\right\}\left\{\hat{Z}(\boldsymbol{x}_0)-Z(\boldsymbol{x}_0)\right\}\right]\,.\end{aligned}$$

By Theorem 2.4, the first expectation is zero. Replacing the estimate by its Definition 2.1 and expanding again,

$$\begin{aligned}
&\mathrm{E}\left[\hat{Z}(\boldsymbol{x}_0+\boldsymbol{h})\left\{\hat{Z}(\boldsymbol{x}_0)-Z(\boldsymbol{x}_0)\right\}\right] \\
&\quad=\mathrm{E}\left[\left\{\sum_{i=1}^{k}\lambda_i\left(Z(\boldsymbol{x}_i)-m\right)\right\}\left\{\sum_{j=1}^{k}\lambda_j\left(Z(\boldsymbol{x}_j)-m\right)\right\}\right] \\
&\qquad-\mathrm{E}\left[\left\{\sum_{i=1}^{k}\lambda_i\left(Z(\boldsymbol{x}_i)-m\right)\right\}\left\{Z(\boldsymbol{x}_0)-m\right\}\right].
\end{aligned}$$

By the commutative property of summation and expectation

$$\begin{aligned}
&\mathrm{E}\left[\hat{Z}(\boldsymbol{x}_0+\boldsymbol{h})\left\{\hat{Z}(\boldsymbol{x}_0)-Z(\boldsymbol{x}_0)\right\}\right] \\
&\quad=\sum_{i=1}^{k}\sum_{j=1}^{k}\lambda_i\lambda_j\mathrm{E}\left[\left(Z(\boldsymbol{x}_i)-m\right)\left(Z(\boldsymbol{x}_j)-m\right)\right] \\
&\qquad-\sum_{i=1}^{k}\lambda_i\mathrm{E}\left[\left(Z(\boldsymbol{x}_i)-m\right)\left(Z(\boldsymbol{x}_0)-m\right)\right],
\end{aligned}$$

which by Assumption 2.2 is equal to

$$\begin{aligned}
&\mathrm{E}\left[\hat{Z}(\boldsymbol{x}_0+\boldsymbol{h})\left\{\hat{Z}(\boldsymbol{x}_0)-Z(\boldsymbol{x}_0)\right\}\right] \\
&\quad=\sum_{i=1}^{k}\sum_{j=1}^{k}\lambda_i\lambda_j\mathrm{Cov}(\boldsymbol{x}_i,\boldsymbol{x}_j)-\sum_{i=1}^{k}\lambda_i\mathrm{Cov}(\boldsymbol{x}_i,\boldsymbol{x}_0).
\end{aligned}$$

Finally, Lemma 2.7 proves the theorem because the terms in the right-hand side of the equation are equal. □

Corollary 2.1

The estimation variance in an exact interpolation is zero.

Proof

By Theorems 2.3 and 2.5,

$$\sigma^2_{SK}(\boldsymbol{x}_0)=\mathrm{Cov}(0)-\mathrm{Cov}(\boldsymbol{x}_0,\boldsymbol{x}_1)$$

and the proof follows because as $\boldsymbol{x}_0=\boldsymbol{x}_1$, then $\mathrm{Cov}(\boldsymbol{x}_0,\boldsymbol{x}_1)=\mathrm{Cov}(0)$. □

Corollary 2.2

It is not possible to run simple kriging when the sampling includes observations at duplicate sites.

Proof

From Theorem 2.2, a duplication in sites will imply duplication in equations. For example, if $\boldsymbol{x}_1=\boldsymbol{x}_2$, the second equation will be identical to the first one. Duplicate equations lead to unsolvable singular matrices. □

Therefore, in practice, two necessary conditions to run simple kriging successfully are:

(a) By Theorem 2.2, on the theoretical side, the covariance must be positive definite.

(b) On practical grounds, the sampling must not include duplicate sampling sites.

Corollary 2.3

Simple kriging is independent of translations of the Cartesian system.

Proof

From Steps 4 and 5 in Algorithm 2.1, both the estimate and the estimation variance depend only on covariance values. By Assumption 2.2, the covariance is independent of location and dependent only on distance, which is insensitive to translations of the reference axes. □

Corollary 2.4

Multiplication of the covariance by a factor does not change the estimate.

Proof

From Theorem 2.2, multiplication of the original covariance by a constant will produce a new system of equations in which both sides will be the sides in the original system of equations multiplied by the constant. Proof of the corollary follows because multiplication of both sides of one or more equations in a system of equations does not change its solution and if the weights are the same, so is the estimate in Definition 2.1. □

Corollary 2.5

Multiplication of the covariance by a factor will change the estimation variance by the same factor.

Proof

Let $\sigma_{SK}^2(\boldsymbol{x}_0)$ be the original estimation variance and let α be the factor. By Corollary 2.4, the weights are invariant to a change in the covariance by a factor. Then, by Theorem 2.3, the new estimation variance $\sigma_{SK}^2{}'(\boldsymbol{x}_0)$ is:

$$\begin{aligned}\sigma_{SK}^2{}'(\boldsymbol{x}_0) &= \alpha\text{Cov}(0) - \sum_{i=i}^{k} \lambda_i \alpha \text{Cov}(\boldsymbol{x}_0, \boldsymbol{x}_i) \\ &= \alpha\sigma_{SK}^2(\boldsymbol{x}_0)\,. \quad \square\end{aligned}$$

Corollary 2.6

The estimation variance depends on the data configuration.

Proof

From Step 5 in Algorithm 2.1, the estimation variance is a function of the covariance matrix $\mathbf{C}$. By Assumption 2.2, the covariance is solely a function of distance, which depends on sampling configuration. Hence the estimation variance depends on the sampling pattern. □

For a fixed sampling size, different sampling patterns can produce significantly different estimation variances. In two dimensions, the regular patterns in Figure 2.5 are at the top of the efficiency scale in terms of achieving a given average estimation variance with the minimum number of observations, while clustered sampling is the most inefficient (Olea, 1984).

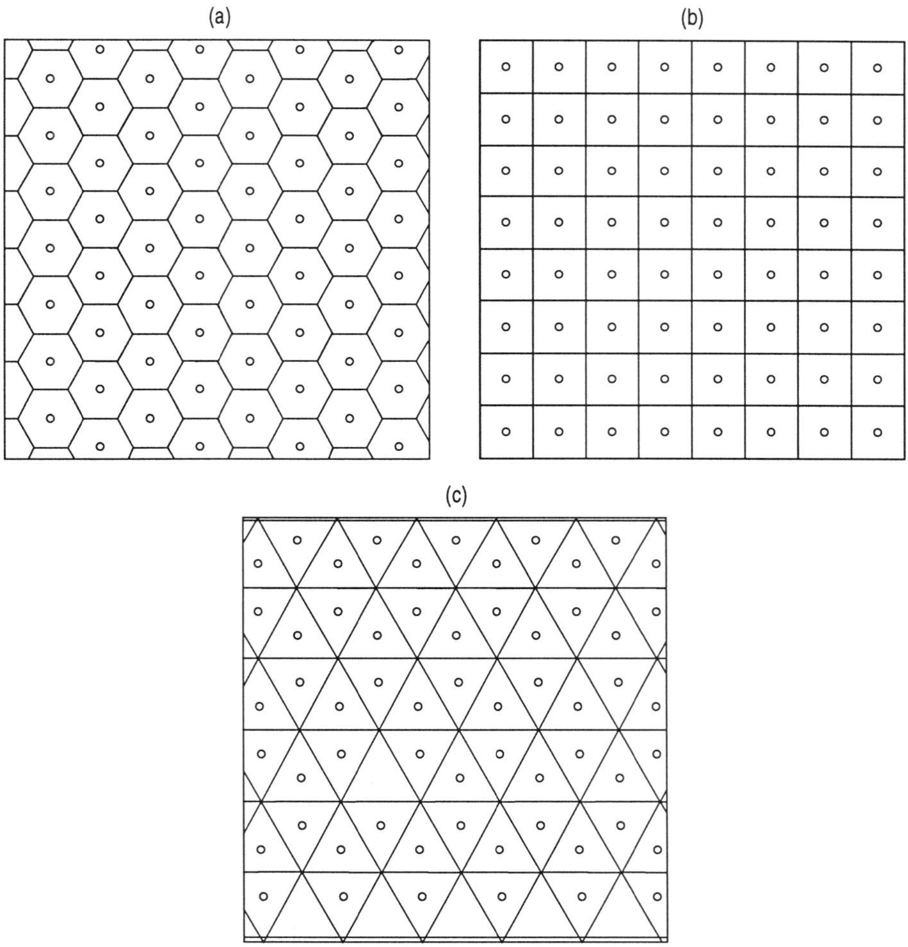

Figure 2.5. Two-dimensional regular patterns of point sampling: (a) hexagonal, (b) square, (c) triangular.

Corollary 2.7

The estimation variance does not depend directly on the data.

Proof

By simple inspection of Step 5 in Algorithm 2.1, one can see that the estimation variance depends exclusively on covariance terms—not on any particular observation $z(\boldsymbol{x}_i)$. □

The independence of the estimation variance from the observation $z(\boldsymbol{x}_i)$ used by the estimator in Definition 2.1 has been employed to downplay the importance that the estimation variance may have in assessing the reliability of the estimation (Journel, 1986).

Corollary 2.8

Consider a sampling where all random variables are independent from one another, but not necessarily from the one at the estimation location $\boldsymbol{x}_0$. Let λ_i be the simple kriging weight in Definition 2.1 for a random function with covariance $\operatorname{Cov}(\cdot)$. Then,

$$\lambda_i = \frac{\operatorname{Cov}(\boldsymbol{x}_0, \boldsymbol{x}_i)}{\operatorname{Cov}(0)}, \text{ for } i = 1, 2, \cdots, k.$$

Proof

Because the variables are independent, $\operatorname{Cov}(\boldsymbol{x}_i, \boldsymbol{x}_j) = 0$, for $i > 0$ and $i \neq j$. From Theorem 2.2

$$\operatorname{Cov}(\boldsymbol{x}_i, \boldsymbol{x}_i)\,\lambda_i = \operatorname{Cov}(\boldsymbol{x}_0, \boldsymbol{x}_i), \text{ for } i = 1, 2, \cdots, k$$

and because $\operatorname{Cov}(\boldsymbol{x}_i, \boldsymbol{x}_i) = \operatorname{Cov}(0)$, the corollary follows by dividing both members by $\operatorname{Cov}(0)$. □

Notice that $\operatorname{Cov}(\boldsymbol{x}_0, \boldsymbol{x}_i)/\operatorname{Cov}(0)$ represents the correlation coefficient of classical statistics.

Corollary 2.9

Consider a sampling where all random variables are independent from one another and with respect to the estimator. Then the estimate $\hat{z}(\boldsymbol{x}_0)$ is equal to the mean m.

Proof

From Corollary 2.8,

$$\lambda_i = \frac{\operatorname{Cov}(\boldsymbol{x}_0, \boldsymbol{x}_i)}{\operatorname{Cov}(0)}, \quad i = 1, 2, \cdots, k,$$

where λ_i is a simple kriging weight in Definition 2.1 for a random function with covariance $\operatorname{Cov}(\cdot)$. If, in addition, the random variables are independent from the estimate, $\text{Cov}(\boldsymbol{x}_0, \boldsymbol{x}_i)$ is zero and all weights are zero too. Then, from Definition 2.1,

$$\hat{z}(\boldsymbol{x}_0) = m. \quad \square$$

Corollary 2.9 explains the smoothing property of simple kriging. As spatial dependency in the estimation decreases, estimates tend toward the constant mean regardless of the sampling and the true values, generating a featureless surface. In the two-dimensional case, areas with perfect spatial independence will map like the surface of a lake, with all "lakes" having the same elevation m. Stochastic simulation in Chapter 9 has been proposed as an alternative to kriging when smoothing is a drawback.

Corollary 2.10

The mean can be regarded as an additional observation in the estimation of a random function.

Proof

Rearranging Definition 2.1,

$$\hat{Z}(\boldsymbol{x}_0) = \sum_{i=1}^{k} \lambda_i Z(\boldsymbol{x}_i) + m\left(1 - \sum_{i=1}^{k} \lambda_i\right),$$

which can be regarded as a linear estimator involving an extra $(k+1)$th observation m taking the weight $\sum_{i=1}^{k} \lambda_i - 1$. $\square$

From the corollary it follows that the closer the sum of the weights is to 1, the smaller is the influence that the mean has in the estimation.

Corollary 2.11

If the simple kriging weights sum to 1, the estimator is independent from the value of the mean.

Proof

From Corollary 2.10, if the weights sum to 1, the weights of the mean equal zero and the estimator depends only on the sampling. $\square$

EXERCISE 2.2

Given the sampling

Index	Easting	Northing
1	10	20
2	30	280
3	387	72
4	392	81
5	388	56

if the mean of the attribute is 110 and its covariance $\operatorname{Cov}(\boldsymbol{h}) = 2000\mathrm{e}^{-\frac{h}{350}}$ and the estimation location is $\boldsymbol{x}_0 = (180, 120)$, then employing simple kriging:

1. Calculate the estimation weights.
2. Make remarks about the estimation weights.

Solution

1. The distance matrix is

$$\mathbf{D} = \begin{bmatrix} 0 & & & & \\ 260.8 & 0 & & & \\ 380.6 & 413.2 & 0 & & \\ 386.8 & 413.1 & 10.3 & 0 & \\ 379.7 & 422.3 & 16.3 & 25.3 & 0 \end{bmatrix}$$

and the distance vector is

$$\mathbf{d} = [197.2 \quad 219.3 \quad 212.5 \quad 215.6 \quad 217.6]'.$$

The corresponding covariance matrices in Algorithm 2.1 are

$$\mathbf{C} = \begin{bmatrix} 2000 & & & & \\ 704.8 & 2000 & & & \\ 436.2 & 383.0 & 2000 & & \\ 425.6 & 383.2 & 1919.4 & 2000 & \\ 438.0 & 369.4 & 1875.8 & 1807.4 & 2000 \end{bmatrix} \quad \mathbf{c} = \begin{bmatrix} 908.7 \\ 831.8 \\ 854.9 \\ 844.4 \\ 837.5 \end{bmatrix},$$

which implies that the optimal weights are

$$\mathbf{X} = \begin{bmatrix} 2000 & & & & \\ 704.8 & 2000 & & & \\ 436.2 & 383.0 & 2000 & & \\ 425.6 & 383.2 & 1919.4 & 2000 & \\ 438.0 & 369.4 & 1875.8 & 1807.4 & 2000 \end{bmatrix}^{-1} \begin{bmatrix} 908.7 \\ 831.8 \\ 854.9 \\ 844.4 \\ 837.5 \end{bmatrix}$$

$$= [0.296 \quad 0.250 \quad 0.116 \quad 0.110 \quad 0.100]'.$$

2. (a) In Exercise 2.1 the sum of the weights was 0.958 and now it is 0.872. These two results prove that the weights do not sum to a constant.

(b) As can be seen in Figure 2.6, again the weights do not vary proportionally to the distances between the estimator and sampling sites. The distance $\boldsymbol{x}_0 - \boldsymbol{x}_2$ is slightly larger than the distance to any of the three eastern observations, yet the weight λ_2 is more than twice as large as any λ_3, λ_4, or λ_5, and λ_1 is almost as large as $\lambda_3 + \lambda_4 + \lambda_5$ despite the fact that the distance $\boldsymbol{x}_0 - \boldsymbol{x}_1$ is only slightly more than 90% of any of $\boldsymbol{x}_0 - \boldsymbol{x}_3$, $\boldsymbol{x}_0 - \boldsymbol{x}_4$, or $\boldsymbol{x}_0 - \boldsymbol{x}_5$.

What is special about the eastern weights is that they are in a cluster. Simple kriging regards observations in a cluster as redundant measurements and automatically assigns to such observations weights that collectively are

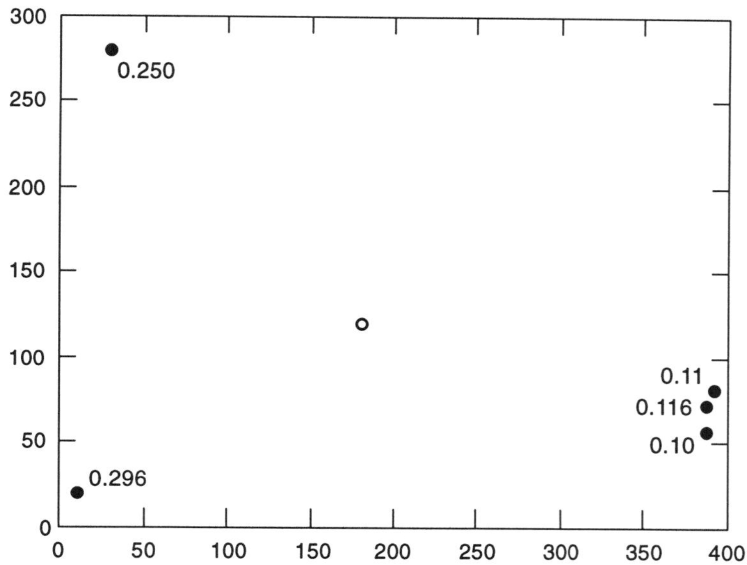

Figure 2.6. Weights for the original sampling in Exercise 2.2.

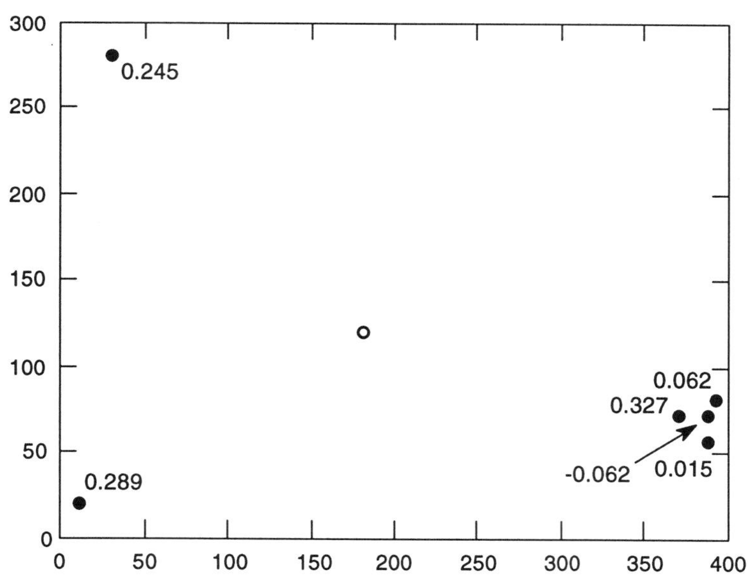

Figure 2.7. Weights resulting from the addition of one observation to the sampling in Exercise 2.2.

close to the weight it would assign to a single observation at the centroid of the cluster. In this instance, the sum of the weights is 0.326 and the weight that a single observation at the centroid of the cluster would receive is 0.311.

This characteristic of the estimator is called the *declustering property* of kriging.

(c) The declustering property and the screen effect—like any of the simple kriging properties—operate simultaneously. For example, addition of a sixth observation at $\mathbf{X} = [0.289 \quad 0.245 \quad -0.062 \quad 0.062 \quad 0.015 \quad 0.327]'$, as displayed in Figure 2.7. □

Chapter 3
NORMALIZATION

We have seen in the previous chapter that simple kriging is the best linear unbiased estimator in a minimum mean square error sense. No distributional assumptions were necessary to reach that level of optimality, but if one is willing to assume multinormality, the method can be taken to new heights.

If the joint distribution of the random function is multivariate normal, simple kriging is the best of all possible mean square estimators—linear or nonlinear, biased or unbiased (see, *e.g.*, Cressie, 1991, p. 110). In practical terms though, multinormality is a formidable requirement to fulfill and one that is impossible to verify.

In practice, if the univariate distribution approximates a normal distribution, the strong but not uncommon working hypothesis is that the data were generated by a multivariate normal distribution.

Comparing Two Distributions

Let $F(z)$ be the theoretical univariate cumulative distribution for attribute Z and let $F_n(z)$ be the observed step function for a sample of size n (Fig. 3.1). Different distribution functions give different cumulative distribution functions. All cumulative distribution functions, however, coincide below the smallest possible value—for which all cumulative distributions are zero—and above the largest one—for which they are 1. So it is the behavior between the extreme values that makes the difference.

The statistic

$$D = \max_{-\infty<z<\infty} |F(z) - F_n(z)|,$$

illustrated in Figure 3.1, measures the overall fit between two cumulative distributions. It has the appeal of being insensitive to a linear transformation of Z. In classical statistics, assuming a random sampling, one can calculate D's distribution in the case of the so-called two-tailed null hypothesis

$$H_0 : F(z) = F_n(z), \text{ for all } z$$

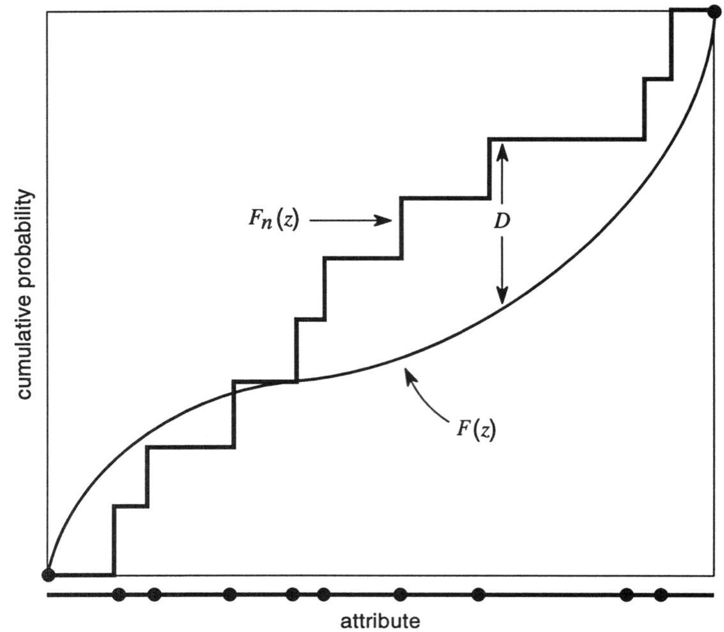

Figure 3.1. Maximum discrepancy D between two distributions. In this example the statistic relates to the continuous cumulative distribution $F(z)$ and the step cumulative distribution $F_n(z)$ for the sampling denoted by the dots along the abscissa. (Modified from Press *et al.*, 1992, p. 618.)

with the alternative

$$H_1 : H_0 \text{ is false ,}$$

thus making it possible to investigate the significance of any nonzero value of D. This is the Kolmogorov–Smirnov test.

In geostatistics, because of the presence of spatial dependence among the sampling, one cannot calculate levels of significance. The magnitude of the statistic D is all that can be used as a guide to accept or reject the presumption of similarity of two distributions. The empirical rule is to consider that the two distributions are similar when $D \leq 0.1$. Although this is not completely satisfactory, it is better than the alternative of making a subjective call based on graphical appearance.

Most statistical and mathematical packages—IMSL Library (IMSL, 1987), Press *et al.* (1992)—have programs that can be used to calculate the Kolmogorov–Smirnov statistic D.

EXERCISE 3.1

Using the sampling in Appendix A:

1. Calculate the sample mean and the sample variance. Display the histogram and the cumulative distribution for the attribute using a normal probability scale.

2. For a maximum discrepancy of 0.1, does the sampling approximate a normal distribution?

Solution

1. The sample mean is 11.92% and the variance is 2.97%^{2}. Figure 3.2 displays the histogram and Figure 3.3 shows the cumulative distribution on a normal probability scale.

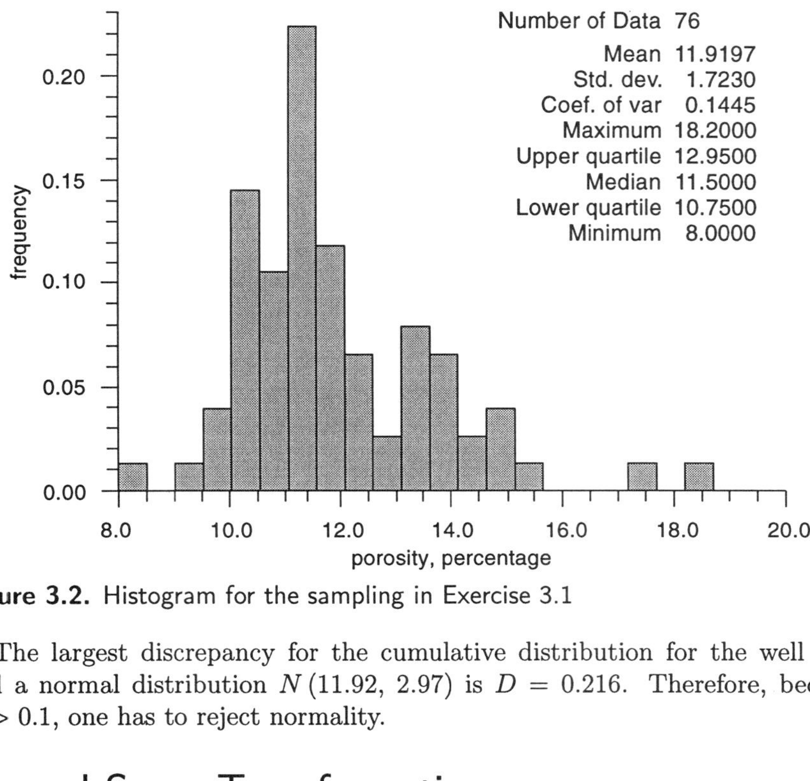

Figure 3.2. Histogram for the sampling in Exercise 3.1

2. The largest discrepancy for the cumulative distribution for the well data and a normal distribution $N(11.92,\ 2.97)$ is $D = 0.216$. Therefore, because $D > 0.1$, one has to reject normality.

Normal Score Transformation

If a sampling is not univariate normal, it can be made so through a nonlinear transformation referred to as either *Gaussian anamorphosis* or *normal score transformation.* The terms are indistinguishable, although the latter has gained more support recently. A property of the transformation is that it preserves the rank.

The idea behind normal score transformation and its implementation are fairly straightforward. First one ranks the data and then assigns to every ith observation the value of the ith percentile in a standardized normal distribution $N(0, 1)$. This percentile is the transformed value of the observation (Figure 3.4). In the presence of preferential sampling, it is recommended to preprocess the data to avoid distortions that may result from clustering in the sampling sites (Isaaks and Srivastava, 1989, chap. 10; Deutsch and Journel, 1992, p. 207–209).

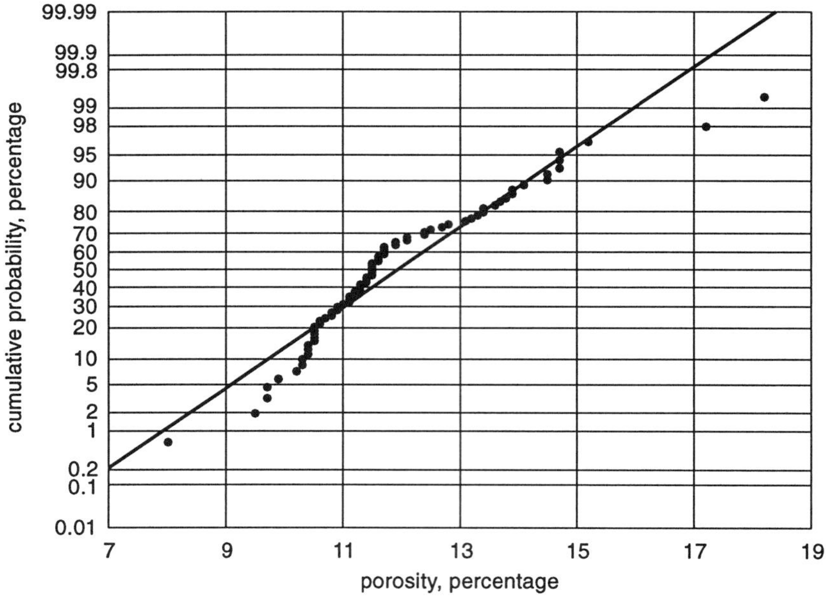

Figure 3.3. Cumulative distribution for $N(11.92,\ 2.97)$ and for the sampling in Exercise 3.1.

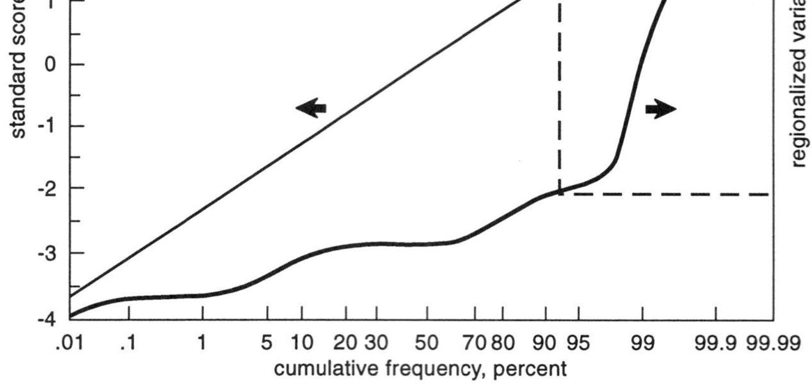

Figure 3.4. Normal score forward and backward transformation. The dashed line relates a normal score value with a value for the attribute with the same cumulative probability.

In the absence of declustering compensations, the new sampling, by construction, exactly follows a standardized normal distribution when there are no duplicated values in the original sampling. In case of duplications, different solutions handle the ties differently, resulting in minor deviations from perfection.

A special case that has received great attention is the lognormal distribution (Rivoirard, 1990), for which one can obtain normality either numerically through normal scores or analytically by simply considering the logarithm of the variable instead of the variable itself.

Simple Kriging of Normal Scores

The strong requirement of simple kriging that the true mean of the attribute must be known severely limits the applicability of the method. The most important exception is the case of the data that results from a normal score transformation without declustering, in which situation the mean is known to be zero.

Simple kriging is first applied to the normal scores and then a back transformation is employed to bring the estimated normal scores to the domain of the data, a procedure also known as multigaussian kriging. Except for the tails, the back transformation is a trivial interpolation using the reference points provided by the initial transformation of the data. For example let us assume that the simple kriging estimate of the normal score is 1.4136. Examining the table of normal scores for the sampling, if the value 1.4136 is exactly in the middle of the interval defined by the values 1.3688 for an original observation of 14.5 and 1.4584 for 14.7, the straight linear back transformed value of the estimate is 14.6. The software package GSLIB (Deutsch and Journel, 1992) has routines for the preparation and back transformation of normal scores.

EXERCISE 3.2

For the sampling in Exercise 3.1:

1. Compute the normal scores.
2. If the normal score covariance is $\mathrm{Cov}(\boldsymbol{h}) = \mathbf{e}^{-\frac{h}{0.8}}$, using the five closest wells and simple kriging, estimate the value of the normal score at (2, 4.3).
3. Calculate the back transformed value of the estimate.

Solution

1. Data in this sampling are arranged in an almost perfect grid. Thus no declustering is necessary. The normal scores are:

Well	Easting	Northing	Porosity	Normal Score	Rank
3	2.097	4.611	12.7	0.6136	56
5	1.863	4.370	14.1	1.2168	68
6	2.112	4.371	11.6	0.1824	42
7	2.175	4.317	11.2	−0.3186	28

(cont.)

Well	Easting	Northing	Porosity	Normal Score	Rank
10	1.634	4.136	15.2	1.8399	74
11	1.880	4.126	14.7	1.5614	71
12	2.096	4.129	13.3	0.7820	60
14	1.613	3.885	17.2	2.0592	75
15	1.841	3.883	14.7	1.6844	72
16	2.113	3.881	12.4	0.5358	53
18	1.610	3.628	13.4	0.8276	61
19	1.861	3.627	13.7	0.9762	64
20	2.088	3.627	13.9	1.0886	66
21	2.358	3.627	14.5	1.2891	69
23	1.365	3.381	13.1	0.6953	58
24	1.614	3.382	14.5	1.3688	70
25	1.863	3.380	13.2	0.7380	59
26	2.088	3.380	11.9	0.3535	49
29	1.342	3.270	13.4	0.8750	62
30	1.177	3.131	12.1	0.4247	51
31	1.363	3.131	13.8	1.0308	65
32	1.614	3.130	13.6	0.9244	63
33	1.862	3.132	11.5	0.0165	36
34	2.117	3.129	12.5	0.5743	55
36	1.118	2.883	12.4	0.4981	54
37	1.363	2.880	14.7	1.4584	73
38	1.614	2.881	11.7	0.2161	45
39	1.861	2.882	11.5	−0.0165	37
40	1.924	2.943	11.5	−0.0495	38
41	2.113	2.878	11.4	−0.1824	33
44	1.151	2.628	12.8	0.6539	57
45	1.362	2.628	12.1	0.4611	52
46	1.615	2.631	10.6	−0.7380	17
47	1.862	2.629	11.5	0.0495	39
48	2.088	2.626	10.8	−0.6136	20
49	1.140	2.381	11.4	−0.1157	34
50	1.373	2.380	11.4	−0.1490	35
51	1.615	2.378	11.7	0.3186	46
52	1.862	2.379	11.3	−0.2499	30
53	2.127	2.388	10.5	−0.8750	12
54	2.363	2.377	10.3	−1.3688	7
57	1.142	2.127	10.7	−0.6953	18
58	1.364	2.129	11.1	−0.4611	25
59	1.616	2.127	10.5	−0.9244	13

(*cont.*)

Well	Easting	Northing	Porosity	Normal Score	Rank
60	1.864	2.127	13.9	1.1504	67
61	2.092	2.124	10.6	−0.7820	18
62	2.363	2.125	9.9	−1.5614	5
63	2.612	2.124	10.5	−0.9762	14
65	1.121	1.878	10.9	−0.5743	23
66	1.369	1.880	11.7	0.2841	47
67	1.676	1.937	11.2	−0.3535	29
68	1.866	1.880	11.7	0.2499	48
69	2.116	1.881	10.4	−1.1503	9
70	2.339	1.875	10.4	−1.2168	10
73	1.123	1.626	11.5	0.0825	40
74	1.369	1.627	11.6	0.1490	43
75	1.614	1.628	11.6	0.1157	44
76	1.867	1.627	11.1	−0.3889	26
77	2.117	1.626	10.3	−1.2891	8
80	0.887	1.396	18.2	2.4795	76
81	1.157	1.371	11.3	−0.2841	31
82	1.372	1.373	11.5	−0.0825	41
83	1.616	1.374	10.9	−0.5358	22
84	1.868	1.386	11.3	−0.2161	32
86	1.137	1.133	10.8	−0.6539	21
87	1.373	1.122	11.0	−0.4981	24
88	1.616	1.123	10.5	−1.0308	15
89	1.867	1.121	11.9	0.3889	50
91	1.121	0.872	8.0	−2.4795	1
92	1.400	0.874	11.1	−0.4247	27
93	1.592	0.899	10.5	−0.8276	26
94	1.849	0.872	10.4	−1.0886	11
95	1.120	0.616	9.5	−2.0592	2
96	1.397	0.619	9.7	−1.6844	4
97	1.623	0.639	10.2	−1.4584	6
99	1.374	0.371	9.7	−1.8399	3

2. The five closest measurements to (2, 4.3) are:

6	2.112	4.371	11.6	0.182
5	1.863	4.370	14.1	1.217
7	2.175	4.317	11.2	−0.319
12	2.096	4.129	13.3	0.782
11	1.880	4.126	14.7	1.561

From Algorithm 2.1

$$\mathbf{X} = \begin{bmatrix} 1 & & & & \\ 0.733 & 1 & & & \\ 0.902 & 0.673 & 1 & & \\ 0.739 & 0.658 & 0.775 & 1 & \\ 0.656 & 0.737 & 0.644 & 0.763 & 1 \end{bmatrix}^{-1} \begin{bmatrix} 0.847 \\ 0.825 \\ 0.803 \\ 0.783 \\ 0.768 \end{bmatrix}$$

$$= \begin{bmatrix} 0.354 & 0.319 & 0.050 & 0.163 & 0.145 \end{bmatrix}'$$

Because the mean is zero, the residuals are equal to the normal scores,

$$\begin{aligned} \hat{z}_{SK}(2,\ 4.3) &= \begin{bmatrix} 0.182 & 1.217 & -0.319 & 0.782 & 1.561 \end{bmatrix}' \\ &\quad \begin{bmatrix} 0.354 & 0.319 & 0.050 & 0.163 & 0.145 \end{bmatrix} \\ &= 0.790. \end{aligned}$$

3. Sorting the data by the ranking, one obtains

		⋮		
25	1.863	3.380	13.2	0.738
12	2.096	4.129	13.3	0.782
18	1.610	3.628	13.4	0.828
29	1.342	3.270	13.4	0.875
		⋮		

By linear interpolation, the back transformed estimate is

$$\begin{aligned} \hat{z}(2,\ 4.3) &= 13.3 + 0.1 \frac{0.79 - 0.782}{0.828 - 0.782} \\ &= 13.32. \quad \square \end{aligned}$$

Chapter 4
ORDINARY KRIGING

Historically, ordinary kriging was the first improvement of simple kriging, whose roots predate geostatistics (Goldberg, 1962; Cressie, 1990). We have seen that simple kriging requires knowledge of the mean to solve the problem of finding weights that minimize the variance of the estimation error. Ordinary kriging elegantly discards the requirement by filtering out the mean, taking advantage of Corollary 2.11. And by removing the mean from the estimator, the entire formulation becomes independent from the mean, as both the weights and the estimation variance in the simple kriging Algorithm 2.1 were already independent from the mean.

In terms of the formulation of the method, the main difference relative to simple kriging is that the filtering changes the unconstrained optimization into a constrained optimization that is solved through the Lagrange method of multipliers. As shown in this chapter, depending on the assumptions, ordinary kriging is really four different methods to which the same name is applied.

Assumptions

There are two forms of stationary ordinary kriging. The first formulation is a straight generalization of the assumptions of simple kriging, dropping Assumption 2.3 which is no longer necessary.

Assumption 4.1

The sampling is a partial realization of a random function $Z(\boldsymbol{x})$, where $\boldsymbol{x}$ denotes spatial location. □

Assumption 4.2

The random function is second order stationary over the sampling domain, which implies that

$$\mathrm{E}[Z(\boldsymbol{x})] = m$$

$$\begin{aligned}\mathrm{E}\left[\left(Z\left(\boldsymbol{x}\right)-m\right)\left(Z\left(\boldsymbol{x}+\boldsymbol{h}\right)-m\right)\right] &= \mathrm{E}\left[Z\left(\boldsymbol{x}\right)Z\left(\boldsymbol{x}+\boldsymbol{h}\right)\right]-m^2\\ &= \mathrm{Cov}\left[Z\left(\boldsymbol{x}\right),Z\left(\boldsymbol{x}+\boldsymbol{h}\right)\right]\\ &= \mathrm{Cov}\left(\boldsymbol{x},\boldsymbol{x}+\boldsymbol{h}\right)=\mathrm{Cov}\left(\boldsymbol{h}\right),\end{aligned}$$

where $\mathrm{E}\left[\cdot\right]$ denotes expected value, m is a constant mean, $\boldsymbol{h}$ is a vectorial distance in the sampling space, and $\mathrm{Cov}\left[\cdot\right]$ is the covariance of the random function. □

Notice that although one does not need to know the value of the mean, the mean still remains a constant but now unknown value. Considering that the second order stationarity approach of simple kriging requires a constant covariance under translation plus the existence of a random function variance, geostatistics has provided an alternative set of assumptions that relaxes such additional requirements.

Assumption 4.3

The sampling is a partial realization of a random function $Z\left(\boldsymbol{x}\right)$, where $\boldsymbol{x}$ denotes spatial location. □

Assumption 4.4

The random function honors the intrinsic hypothesis over the sampling domain

$$\begin{aligned}\mathrm{E}\left[Z\left(\boldsymbol{x}\right)\right] &= m\\ \mathrm{Var}\left[Z\left(\boldsymbol{x}\right)-Z\left(\boldsymbol{x}+\boldsymbol{h}\right)\right] &= 2\gamma\left(\boldsymbol{h}\right),\end{aligned}$$

where $\gamma\left(\cdot\right)$ is the semivariogram of the random function. □

Every random function that is second order stationary is also intrinsic, but the converse is not true.

Important Relationships

Let us demonstrate some interesting relationships that follow from the two sets of assumptions.

Lemma 4.1

Let Z be a random function that is either second order stationary or intrinsic. Then

$$\mathrm{E}\left[Z\left(\boldsymbol{x}\right)-Z\left(\boldsymbol{x}+\boldsymbol{h}\right)\right]=0.$$

Proof

Expanding the expectation,

$$\mathrm{E}\left[Z\left(\boldsymbol{x}\right)-Z\left(\boldsymbol{x}+\boldsymbol{h}\right)\right]=\mathrm{E}\left[Z\left(\boldsymbol{x}\right)\right]-\mathrm{E}\left[Z\left(\boldsymbol{x}+\boldsymbol{h}\right)\right].$$

But, by Assumption 4.2 or 4.4, the expected value of any variable is a constant and the difference cancels out. □

Lemma 4.2

Let Z be an intrinsic random function. Then

$$\operatorname{Var}\left[Z(\boldsymbol{x})-Z(\boldsymbol{x}+\boldsymbol{h})\right]=\operatorname{E}\left[\{Z(\boldsymbol{x})-Z(\boldsymbol{x}+\boldsymbol{h})\}^2\right].$$

Proof

Because the mean is constant, from the definition of variance,

$$\operatorname{Var}\left[Z(\boldsymbol{x})-Z(\boldsymbol{x}+\boldsymbol{h})\right]=\operatorname{E}\left[\{Z(\boldsymbol{x})-Z(\boldsymbol{x}+\boldsymbol{h})\}^2\right]-\operatorname{E}\left[Z(\boldsymbol{x})-Z(\boldsymbol{x}+\boldsymbol{h})\right]^2$$

and the proof follows because the second term in the right-hand side of the equation is zero according to Lemma 4.1. □

Therefore, provided that the mean is constant, the semivariogram $\gamma(\boldsymbol{h})$ is equal to half the mean of the squared difference of variates a distance $\boldsymbol{h}$ apart.

Lemma 4.3

If Z is an intrinsic random function. Then

$$2\gamma(\boldsymbol{h})=\operatorname{Var}\left[Z(\boldsymbol{x})\right]+\operatorname{Var}\left[Z(\boldsymbol{x}+\boldsymbol{h})\right]-2\operatorname{Cov}(\boldsymbol{x},\ \boldsymbol{x}+\boldsymbol{h}),$$

where $\boldsymbol{x}$ and $\boldsymbol{x}+\boldsymbol{h}$ are any two locations in the sample domain.

Proof

By its definition in Assumption 4.4 and by Lemma 4.2,

$$2\gamma(\boldsymbol{h})=\operatorname{E}\left[\{Z(\boldsymbol{x})-Z(\boldsymbol{x}+\boldsymbol{h})\}^2\right].$$

Expanding and then adding and subtracting the term $2m^2$,

$$2\gamma(\boldsymbol{h})=\operatorname{E}\left[Z^2(\boldsymbol{x})\right]-m^2+\operatorname{E}\left[Z^2(\boldsymbol{x}+\boldsymbol{h})\right]-m^2-2\operatorname{E}\left[Z(\boldsymbol{x})\,Z(\boldsymbol{x}+\boldsymbol{h})\right]+2m^2$$

and the proof follows by Assumption 4.4 that the mean m is constant, and by the definitions of variance and covariance. □

Corollary 4.1

If Z is a second order stationary random function, then

$$\gamma(\boldsymbol{h})=\operatorname{Cov}(0)-\operatorname{Cov}(\boldsymbol{h}).$$

Proof

By Lemma 4.3

$$2\gamma\left(\boldsymbol{h}\right)=\mathrm{Var}\left[Z\left(\boldsymbol{x}\right)\right]+\mathrm{Var}\left[Z\left(\boldsymbol{x}+\boldsymbol{h}\right)\right]-2\mathrm{Cov}\left(\boldsymbol{x},\ \boldsymbol{x}+\boldsymbol{h}\right),$$

which is equivalent to

$$2\gamma\left(\boldsymbol{h}\right)=\mathrm{Cov}\left(\boldsymbol{x},\ \boldsymbol{x}\right)+\mathrm{Cov}\left(\boldsymbol{x}+\boldsymbol{h},\ \boldsymbol{x}+\boldsymbol{h}\right)-2\mathrm{Cov}\left(\boldsymbol{x},\ \boldsymbol{x}+\boldsymbol{h}\right).$$

If the random function is second order stationary, the covariance is independent of location and depends only on the distance $\boldsymbol{h}$ between the two variates:

$$2\gamma\left(\boldsymbol{h}\right)=2\mathrm{Cov}\left(0\right)-2\mathrm{Cov}\left(\boldsymbol{h}\right).$$

The proof follows by cancellation of the factor 2. □

Lemma 4.4

Let Z be an intrinsic random function. Then

$$\gamma\left(\boldsymbol{x}_i-\boldsymbol{x}_j\right)=\gamma\left(\boldsymbol{x}_i-\boldsymbol{x}\right)+\gamma\left(\boldsymbol{x}_j-x\right)-\mathrm{Cov}\big(Z\left(\boldsymbol{x}_i\right)-Z\left(\boldsymbol{x}\right),Z\left(\boldsymbol{x}_j\right)-Z\left(\boldsymbol{x}\right)\big).$$

Proof

By its definition in Assumption 4.4 and Lemma 4.2

$$\gamma\left(\boldsymbol{x}_i-\boldsymbol{x}_j\right)=\frac{1}{2}\mathrm{E}\left[\left\{Z\left(\boldsymbol{x}_i\right)-Z\left(\boldsymbol{x}_j\right)\right\}^2\right],$$

which does not change by adding and subtracting a third variate

$$\gamma\left(\boldsymbol{x}_i-\boldsymbol{x}_j\right)=\frac{1}{2}\mathrm{E}\left[\left\{\left\{Z\left(\boldsymbol{x}_i\right)-Z\left(\boldsymbol{x}\right)\right\}-\left\{Z\left(\boldsymbol{x}_j\right)-Z\left(\boldsymbol{x}\right)\right\}\right\}^2\right].$$

Expanding,

$$\begin{aligned}\gamma\left(\boldsymbol{x}_i-\boldsymbol{x}_j\right)=&\frac{1}{2}\mathrm{E}\left[\left\{Z\left(\boldsymbol{x}_i\right)-Z\left(\boldsymbol{x}\right)\right\}^2\right]+\frac{1}{2}\mathrm{E}\left[\left\{Z\left(\boldsymbol{x}_j\right)-Z\left(\boldsymbol{x}\right)\right\}^2\right]\\&-\mathrm{E}\left[Z\left(\boldsymbol{x}_i\right)-Z\left(\boldsymbol{x}\right),Z\left(\boldsymbol{x}_j\right)-Z\left(\boldsymbol{x}\right)\right]\end{aligned}$$

by Lemma 4.2 and the definition in Assumption 4.4

$$\gamma\left(\boldsymbol{x}_i-\boldsymbol{x}_j\right)=\gamma\left(\boldsymbol{x}_i-\boldsymbol{x}\right)+\gamma\left(\boldsymbol{x}_j-\boldsymbol{x}\right)-\mathrm{E}\left[Z\left(\boldsymbol{x}_i\right)-Z\left(\boldsymbol{x}\right),Z\left(\boldsymbol{x}_j\right)-Z\left(\boldsymbol{x}\right)\right],$$

and the proof follows from the definition of covariance because in this case, by Lemma 4.1, the expected value of the differences is zero. □

The Estimator

The ordinary kriging estimator is nothing other than a simple kriging estimator under the constraint that the weights must sum to one, from which results the following estimator that is mean-free and more compact than the one in Definition 2.1.

Definition 4.1

Let Z be a second order stationary or an intrinsic random function. The estimator $\hat{Z}(\boldsymbol{x}_0)$ at site $\boldsymbol{x}_0$ is given by the following linear combination of random variables at sites $\boldsymbol{x}_i$ considered in the sampling:

$$\hat{Z}(\boldsymbol{x}_0) = \sum_{i=1}^{k} \lambda_i Z(\boldsymbol{x}_i)$$

subject to $\sum_{i=1}^{k} \lambda_i = 1$. □

The Estimation Variance

The estimation variance is the objective function in our problem of optimizing the weights. Assumption of the intrinsic hypothesis is a more efficient alternative in terms of the derivation of a solution. All results obtained for an intrinsic random function can be extended to the particular case of second order stationary random functions by taking advantage of Corollary 4.1. The converse is not possible.

Theorem 4.1

Let $\hat{Z}(\boldsymbol{x}_0)$ be the linear unbiased estimator in Definition 4.1 and let $\gamma(\boldsymbol{h})$ be the semivariogram for the random function. Then the estimation variance $\sigma^2(\boldsymbol{x}_0)$ is equal to

$$\sigma^2(\boldsymbol{x}_0) = 2\sum_{i=1}^{k} \lambda_i \gamma(\boldsymbol{x}_i,\ \boldsymbol{x}_0) - \sum_{i=1}^{k}\sum_{j=1}^{k} \lambda_i \lambda_j \gamma(\boldsymbol{x}_i,\ \boldsymbol{x}_j).$$

Proof

The variance of the estimation error is

$$\sigma^2(\boldsymbol{x}_0) = \mathrm{Var}\left[\hat{Z}(\boldsymbol{x}_0) - Z(\boldsymbol{x}_0)\right].$$

By Definition 4.1

$$\sigma^2(\boldsymbol{x}_0) = \mathrm{Var}\left[\sum_{i=1}^{k} \lambda_i \big(Z(\boldsymbol{x}_i) - Z(\boldsymbol{x}_0)\big)\right]$$

and by Lemma 2.4

$$\sigma^2(\boldsymbol{x}_0) = \sum_{i=1}^{k}\sum_{j=1}^{k}\lambda_i\lambda_j \mathrm{Cov}\big(Z(\boldsymbol{x}_i) - Z(\boldsymbol{x}_0), Z(\boldsymbol{x}_j) - Z(\boldsymbol{x}_0)\big),$$

which by Lemma 4.4 is equivalent to

$$\sigma^2(\boldsymbol{x}_0) = \sum_{i=1}^{k}\sum_{j=1}^{k}\lambda_i\lambda_j\{\gamma(\boldsymbol{x}_i - \boldsymbol{x}_0) + \gamma(\boldsymbol{x}_j - \boldsymbol{x}_0) - \gamma(\boldsymbol{x}_i - \boldsymbol{x}_j)\}.$$

Expanding, because of the constraint $\sum_{i=1}^{k}\lambda_i = 1$,

$$\sigma^2(\boldsymbol{x}_0) = \sum_{i=1}^{k}\lambda_i\gamma(\boldsymbol{x}_i - \boldsymbol{x}_0) + \sum_{j=1}^{k}\lambda_j\gamma(\boldsymbol{x}_j - \boldsymbol{x}_0) - \sum_{i=1}^{k}\sum_{j=1}^{k}\lambda_i\lambda_j\gamma(\boldsymbol{x}_i - \boldsymbol{x}_j).$$

The first two terms in the right-hand side of the equation are the same, which proves the theorem. □

The Optimization Problem

Calculation of the optimal weights is an example of constrained optimization—the objective function is the mean square error in Theorem 4.1 and the constraint is the restriction on the weights in Definition 4.1.

The Lagrange method of multipliers allows the problem to be solved by reducing it to an unconstrained problem by means of a new objective function called the *Lagrangian function* (Hillier and Lieberman, 1995, p. 961). The Lagrangian function results from the addition of a weighted sum of the constraints to the original objective function.

Definition 4.2

Let N be a positive integer, let $\{\alpha_1, \alpha_2, \ldots, \alpha_N\}$ be a set of real or complex numbers, and let $\{\boldsymbol{x}_1, \boldsymbol{x}_2, \ldots, \boldsymbol{x}_N\}$ be a set of points in an n-dimensional Euclidean space. Then the continuous function $\phi(\boldsymbol{x}_i, \boldsymbol{x}_j)$ is said to be negative definite (Itô, 1987, p. 1460) if

$$-\sum_{i=1}^{N}\sum_{j=1}^{N}\alpha_i\alpha_j\phi(\boldsymbol{x}_i, \boldsymbol{x}_j) \geq 0. \quad \square$$

Definition 4.3

Let $\sigma^2(\boldsymbol{x}_0)$ be the estimation variance in Theorem 4.1 for an estimate at $\boldsymbol{x}_0$, let λ_i be the weights in Definition 4.1, and let μ be a Lagrange multiplier. Then

$$L(\lambda_1, \lambda_2, \cdots \lambda_k; \mu) = \sigma^2(\boldsymbol{x}_0) + 2\mu\left(\sum_{i=1}^{k}\lambda_i - 1\right)$$

is the Lagrangian function for ordinary kriging. □

The factor 2 is selected for convenience in the final solution.

Theorem 4.2

Let λ_i be the weights in Definition 4.1, let μ be the Lagrange multiplier in Definition 4.3, and let $\gamma(\boldsymbol{h})$ be the semivariogram for the random function of interest. If the semivariogram is negative definite, then (a) the weights that produce the unique minimum estimation variance are the solution to

$$\left.\begin{array}{l}\sum_{i=1}^{k} \lambda_i \gamma(\boldsymbol{x}_i,\ \boldsymbol{x}_1) - \mu = \gamma(\boldsymbol{x}_1,\ \boldsymbol{x}_0) \\ \sum_{i=1}^{k} \lambda_i \gamma(\boldsymbol{x}_i,\ \boldsymbol{x}_2) - \mu = \gamma(\boldsymbol{x}_2,\ \boldsymbol{x}_0) \\ \cdots\cdots\cdots\cdots\cdots\cdots\cdots\cdots \\ \sum_{i=1}^{k} \lambda_i \gamma(\boldsymbol{x}_k,\ \boldsymbol{x}_2) - \mu = \gamma(\boldsymbol{x}_k,\ \boldsymbol{x}_0) \\ \qquad \sum_{i=1}^{k} \lambda_i = 1 \end{array}\right\}$$

and (b) the kriging variance is never negative.

Proof

(a) Replacing the estimation variance by its value in Theorem 4.1,

$$L(\lambda_1, \lambda_2, \cdots \lambda_k; \mu) = 2 \sum_{i=1}^{k} \lambda_i \gamma(\boldsymbol{x}_i,\ \boldsymbol{x}_0)$$
$$- \sum_{i=1}^{k} \sum_{j=1}^{k} \lambda_i \lambda_j \gamma(\boldsymbol{x}_i,\ \boldsymbol{x}_j) + 2\mu \left(\sum_{i=1}^{k} \lambda_i - 1 \right),$$

one can see that the Lagrangian function is a quadratic expression in the unknown weights and the constraint on the weights is linear. In such a case, the necessary and sufficient condition to have a unique global minimum is that the term

$$- \sum_{i=1}^{k} \sum_{j=1}^{k} \lambda_i \lambda_j \gamma(\boldsymbol{x}_i,\ \boldsymbol{x}_j)$$

must be larger than or equal to zero (Cooper, 1974, p. 65), which by Definition 4.2 is mathematically equivalent to saying that the semivariogram must be negative definite.

The minimum mean square error is given by those weights that make zero all first derivatives of the Lagrangian function with respect to the unknowns:

$$\frac{\partial L(\lambda_1, \lambda_2, \cdots \lambda_k; \mu)}{\partial \lambda_i} = 2\gamma(\boldsymbol{x}_i,\ \boldsymbol{x}_0) - 2\gamma(\boldsymbol{x}_i,\ \boldsymbol{x}_j) + 2\mu, \quad \text{for } i = 1,\ 2,\ \cdots,\ k$$

$$\frac{\partial L(\lambda_1, \lambda_2, \cdots \lambda_k; \mu)}{\partial \mu} = \left(\sum_{i=1}^{k} \lambda_i - 1 \right).$$

The proof follows after making each equation equal to zero, plus, in the first equations, dividing both sides by -2 and adding $\gamma(\boldsymbol{x}_i, \boldsymbol{x}_0)$.

(b) One can rewrite Theorem 4.1 as

$$\sigma^2(\boldsymbol{x}_0) = -\sum_{i=0}^{k}\sum_{j=0}^{k} \lambda_i \lambda_j \gamma(\boldsymbol{x}_i,\ \boldsymbol{x}_j),$$

with $\lambda_0 = -1$. The proof follows directly from the negative definite condition. □

Minimum Mean Square Error

The expression for the estimation variance given by Theorem 4.1 is valid for any weights λ_i. The next lemma and theorem provide the value of the estimation variance for the special case of the optimal weights.

Lemma 4.5

Let λ_i be the optimal weights for the ordinary kriging estimator in Definition 4.1 and let μ be the Lagrange multiplier in Definition 4.3. If $\gamma(\boldsymbol{h})$ is the semivariogram for the random function of interest, then

$$\sum_{i=1}^{k}\sum_{j=1}^{k} \lambda_i \lambda_j \gamma(\boldsymbol{x}_i,\ \boldsymbol{x}_j) = \sum_{i=1}^{k} \lambda_i \gamma(\boldsymbol{x}_i,\ \boldsymbol{x}_0) + \mu\ .$$

Proof

From Theorem 4.2,

$$\sum_{j=1}^{k} \lambda_j \gamma(\boldsymbol{x}_i,\ \boldsymbol{x}_j) = \gamma(\boldsymbol{x}_i,\ \boldsymbol{x}_0) + \mu\,, \quad \text{for } i = 1,\ 2,\ \ldots,\ k\ .$$

Each of the k expressions is still true if one multiplies both sides of the ith equation by λ_i. Summing them up all together,

$$\sum_{i=1}^{k} \lambda_i \sum_{j=1}^{k} \lambda_j \gamma(\boldsymbol{x}_i,\ \boldsymbol{x}_j) = \sum_{i=1}^{k} \lambda_i \gamma(\boldsymbol{x}_i,\ \boldsymbol{x}_0) + \mu \sum_{i=1}^{k} \lambda_i\ .$$

The proof follows by remembering that by Definition 4.1 the sum of the weights is 1, and by introducing λ_i under the second summation. □

Theorem 4.3

Let λ_i be the optimal weights for the ordinary kriging estimator in Definition 4.1 and let μ be the Lagrange multiplier in Definition 4.3. If $\gamma(\boldsymbol{h})$ is the semivariogram for the random function of interest, then the minimum mean square error for ordinary kriging $\sigma^2_{OKI}(\boldsymbol{x}_0)$ is:

$$\sigma^2_{OKI}(\boldsymbol{x}_0) = \sum_{i=i}^{k} \lambda_i \gamma(\boldsymbol{x}_0,\, \boldsymbol{x}_i) - \mu\,.$$

Proof

From Theorem 4.1,

$$\sigma^2(\boldsymbol{x}_0) = 2\sum_{i=1}^{k} \lambda_i \gamma(\boldsymbol{x}_i,\, \boldsymbol{x}_0) - \sum_{i=1}^{k}\sum_{j=1}^{k} \lambda_i \lambda_j \gamma(\boldsymbol{x}_i,\, \boldsymbol{x}_j),$$

but by Lemma 4.5,

$$\sigma^2_{OKI}(\boldsymbol{x}_0) = 2\sum_{i=1}^{k} \lambda_i \gamma(\boldsymbol{x}_i,\, \boldsymbol{x}_0) - \sum_{i=1}^{k} \lambda_i \gamma(\boldsymbol{x}_i,\, \boldsymbol{x}_0) - \mu\,.$$

The addition of the first two terms on the right-hand side of the equation completes the proof. □

Algorithm for Intrinsic Random Functions

For more compact display of the results let us define some matrices.

Definition 4.4

Let $\boldsymbol{x}_i$'s be the sampling sites of a sample subset of size k, $i = 1,\, 2,\, \ldots,\, k$ and let $\gamma(\boldsymbol{x}_i,\, \boldsymbol{x}_j)$'s be the semivariogram. Then $\mathbf{G}$ is the matrix

$$\mathbf{G} = \begin{bmatrix} \gamma(\boldsymbol{x}_1,\, \boldsymbol{x}_1) & \gamma(\boldsymbol{x}_2,\, \boldsymbol{x}_1) & \cdots & \gamma(\boldsymbol{x}_k,\, \boldsymbol{x}_1) & 1 \\ \gamma(\boldsymbol{x}_1,\, \boldsymbol{x}_2) & \gamma(\boldsymbol{x}_2,\, \boldsymbol{x}_2) & \cdots & \gamma(\boldsymbol{x}_k,\, \boldsymbol{x}_2) & 1 \\ \cdots & \cdots & \cdots & \cdots & \cdots \\ \gamma(\boldsymbol{x}_1,\, \boldsymbol{x}_k) & \gamma(\boldsymbol{x}_2,\, \boldsymbol{x}_k) & \cdots & \gamma(\boldsymbol{x}_k,\, \boldsymbol{x}_k) & 1 \\ 1 & 1 & \cdots & 1 & 0 \end{bmatrix}. \quad \square$$

Definition 4.5

Let λ_i's be the optimal weights for the estimator in Definition 2.1 and let the prime stand for the transpose of the matrix. Then $\mathbf{W}$ is the matrix

$$\mathbf{W} = [\, \lambda_1 \quad \lambda_2 \quad \cdots \quad \lambda_k \quad -\mu\,]'. \quad \square$$

Definition 4.6

Let $\gamma(\cdot)$ be the semivariogram of the random function, $\boldsymbol{x}_0$ be the estimation location, and the x_i's be sampling sites of a sample subset of size k, $i = 1,\, 2,\, \ldots,\, k$. Then $\mathbf{g}$ is the matrix

$$\mathbf{g} = [\, \gamma(\boldsymbol{x}_0,\, \boldsymbol{x}_1) \quad \gamma(\boldsymbol{x}_0,\, \boldsymbol{x}_2) \quad \cdots \quad \gamma(\boldsymbol{x}_0,\, \boldsymbol{x}_k) \quad 1\,]'. \quad \square$$

Definition 4.7

Let $Z(\boldsymbol{x}_i)$ be random variables of a random function sampled at sites $\boldsymbol{x}_i$, $i = 1, 2, \ldots, k$. Then the matrix $\mathbf{Z}$ is

$$\mathbf{Z} = [\, Z(\boldsymbol{x}_1) \quad Z(\boldsymbol{x}_2) \quad \cdots \quad Z(\boldsymbol{x}_k) \quad 0]' . \ \square$$

Algorithm 4.1

This is an algorithm for ordinary kriging estimation of an attribute honoring Assumptions 4.3 and 4.4.

1. Calculate each term in matrix $\mathbf{G}$ in Definition 4.4.
2. Calculate each term in matrix g in Definition 4.6.
3. Solve the system of equations

$$\mathbf{GW} = \mathbf{g},$$

where $\mathbf{W}$ is the matrix in Definition 4.5.

4. Compute the estimate

$$\hat{Z}_{OKI}(\boldsymbol{x}_0) = \mathbf{Z}'\mathbf{W} = \mathbf{Z}'\mathbf{G}^{-1}\mathbf{g},$$

where $\mathbf{Z}$ is the matrix in Definition 4.7.

5. Calculate the ordinary kriging estimation variance

$$\sigma^2_{OKI}(\boldsymbol{x}_0) = \mathbf{g}'\mathbf{W} = \mathbf{g}'\mathbf{G}^{-1}\mathbf{g}. \ \square$$

Second Order Stationary Ordinary Kriging

Using Corollary 4.1, findings obtained assuming the intrinsic hypothesis can be translated into equivalent results assuming second order stationarity. I will limit the conversion to the final and most relevant results only.

From Definition 4.3, Theorem 4.1, and Corollary 4.1, the Lagrangian function changes to:

$$L(\lambda_1, \lambda_2, \cdots \lambda_k; \mu) = \mathrm{Cov}(0) + \sum_{i=1}^{k}\sum_{j=1}^{k} \lambda_i\lambda_j \mathrm{Cov}(\boldsymbol{x}_i, \boldsymbol{x}_j)$$
$$- 2\sum_{i=1}^{k} \lambda_i \mathrm{Cov}(\boldsymbol{x}_i,\ \boldsymbol{x}_0) + 2\mu\left(\sum_{i=1}^{k}\lambda_i - 1\right).$$

$\mathrm{Cov}(\cdot)$ must be positive definite (Definition 2.3) for the system to have all the favorable characteristics of a quadratic system.

Theorem 4.4

Let λ_i be the weights in Definition 4.1, let μ be the Lagrange multiplier in Definition 4.3, and let $\operatorname{Cov}(\boldsymbol{h})$ be the covariance for a second order stationary random function. Then the weights that produce the minimum estimation variance are the solution to

$$\left.\begin{array}{r}\sum_{i=1}^{k}\lambda_i \operatorname{Cov}(\boldsymbol{x}_i,\ \boldsymbol{x}_1)+\mu\ =\operatorname{Cov}(\boldsymbol{x}_1,\ \boldsymbol{x}_0)\\ \sum_{i=1}^{k}\lambda_i \operatorname{Cov}(\boldsymbol{x}_i,\ \boldsymbol{x}_2)+\mu\ =\operatorname{Cov}(\boldsymbol{x}_2,\ \boldsymbol{x}_0)\\ \cdots\cdots\cdots\cdots\cdots\cdots\cdots\cdots\\ \sum_{i=1}^{k}\lambda_i \operatorname{Cov}(\boldsymbol{x}_i,\ \boldsymbol{x}_k)+\mu\ =\operatorname{Cov}(\boldsymbol{x}_k,\ \boldsymbol{x}_0)\\ \sum_{i=1}^{k}\lambda_i=1\end{array}\right\}.$$

Proof

From Theorem 4.2,

$$\left.\begin{array}{r}\sum_{i=1}^{k}\lambda_i \gamma(\boldsymbol{x}_i,\ \boldsymbol{x}_1)-\mu\ =\gamma(\boldsymbol{x}_1,\ \boldsymbol{x}_0)\\ \sum_{i=1}^{k}\lambda_i \gamma(\boldsymbol{x}_i,\ \boldsymbol{x}_2)-\mu\ =\gamma(\boldsymbol{x}_2,\ \boldsymbol{x}_0)\\ \cdots\cdots\cdots\cdots\cdots\cdots\cdots\cdots\\ \sum_{i=1}^{k}\lambda_i \gamma(\boldsymbol{x}_i,\ \boldsymbol{x}_k)-\mu\ =\gamma(\boldsymbol{x}_k,\ \boldsymbol{x}_0)\\ \sum_{i=1}^{k}\lambda_i=1\end{array}\right\}$$

and from Corollary 4.1,

$$\left.\begin{array}{r}\sum_{i=1}^{k}\lambda_i\left\{\operatorname{Cov}(0)-\operatorname{Cov}(\boldsymbol{x}_i,\ \boldsymbol{x}_1)\right\}-\mu\ =\operatorname{Cov}(0)-\operatorname{Cov}(\boldsymbol{x}_1,\ \boldsymbol{x}_0)\\ \sum_{i=1}^{k}\lambda_i\left\{\operatorname{Cov}(0)-\operatorname{Cov}(\boldsymbol{x}_i,\boldsymbol{x}_1)\right\}-\mu\ =\operatorname{Cov}(0)-\operatorname{Cov}(\boldsymbol{x}_2,\ \boldsymbol{x}_0)\\ \cdots\cdots\cdots\cdots\cdots\cdots\cdots\cdots\cdots\cdots\cdots\cdots\\ \sum_{i=1}^{k}\lambda_i\left\{\operatorname{Cov}(0)-\operatorname{Cov}(\boldsymbol{x}_i,\boldsymbol{x}_1)\right\}-\mu\ =\operatorname{Cov}(0)-\operatorname{Cov}(\boldsymbol{x}_k,\ \boldsymbol{x}_0)\\ \sum_{i=1}^{k}\lambda_i=1\end{array}\right\};$$

expanding the summation:

$$\left.\begin{array}{r}\sum_{i=1}^{k}\lambda_i \operatorname{Cov}(0)-\sum_{i=1}^{k}\lambda_i \operatorname{Cov}(\boldsymbol{x}_i,\ \boldsymbol{x}_1)-\mu\ =\operatorname{Cov}(0)-\operatorname{Cov}(\boldsymbol{x}_1,\ \boldsymbol{x}_0)\\ \sum_{i=1}^{k}\lambda_i \operatorname{Cov}(0)-\sum_{i=1}^{k}\lambda_i \operatorname{Cov}(\boldsymbol{x}_i,\ \boldsymbol{x}_2)-\mu\ =\operatorname{Cov}(0)-\operatorname{Cov}(\boldsymbol{x}_2,\ \boldsymbol{x}_0)\\ \cdots\cdots\cdots\cdots\cdots\cdots\cdots\cdots\cdots\cdots\cdots\cdots\\ \sum_{i=1}^{k}\lambda_i \operatorname{Cov}(0)-\sum_{i=1}^{k}\lambda_i \operatorname{Cov}(\boldsymbol{x}_i,\ \boldsymbol{x}_k)-\mu\ =\operatorname{Cov}(0)-\operatorname{Cov}(\boldsymbol{x}_k,\ \boldsymbol{x}_0)\\ \sum_{i=1}^{k}\lambda_i=1\end{array}\right\}.$$

According to the constraint

$$\sum_{i=1}^{k} \lambda_i = 1$$

in Definition 4.1, in the first k equations the first terms on each side of the equations are equal, thus they cancel out. The proof follows after multiplying both sides of the first k equations by -1. □

Theorem 4.5

Let λ_i be the optimal weights for the ordinary kriging estimator in Definition 4.1 and let μ be the Lagrange multiplier in Definition 4.3. If $\mathrm{Cov}(\boldsymbol{h})$ is the covariance for a second order stationary random function, then the minimum mean square error for ordinary kriging $\sigma^2_{OK}(\boldsymbol{x}_0)$ is

$$\sigma^2_{OK}(\boldsymbol{x}_0) = \mathrm{Cov}(0) - \sum_{i=i}^{k} \lambda_i \, \mathrm{Cov}(\boldsymbol{x}_0,\, \boldsymbol{x}_i) - \mu \, .$$

Proof

From Theorem 4.3

$$\sigma^2_{OK}(\boldsymbol{x}_0) = \sum_{i=i}^{k} \lambda_i \, \gamma(\boldsymbol{x}_0,\, \boldsymbol{x}_i) - \mu,$$

but by Corollary 4.1,

$$\sigma^2_{OK}(\boldsymbol{x}_0) = \sum_{i=i}^{k} \lambda_i \{\mathrm{Cov}(0) - \mathrm{Cov}(\boldsymbol{x}_0,\, \boldsymbol{x}_i)\} - \mu \, .$$

Expanding the summation,

$$\sigma^2_{OK}(\boldsymbol{x}_0) = \sum_{i=i}^{k} \lambda_i \, \mathrm{Cov}(0) - \sum_{i=i}^{k} \lambda_i \, \mathrm{Cov}(\boldsymbol{x}_0,\, \boldsymbol{x}_i) - \mu$$

and the proof follows by the constraint on the weights in Definition 4.1. □

If, in addition, one defines three more matrices, a second algorithm can be formulated.

Definition 4.8

Let $\boldsymbol{x}_i$'s be the sampling sites of a sample subset of size k, $i = 1, 2, \ldots, k$ and let $\mathrm{Cov}(\boldsymbol{x}_i,\, \boldsymbol{x}_j)$ be the covariance. Then $\mathbf{V}$ is the matrix

$$\mathbf{V} = \begin{bmatrix} \mathrm{Cov}(\boldsymbol{x}_1,\, \boldsymbol{x}_1) & \mathrm{Cov}(\boldsymbol{x}_2,\, \boldsymbol{x}_1) & \cdots & \mathrm{Cov}(\boldsymbol{x}_k,\, \boldsymbol{x}_1) & 1 \\ \mathrm{Cov}(\boldsymbol{x}_1,\, \boldsymbol{x}_2) & \mathrm{Cov}(\boldsymbol{x}_2,\, \boldsymbol{x}_2) & \cdots & \mathrm{Cov}(\boldsymbol{x}_k,\, \boldsymbol{x}_2) & 1 \\ \cdots & \cdots & \cdots & \cdots & \cdots \\ \mathrm{Cov}(\boldsymbol{x}_1,\, \boldsymbol{x}_k) & \mathrm{Cov}(\boldsymbol{x}_2,\, \boldsymbol{x}_k) & \cdots & \mathrm{Cov}(\boldsymbol{x}_k,\, \boldsymbol{x}_k) & 1 \\ 1 & 1 & \cdots & 1 & 0 \end{bmatrix} . \quad \square$$

Definition 4.9

Let λ_i's be the optimal weights for the estimator in Definition 2.1 and let the prime stand for the transpose of the matrix. Then **L** is the matrix

$$\mathbf{L} = [\ \lambda_1 \quad \lambda_2 \quad \cdots \quad \lambda_k \quad \mu\,]'. \ \square$$

Definition 4.10

Let $\text{Cov}(\boldsymbol{x}_i,\ \boldsymbol{x}_j)$ be the covariance of a second order stationary random function, $\boldsymbol{x}_0$ be the estimation location, and the $\boldsymbol{x}_i$'s be sampling sites of a sample subset of size k, $i = 1,\ 2,\ \ldots,\ k$. Then **v** is the matrix

$$\mathbf{v} = [\ \text{Cov}(\boldsymbol{x}_0,\ \boldsymbol{x}_1) \quad \text{Cov}(\boldsymbol{x}_0,\ \boldsymbol{x}_2) \quad \cdots \quad \text{Cov}(\boldsymbol{x}_0,\ \boldsymbol{x}_k) \quad 1\,]'. \ \square$$

Algorithm 4.2

This is an algorithm for the ordinary kriging estimation of attributes honoring Assumptions 4.1 and 4.2.

1. Calculate each term in matrix **V** in Definition 4.8.
2. Calculate each term in matrix **v** in Definition 4.10.
3. Solve the system of equations

$$\mathbf{VL} = \mathbf{v}$$

where **L** is the matrix in Definition 4.9.

4. Compute the estimate

$$\hat{Z}_{OK}(\boldsymbol{x}_0) = \mathbf{Z}'\mathbf{L} = \mathbf{Z}'\mathbf{V}^{-1}\mathbf{v}$$

where **Z** is the matrix in Definition 4.7.

5. Calculate the ordinary kriging estimation variance

$$\sigma^2_{OK}(\boldsymbol{x}_0) = \text{Cov}(0) - \mathbf{v}'\mathbf{L} = \text{Cov}(0) - \mathbf{v}'\mathbf{V}^{-1}\mathbf{v}. \ \square$$

In practice it may be seen that Algorithm 4.2 is implemented more widely than Algorithm 4.1, despite the fact that the latter is more general. There are two reasons for this:

i) processes for which there is a semivariogram and no covariance are of rare and limited interest, and

ii) the system of equations in Algorithm 4.2 is better behaved than the algorithm for intrinsic random functions. In Algorithm 4.2, except for the last row, the largest elements in matrix **V** are along the main diagonal, while all the elements in matrix **G** of Algorithm 4.1 are zero.

EXERCISE 4.1

Repeat Exercise 2.1 using ordinary kriging and make comparisons. Given the sampling

Index	Easting	Northing	Measurement
1	10	20	40
2	30	280	130
3	250	130	90
4	360	120	160

and an estimation location $\boldsymbol{x}_0 = (180, 120)$, and covariance $\text{Cov}\,(\boldsymbol{h}) = 2000e^{-\frac{h}{250}}$:

1. Calculate the estimation weights.
2. What comment can you make about the estimation weights?
3. Find the estimate.
4. Compute the kriging estimation variance.

Solution

1. The matrix of distances $\mathbf{D}$ among the observations is

$$\mathbf{D} = \begin{bmatrix} 0 & & & \\ 260.8 & 0 & & \\ 264.0 & 266.3 & 0 & \\ 364.0 & 366.7 & 110.4 & 0 \end{bmatrix},$$

thus, from Algorithm 4.2, the covariance matrix is

$$\mathbf{V} = \begin{bmatrix} 2000 & & & & \\ 704.8 & 2000 & & & \\ 695.6 & 689.4 & 2000 & & \\ 466.4 & 461.2 & 1285.8 & 2000 & \\ 1 & 1 & 1 & 1 & 0 \end{bmatrix}.$$

The distance vector from the sampling sites to the estimation location is

$$\mathbf{d} = [\,197.2 \quad 219.3 \quad 70.7 \quad 180.0\,]',$$

hence,

$$\mathbf{v} = [\,908.7 \quad 831.8 \quad 1507.2 \quad 973.6 \quad 1\,]'.$$

Therefore, the weights are

$$\mathbf{L} = \begin{bmatrix} 2000 & & & & \\ 704.8 & 2000 & & & \\ 695.6 & 689.4 & 2000 & & \\ 466.4 & 461.2 & 1285.8 & 2000 & \\ 1 & 1 & 1 & 1 & 0 \end{bmatrix}^{-1} \begin{bmatrix} 908.7 \\ 831.8 \\ 1507.2 \\ 973.6 \\ 1 \end{bmatrix}$$

$$= [\,0.197 \quad 0.141 \quad 0.650 \quad 0.011 \quad -42.714\,]'.$$

2. (a) The weights are similar in value and have the same rank as those obtained in Exercise 2.1 using simple kriging. Their sum, however, is 0.999 instead of 0.958 as in Exercise 2.1. The small deviation from the theoretical value of 1.0 is due to truncations in the calculations.

(b) The screen effect, although less pronounced, is still present.

3. From Step 4 in Algorithm 4.2,

$$\hat{z}_{OK}(\boldsymbol{x}_0) = [40 \quad 130 \quad 90 \quad 160 \quad 0][0.197 \quad 0.141 \quad 0.650 \quad 0.011 \quad -42.714]'$$

$$= 86.6,$$

which is a variation of approximately 0.1% with respect to the estimator in Exercise 2.1.

4. From Step 5 in Algorithm 4.2,

$$\sigma^2_{OK}(\boldsymbol{x}_0) = 2000$$

$$- [908.6 \quad 831.8 \quad 1507.2 \quad 973.6 \quad 1]' [0.197 \quad 0.141 \quad 0.650 \quad 0.011 \quad -42.714]$$

$$= 754.8,$$

which is 1.9 units larger than $\sigma^2_{SK}(\boldsymbol{x}_0)$ in Exercise 2.1.

Properties

Simple kriging shares most of its properties with ordinary kriging. It goes without proof that ordinary kriging has the following properties:

i) minimum mean square error,
ii) estimation interval not restricted to the data interval,
iii) screen effect,
iv) declustering ability,
v) exact interpolation with zero kriging variance,
vi) inability to handle duplicate sampling sites,
vii) independence from translation of the reference axes,
viii) dependence upon the sampling pattern, and
ix) independence of the kriging variance from individual observations.

The first two properties do not need proof—they follow from the design of the estimator. As proved in Theorem 4.2, ordinary kriging is the minimum estimation variance estimator. The unbounded nature of the estimate derives from Definition 4.1, which does not restrict the value of the weights to the interval (0, 1). The screen effect and the declustering ability are only illustrated through numerical examples. Proofs of properties (v) through (ix) are similar to those for simple kriging and are left as an exercise for the reader. The proofs for other, less evident properties are given below.

Lemma 4.6

Let λ_i be the weights for the estimator in Definition 4.1. Then the estimator is unbiased if and only if

$$\sum_{i=1}^{k} \lambda_i = 1\,.$$

Proof

(a) If the estimator is unbiased,

$$\mathrm{E}\big[\hat{Z}\left(\boldsymbol{x}_0\right)\big] = \mathrm{E}\big[Z\left(\boldsymbol{x}_0\right)\big].$$

By Definition 4.1, exchanging summation and expectation,

$$\sum_{i=1}^{k} \mathrm{E}\big[Z\left(\boldsymbol{x}_i\right)\big]\lambda_i = \mathrm{E}\big[Z\left(\boldsymbol{x}_0\right)\big]$$

and the expected values cancel out because of Assumption 4.4.

(b) The constraint $\sum_{i=1}^{k} \lambda_i = 1$ does not change when both sides of the equation are multiplied by the constant value of the mean

$$\sum_{i=1}^{k} m\lambda_i = m$$

in Assumption 4.4, which by the same assumption is equal to

$$\sum_{i=1}^{k} \mathrm{E}\big[Z\left(\boldsymbol{x}_i\right)\big]\lambda_i = \mathrm{E}\big[Z\left(\boldsymbol{x}_0\right)\big].$$

Introducing the weight inside the expectation and exchanging summation and expectation,

$$\mathrm{E}\left[\sum_{i=1}^{k} \lambda_i Z\left(\boldsymbol{x}_i\right)\right] = \mathrm{E}\big[Z\left(\boldsymbol{x}_0\right)\big]$$

but, by Definition 4.1, the expression in the first expectation is the ordinary kriging estimator

$$\mathrm{E}\big[\hat{Z}\left(\boldsymbol{x}_0\right)\big] = \mathrm{E}\big[Z\left(\boldsymbol{x}_0\right)\big],$$

which is the condition for an estimator to be unbiased. The two implications prove the lemma. □

Hence, the constraint on the weights serves the double purpose of making the estimator in Definition 2.1 independent of the mean and making the new estimator unbiased.

Corollary 4.2

Changes of scale in the semivariogram do not change the estimate.

Proof

From Theorem 4.2, multiplication of the original semivariogram by a constant α will produce the following new system of equations:

$$\left.\begin{array}{c}\sum\limits_{i=1}^{k}\lambda_i\alpha\gamma(\boldsymbol{x}_i,\ \boldsymbol{x}_1)-\mu'=\alpha\gamma(\boldsymbol{x}_1,\ \boldsymbol{x}_0)\\ \sum\limits_{i=1}^{k}\lambda_i\alpha\gamma(\boldsymbol{x}_i,\ \boldsymbol{x}_2)-\mu'=\alpha\gamma(\boldsymbol{x}_2,\ \boldsymbol{x}_0)\\ \cdots\cdots\cdots\cdots\cdots\cdots\cdots\cdots\cdots\\ \sum\limits_{i=1}^{k}\lambda_i\alpha\gamma(\boldsymbol{x}_k,\ \boldsymbol{x}_2)-\mu'=\alpha\gamma(\boldsymbol{x}_k,\ \boldsymbol{x}_0)\\ \sum\limits_{i=1}^{k}\lambda_i=1\end{array}\right\}$$

which, after dividing the first k equations by α and defining $\mu=\mu'/\alpha$, turns into

$$\left.\begin{array}{c}\sum\limits_{i=1}^{k}\lambda_i\gamma(\boldsymbol{x}_i,\ \boldsymbol{x}_1)-\mu=\gamma(\boldsymbol{x}_1,\ \boldsymbol{x}_0)\\ \sum\limits_{i=1}^{k}\lambda_i\gamma(\boldsymbol{x}_i,\ \boldsymbol{x}_2)-\mu=\gamma(\boldsymbol{x}_2,\ \boldsymbol{x}_0)\\ \cdots\cdots\cdots\cdots\cdots\cdots\cdots\cdots\cdots\\ \sum\limits_{i=1}^{k}\lambda_i\gamma(\boldsymbol{x}_k,\ \boldsymbol{x}_2)-\mu=\gamma(\boldsymbol{x}_k,\ \boldsymbol{x}_0)\\ \sum\limits_{i=1}^{k}\lambda_i=1\end{array}\right\},$$

which is exactly the same original system of equations and, hence, its solution. If the weights are invariable to a scale change in the semivariogram, so is the estimator in Definition 4.1. □

Corollary 4.3

Multiplication of the semivariogram by a factor will change the estimation variance by the same factor.

Proof

Let $\sigma^2_{OK}(\boldsymbol{x}_0)$ be the original estimation variance and let α be the factor. By Corollary 4.2, the weights are invariable to a change in the semivariogram by a factor. Then, by Theorem 4.3, the new estimation variance $\sigma^2_{OK}(\boldsymbol{x}_0)$ is

$$\sigma^2_{OK}{}'(\boldsymbol{x}_0)=\sum_{i=i}^{k}\lambda_i\alpha\gamma(\boldsymbol{x}_0,\ \boldsymbol{x}_i)-\mu'.$$

Factoring by α,

$$\sigma^2_{OK}{}'(\boldsymbol{x}_0)=\alpha\left\{\sum_{i=i}^{k}\lambda_i\gamma(\boldsymbol{x}_0,\ \boldsymbol{x}_i)-\frac{\mu'}{\alpha}\right\}.$$

But, according to Corollary 4.2, μ'/α is exactly equal to the Lagrange multiplier for the original system of equations. Consequently, by Theorem 4.3,

$$\sigma_{OK}^{2}{}'(\boldsymbol{x}_0) = \alpha\sigma_{OK}^2(\boldsymbol{x}_0). \quad \square$$

Corollary 4.4

Consider a sampling where all random variables are independent from one another. If the random function has a semivariogram $\gamma(\cdot)$ that is equal to a constant C for any pair of independent random variables $Z(\boldsymbol{x})$ and $Z(\boldsymbol{x}+\boldsymbol{h})$, then

$$\mu = \frac{(k-1)C - \sum_{i=1}^{k}\gamma(\boldsymbol{x}_i,\, \boldsymbol{x}_0)}{k}.$$

Proof

Because the variables are independent, for $i > 0$,

$$\gamma(\boldsymbol{x}_i,\, \boldsymbol{x}_j) = \begin{cases} 0, & \text{for } i = j \\ C, & \text{otherwise}. \end{cases}$$

In this special case, Theorem 4.2 turns into

$$\left.\begin{array}{l} \sum_{i=2}^{k}\lambda_i C - \mu = \gamma(\boldsymbol{x}_1,\, \boldsymbol{x}_0) \\ C + \sum_{i=3}^{k}\lambda_i C - \mu = \gamma(\boldsymbol{x}_2,\, \boldsymbol{x}_0) \\ \cdots\cdots\cdots\cdots\cdots\cdots\cdots \\ \sum_{i=1}^{k-1}\lambda_i C - \mu = \gamma(\boldsymbol{x}_k,\, \boldsymbol{x}_0) \\ \qquad \sum_{i=1}^{k}\lambda_i = 1 \end{array}\right\}.$$

Adding the first k equations, one has

$$(k-1)C - k\mu = \sum_{i=1}^{k}\gamma(\boldsymbol{x}_i,\, \boldsymbol{x}_0)$$

and the proof follows by solving the equation in terms of the Lagrange multiplier. $\square$

Corollary 4.5

Consider a sampling where all random variables are independent from one another. Let the random function have a semivariogram $\gamma(\cdot)$, let λ_i be an ordinary kriging weight in Definition 4.1, and let μ be the Lagrange multiplier in Corollary 4.4. Then

$$\lambda_i = 1 - \frac{\gamma(\boldsymbol{x}_i,\, \boldsymbol{x}_0) + \mu}{C}, \qquad i = 1,\, 2,\, \cdots,\, k.$$

Proof

Because the variables are independent, for $i > 0$,

$$\gamma(\boldsymbol{x}_i, \boldsymbol{x}_j) = \begin{cases} 0, & \text{for } i = j \\ C, & \text{otherwise .} \end{cases}$$

In this special case, Theorem 4.2 turns into

$$\left.\begin{array}{l} \sum_{i=2}^{k} \lambda_i C - \mu = \gamma(\boldsymbol{x}_1, \boldsymbol{x}_0) \\ C + \sum_{i=3}^{k} \lambda_i C - \mu = \gamma(\boldsymbol{x}_2, \boldsymbol{x}_0) \\ \cdots\cdots\cdots\cdots\cdots\cdots\cdots\cdots \\ \sum_{i=1}^{k-1} \lambda_i C - \mu = \gamma(\boldsymbol{x}_k, \boldsymbol{x}_0) \\ \qquad \sum_{i=1}^{k} \lambda_i = 1 \end{array}\right\} .$$

From the last equation, $1 - \sum_{j=1}^{k;j \neq i} \lambda_j = \lambda_i$. Therefore, from the ith equation,

$$(1 - \lambda_i)\, C - \mu = \gamma(\boldsymbol{x}_i, \boldsymbol{x}_0)$$

and the proof follows by solving the equation in terms of the weight. □

Corollary 4.6

Consider a sampling where all random variables are independent from one another. Let the random function have a semivariogram $\gamma(\cdot)$ and let μ be the Lagrange multiplier in Corollary 4.4. Then

$$\sigma_{OK}^2(\boldsymbol{x}_0) = \sum_{i=i}^{k} \left\{ 1 - \frac{\gamma(\boldsymbol{x}_i, \boldsymbol{x}_0)}{C} \right\} \gamma(\boldsymbol{x}_0, \boldsymbol{x}_i) - \mu \left\{ 1 + \sum_{i=1}^{k} \frac{\gamma(\boldsymbol{x}_0, \boldsymbol{x}_i)}{C} \right\} .$$

Proof

From Corollary 4.5 and Theorem 4.3,

$$\sigma_{OK}^2(\boldsymbol{x}_0) = \sum_{i=i}^{k} \left\{ 1 - \frac{\gamma(\boldsymbol{x}_i, \boldsymbol{x}_0) + \mu}{C} \right\} \gamma(\boldsymbol{x}_0, \boldsymbol{x}_i) - \mu$$

and the proof follows by removing the third term in the first summation and factoring by μ. □

Corollary 4.7

Consider a sampling where all random variables and the estimator are independent from one another. If λ_i is an ordinary kriging weight in Definition 4.1 and k is the number of observations in the same definition, then

$$\lambda_i = \frac{1}{k}, \quad i = 1,\ 2,\ \cdots,\ k\,.$$

Proof

Because all the variables are independent, from Corollary 4.4,

$$\mu = -\frac{C}{k}\,.$$

For the same reason, from Corollary 4.5,

$$\lambda_i = -\frac{\mu}{C}, \quad i = 1,\ 2,\ \cdots,\ k$$

and the proof follows by replacing the value of μ in the expression for λ_i. □

Corollary 4.8

Consider a sampling where all random variables and the estimator are independent from one another. If k is the number of observations in Definition 4.1 and C is the value of the semivariogram for those pairs of independent variables, then

$$\sigma^2_{OK}(\boldsymbol{x}_0) = \frac{(k+1)}{k}\,C\,.$$

Proof

Because all the variables are independent, $\gamma(\boldsymbol{x}_0, \boldsymbol{x}_i) = C$ for $i = 1, 2, \cdots, k$. From Corollary 4.6,

$$\sigma^2_{OK}(\boldsymbol{x}_0) = -\mu\,\{1 + k\}$$

and from Corollary 4.4,

$$\mu = -\frac{C}{k}\,.$$

The proof follows by replacing the value of μ in the expression for $\sigma^2_{OK}(\boldsymbol{x}_0)$. □

Corollary 4.9

Consider a sampling where all random variables are independent from one another and with respect to the estimator. Then the estimator $\hat{Z}(\boldsymbol{x}_0)$ is equal to the estimator for the local mean in the area comprising the k observations.

Proof

From Definition 4.1 and Corollary 4.7,

$$\hat{Z}(\boldsymbol{x}_0) = \frac{1}{k}\sum_{i=1}^{k} Z(\boldsymbol{x}_i)\,,$$

which is an unbiased estimator for the mean of the observations considered in the estimation. □

Relating Simple and Ordinary Kriging

Let us try to quantify how the constraint on the weights makes simple kriging different than ordinary kriging.

The key to all proofs is to put the ordinary kriging Algorithm 4.2 in terms of the matrices defined for Algorithm 2.1.

Lemma 4.7

Let $\mathbf{C}$ be the covariance matrix in Definition 2.4, let $\mathbf{c}$ be the vector in Definition 2.6, and let $\mathbf{1}$ be $[1 \quad 1 \quad \ldots \quad 1]'$. Then the Lagrange multiplier in Definition 4.3 is given by

$$\mu = \left(\mathbf{1}'\mathbf{C}^{-1}\mathbf{c} - 1\right)\left(\mathbf{1}'\mathbf{C}^{-1}\mathbf{1}\right)^{-1}$$

Proof

One can rewrite the system of equations in Theorem 4.4 as

$$\left.\begin{aligned} \mathbf{C}\mathbf{X}_{OK} + \mu\mathbf{1} &= \mathbf{c} \\ \mathbf{1}'\mathbf{X}_{OK} &= 1 \end{aligned}\right\}.$$

Then from the first matrix equation it follows that $\mathbf{X}_{OK} = \mathbf{C}^{-1}\left(\mathbf{c} - \mu\mathbf{1}\right)$. Substituting the value of $\mathbf{X}_{OK}$ back into the last equation,

$$\mathbf{1}'\mathbf{C}^{-1}\left(\mathbf{c} - \mu\mathbf{1}\right) = 1$$

and the proof follows by solving the equation in terms of the multiplier. □

Lemma 4.8

Let $\mathbf{C}$ be the covariance matrix in Definition 2.4 and let $\mathbf{c}$ be the vector in Definition 2.6. Then the ordinary kriging weights $\mathbf{X}_{OK}$ in Definition 4.1 are given by

$$\mathbf{X}_{OK} = \mathbf{C}^{-1}\left\{\mathbf{c} - \left(\mathbf{1}'\mathbf{C}^{-1}\mathbf{c} - 1\right)\left(\mathbf{1}'\mathbf{C}^{-1}\mathbf{1}\right)^{-1}\mathbf{1}\right\}.$$

Proof

One can rewrite the system of equations in Theorem 4.4 as

$$\left.\begin{aligned} \mathbf{C}\mathbf{X}_{OK} + \mu\mathbf{1} &= \mathbf{c} \\ \mathbf{1}'\mathbf{X}_{OK} &= 1 \end{aligned}\right\},$$

but from Lemma 4.7,

$$\mathbf{C}\mathbf{X}_{OK} + \left(\mathbf{1}'\mathbf{C}^{-1}\mathbf{c} - 1\right)\left(\mathbf{1}'\mathbf{C}^{-1}\mathbf{1}\right)^{-1}\mathbf{1} = \mathbf{c}$$

and the proof follows by solving the equation in terms of the weights. □

Theorem 4.6

Given a sampling, the difference between the simple kriging and the ordinary kriging weights is

$$\mathbf{X}_{SK} - \mathbf{X}_{OK} = \mathbf{C}^{-1}\left(\mathbf{1}'\mathbf{C}^{-1}\mathbf{c} - 1\right)\left(\mathbf{1}'C^{-1}\mathbf{1}\right)^{-1}\mathbf{1}.$$

Proof

From Algorithm 2.1, Step 3,

$$\mathbf{X}_{SK} = \mathbf{C}^{-1}\mathbf{c}$$

and from Lemma 4.8,

$$\mathbf{X}_{OK} = \mathbf{C}^{-1}\left\{\mathbf{c} - \left(\mathbf{1}'\mathbf{C}^{-1}\mathbf{c} - 1\right)\left(\mathbf{1}'\mathbf{C}^{-1}\mathbf{1}\right)^{-1}\mathbf{1}\right\}.$$

Consequently,

$$\mathbf{X}_{SK} - \mathbf{X}_{OK} = \mathbf{C}^{-1}\left(\mathbf{1}'\mathbf{C}^{-1}\mathbf{c} - 1\right)\left(\mathbf{1}'C^{-1}\mathbf{1}\right)^{-1}\mathbf{1}. \quad \square$$

In the case of the sampling in Exercise 4.1, this difference is equal to

$$[-0.012 \quad -0.013 \quad -0.004 \quad -0.012]'.$$

Lemma 4.9

Let $\mathbf{C}$ be the covariance matrix in Definition 2.4 and let $\mathbf{c}$ be the vector in Definition 2.6. Then the ordinary kriging estimation variance is

$$\sigma_{OK}^2(\boldsymbol{x}_0) = \mathrm{Cov}(0) - \mathbf{c}'\mathbf{C}^{-1}\mathbf{c} + \left(\mathbf{1}'\mathbf{C}^{-1}\mathbf{c} - 1\right)^2\left(\mathbf{1}'C^{-1}\mathbf{1}\right)^{-1}.$$

Proof

From Step 5 in Algorithm 4.2,

$$\sigma_{OK}^2(\boldsymbol{x}_0) = \mathrm{Cov}(0) - (\mathbf{c}', \mathbf{1})'\left(\mathbf{X}_{OK}', \mu\right).$$

Expanding,

$$\sigma_{OK}^2(\boldsymbol{x}_0) = \mathrm{Cov}(0) - \mathbf{c}'\mathbf{X}_{OK} - \mu$$

and by Lemmas 4.7 and 4.8,

$$\sigma_{OK}^2(\boldsymbol{x}_0) = \mathrm{Cov}(0) - \mathbf{c}'\mathbf{C}^{-1}\left\{\mathbf{c} - \left(\mathbf{1}'\mathbf{C}^{-1}\mathbf{c} - 1\right)\left(\mathbf{1}'\mathbf{C}^{-1}\mathbf{1}\right)^{-1}\mathbf{1}\right\}$$

$$-\left(\mathbf{1}'\mathbf{C}^{-1}\mathbf{c} - 1\right)\left(\mathbf{1}'\mathbf{C}^{-1}\mathbf{1}\right)^{-1}.$$

Expanding,

$$\sigma^2_{OK}(\boldsymbol{x}_0) = \text{Cov}(0) - \mathbf{c}'\mathbf{C}^{-1}\mathbf{c} + \mathbf{c}'\mathbf{C}^{-1}\left(\mathbf{1}'\mathbf{C}^{-1}\mathbf{c} - 1\right)\left(\mathbf{1}'\mathbf{C}^{-1}\mathbf{1}\right)^{-1}\mathbf{1} \\ - \left(\mathbf{1}'\mathbf{C}^{-1}\mathbf{c} - 1\right)\left(\mathbf{1}'\mathbf{C}^{-1}\mathbf{1}\right)^{-1}$$

and factoring by $\left(\mathbf{1}'\mathbf{C}^{-1}\mathbf{c} - 1\right)\left(\mathbf{1}'\mathbf{C}^{-1}\mathbf{1}\right)^{-1}$,

$$\sigma^2_{OK}(\boldsymbol{x}_0) = \text{Cov}(0) - \mathbf{c}'\mathbf{C}^{-1}\mathbf{c} + \left(\mathbf{c}'\mathbf{C}^{-1}\mathbf{1} - 1\right)\left(\mathbf{1}'\mathbf{C}^{-1}\mathbf{c} - 1\right)\left(\mathbf{1}'\mathbf{C}^{-1}\mathbf{1}\right)^{-1}.$$

Because $\mathbf{C}$ is symmetric, $\left(\mathbf{C}^{-1}\right)' = \mathbf{C}^{-1}$ and because, in addition, the term $\mathbf{c}'\mathbf{C}^{-1}\mathbf{1}$ is not a true matrix but a scalar, $\mathbf{c}'\mathbf{C}^{-1}\mathbf{1} = \left(\mathbf{c}'\mathbf{C}^{-1}\mathbf{1}\right)' = \mathbf{1}'\mathbf{C}^{-1}\mathbf{c}$, the expression reduces to

$$\sigma^2_{OK}(\boldsymbol{x}_0) = \text{Cov}(0) - \mathbf{c}'\mathbf{C}^{-1}\mathbf{c} + \left(\mathbf{1}'\mathbf{C}^{-1}\mathbf{c} - 1\right)^2\left(\mathbf{1}'\mathbf{C}^{-1}\mathbf{1}\right)^{-1}. \quad \square$$

Theorem 4.7

Given a sampling, the difference between the estimation variance for simple and ordinary kriging is

$$\sigma^2_{OK}(\boldsymbol{x}_0) - \sigma^2_{SK}(\boldsymbol{x}_0) = \left(\mathbf{1}'\mathbf{C}^{-1}\mathbf{c} - 1\right)^2\left(\mathbf{1}'\mathbf{C}^{-1}\mathbf{1}\right)^{-1}.$$

Proof

From Step 5 of Algorithm 2.1,

$$\sigma^2_{SK}(\boldsymbol{x}_0) = \text{Cov}(0) - \mathbf{c}'\mathbf{C}^{-1}\mathbf{c}$$

and from Lemma 4.9,

$$\sigma^2_{OK}(\boldsymbol{x}_0) = \text{Cov}(0) - \mathbf{c}'\mathbf{C}^{-1}\mathbf{c} + \left(\mathbf{1}'\mathbf{C}^{-1}\mathbf{c} - 1\right)^2\left(\mathbf{1}'\mathbf{C}^{-1}\mathbf{1}\right)^{-1}.$$

Hence, the difference is

$$\sigma^2_{OK}(\boldsymbol{x}_0) - \sigma^2_{SK}(\boldsymbol{x}_0) = \left(\mathbf{1}'\mathbf{C}^{-1}\mathbf{c} - 1\right)^2\left(\mathbf{1}'\mathbf{C}^{-1}\mathbf{1}\right)^{-1}. \quad \square$$

Because the term $\left(\mathbf{1}'\mathbf{C}^{-1}\mathbf{c} - 1\right)^2\left(\mathbf{1}'\mathbf{C}^{-1}\mathbf{1}\right)^{-1}$ is nonnegative, the ordinary kriging estimation variance is never smaller than the one for simple kriging. In the case of Exercise 4.1, $\sigma^2_{OK}(\boldsymbol{x}_0) - \sigma^2_{SK}(\boldsymbol{x}_0) = 1.9$. The term summarizes two things that happen when the mean is filtered out. First, an additional unknown is introduced in the system of equations—the Lagrange multiplier, which like any other Lagrange multiplier measures the sensitivity of the solution to the constraint. Second, in general, the constraint on the weights reduces the space solution, in which case the absolutely minimum error variance of simple kriging is unfeasible.

The term represents an additional contribution to uncertainty if one ignores the mean and uses ordinary kriging instead of simple kriging. If the mean is actually unknown, the term is an indirect penalty for not knowing it. In the latter case the penalty is indirect because the mean is not really computed but filtered out.

Search Neighborhood

Using the whole sampling for kriging has both negative and positive connotations. On the positive side, by using all the observations, the covariance matrix $\mathbf{V}$ in Algorithm 4.2 does not change by moving the estimation location—only matrix $\mathbf{v}$ changes—which means that if several estimations are needed, one may resort to inverting $\mathbf{V}$ only once for the first estimation.

On the negative side, if the sampling size is large, kriging the entire sampling involves the inversion of a huge matrix of an order one unit larger than the sampling size. As the distance estimator observation increases, the weight associated with a distant observation tends to zero because of the combined influence of the screen effect and the small to zero correlation. The result is that, with the exception of a few of the closest observations, the kriging weights are all close to zero. According to Davis and Morris (1997), large normal systems may be unstable for certain combinations of covariances and sampling patterns.

Any sampling that can be employed to produce estimates by simple kriging can also be used to produce estimates using ordinary kriging, which is equivalent to ignoring the known constant mean m over the sampling domain. Table 4.1 contains ordinary kriging results obtained by solving Exercise 3.2 for an increasing number of wells. One can see that beyond 10 wells the estimation is insensitive to the addition of more data.

Table 4.1. Sensitivity of an estimation to size of the sub-sampling.

k	λ_1	λ_2	λ_3	λ_4	λ_5	$\sum_{i=6}^{k} \lambda_i$	$\hat{z}_{OK}(2,\ 4.3)$	$\varepsilon^2_{OK}(2,\ 4.3)$
5	0.351	0.311	0.043	0.157	0.137	—	0.766	0.159
7	0.358	0.328	0.047	0.157	0.157	−0.047	0.750	0.159
8	0.356	0.325	0.050	0.178	0.165	−0.074	0.767	0.158
10	0.355	0.324	0.050	0.178	0.170	−0.077	0.756	0.158
15	0.355	0.324	0.051	0.178	0.169	−0.077	0.753	0.158
20	0.355	0.324	0.052	0.178	0.169	−0.078	0.754	0.158
32	0.355	0.324	0.052	0.178	0.169	−0.078	0.758	0.158

Selection of an optimal search neighborhood is the solution to the inefficiencies and numerical problems that large samplings may generate. The idea is to reduce the size of the covariance matrix by retaining only the relevant observations for each estimation. There are two inconvenient aspects to this simplification. First, unless all estimation locations are fairly closely spaced, the covariance matrix is no longer unique since it comprises different terms for each estimation location. Customarily the advantages of a reduction in the covariance matrix size are presumed to compensate for the repeated inversion of several smaller covariance matrices. If the normal equations are not solved by inverting the covariance matrices, the advantage provided by the uniqueness of the covariance matrix may be immaterial, making its reduction in size even more expedient.

The second and more serious drawback of dealing selectively with large samplings is the lack of tests for determining appropriate neighborhood size. In very general terms, the neighborhood must be large enough to contain three observations at a bare minimum and anything beyond 25 observations is considered more than adequate (Myers, 1991). More precise justification for selection of neighborhood size depends upon the measure of performance achieved and fluctuates according to the nature of the sampling pattern and the covariance. Lacking theoretical criteria, experimentation such as that presented in Table 4.1 is the best alternative. Chapter 7 elaborates more on this matter.

In order to improve the radial data distribution and place a limitation on the maximum number of observations, selection of a search neighborhood can be based on a more elaborate scheme, discarding observations by sectors, usually by quadrants or octants, as demonstrated by the search neighborhood in Figure 4.1.

If the user wishes to introduce a direction of preferential search, the neighborhood can be generalized into an ellipse. An elliptical search is a judicious choice in the presence of anisotropy, wherein the major axis must be oriented in the direction of maximum continuity.

Quasi-Stationary Estimator

A practical implication of using search neighborhoods of fixed sizes is that the estimation always takes place in the vicinity of the estimation location.

Definition 4.11

Let $Z(\boldsymbol{x}_i)$ be k random variables inside a neighborhood $d \subset D$ of the sampling domain D. The quasi-stationary ordinary kriging estimator $\hat{Z}(\boldsymbol{x}_0)$ of the variate $Z(\boldsymbol{x}_0)$ is the linear combination

$$\hat{Z}(\boldsymbol{x}_0) = \sum_{i=1}^{k} \lambda_i Z(\boldsymbol{x}_i)$$

subject to $\sum_{i=1}^{k} \lambda_i = 1$. □

Besides replacing Definition 4.1 by Definition 4.11, the formulation of quasi-stationary ordinary kriging is completed by restricting the stationarity assumptions to the neighborhood d, which implies the replacing of Assumption 4.2 by Assumption 4.5 and Assumption 4.4 by Assumption 4.6.

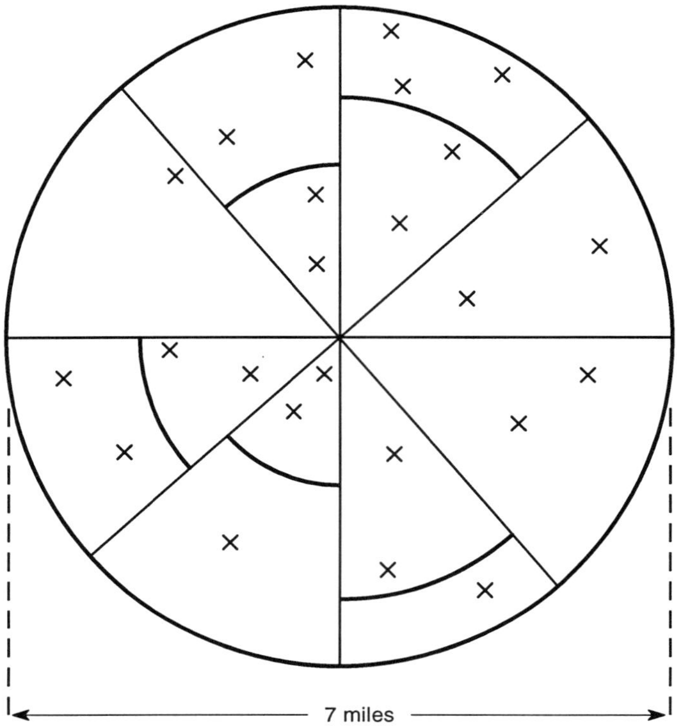

Figure 4.1. Example of search neighborhood in which the sampling is restricted to no more than two observations per octant situated no more than 3.5 miles from the estimation location.

Assumption 4.5

The random function is second order stationary over $d \subset D$, which implies that

$$\left.\begin{aligned} &\mathrm{E}\left[Z\left(\boldsymbol{x}\right)\right] = m \\ &\mathrm{E}\left[\left(Z\left(\boldsymbol{x}\right) - m\right)\left(Z\left(\boldsymbol{x}+\boldsymbol{h}\right) - m\right)\right] = \mathrm{Cov}\left(\boldsymbol{h}\right) \end{aligned}\right\} \text{for all } \boldsymbol{x},\ \boldsymbol{x}+\boldsymbol{h} \in d\,. \quad \square$$

Assumption 4.6

The random function honors the intrinsic hypothesis in $d \subset D$, which implies that

$$\left.\begin{aligned} &\mathrm{E}\left[Z\left(\boldsymbol{x}\right)\right] = m \\ &\mathrm{Var}\left[Z\left(\boldsymbol{x}\right) - Z\left(\boldsymbol{x}+\boldsymbol{h}\right)\right] = 2\gamma\left(\boldsymbol{h}\right) \end{aligned}\right\} \text{for all } \boldsymbol{x},\ \boldsymbol{x}+\boldsymbol{h} \in d\,. \quad \square$$

Formulation of both forms of quasi-stationary ordinary kriging progresses identically to the stationary cases. Because the derivation of stationary ordinary kriging never involved assumptions about the sampling locations, all lemmas, theorems and corollaries are valid for the new estimator and assumptions.

On the other hand, quasi-stationary ordinary kriging is equivalent to universal kriging of order zero, a topic fully covered in Chapter 6.

Beyond the improvement of calculation efficiency, the most remarkable consequence of dealing with restricted neighborhoods is the relaxation of the requirement that an attribute must have a constant mean to qualify for characterization by ordinary kriging. By requiring that the mean be constant by neighborhood instead of being constant over the entire sampling domain, quasi-stationary ordinary kriging goes beyond stationary ordinary kriging. Through a stepwise approximation employing observations close to the estimation location, quasi-stationary ordinary kriging allows a rudimentary modeling of an attribute with a fluctuating mean. Chapter 6 covers an even more general expansion.

These two additional forms of ordinary kriging complete the list of four related but different formulations mentioned at the beginning of the chapter:

i) second order stationary ordinary kriging,
ii) intrinsic ordinary kriging,
iii) second order quasi-stationary ordinary kriging,
iv) quasi-intrinsic ordinary kriging.

Chapter 5
THE SEMIVARIOGRAM

Three good reasons may be cited to explain why the semivariogram is important in geostatistics:

1. The semivariogram is a statistic that assesses the average decrease in similarity between two random variables as the distance between the variables increases, leading to some applications in exploratory data analysis.
2. It has been demonstrated by the foregoing algorithms and exercises that kriging is not possible without knowledge of the semivariogram or the covariance. In the formulation of our exercises, the covariance has been a known analytical expression—which, incidentally, is what the rigorous application of the algorithms demands. Yet, in practice, neither the covariance nor the semivariogram is known. The way in which geostatistics sidesteps this impasse is by use of an estimate of the semivariogram or the covariance instead of the true moments of the random function model, an approximation for which the derivation of the normal equations does not account.
3. In the previous chapter we have also seen that the practice is to solve the kriging system of equations in terms of covariances. This is primarily for convenience in the handling of the square matrices, despite the slight loss in generality. Yet in terms of determining the spatial correlation, the practice continues to be to estimate the semivariogram and then, provided that the covariance exists, to use the following Corollary 5.2 for converting semivariograms into covariances.

In a probabilistic framework, the semivariogram does not require knowledge of the mean of the random function for its estimation. In addition to this advantage, other reasons to prefer estimation of the semivariogram rather than use of the covariance (Isaaks and Srivastava, 1988; Cressie and Grondona, 1992) are:

1. The semivariogram requires a less stringent assumption for its existence than the covariance. The most common example of a random process with a semivariogram and no covariance is the discrete Wiener–Levy process.

2. Estimation of the semivariogram is insensitive to the addition of a constant to the random function, whereas the covariance estimator depends on such a constant.
3. Estimation of the semivariogram is less sensitive than covariance estimation to failure to comply with the requirement that the drift must be constant.

Despite these advantages of estimating the semivariogram compared to using the covariance, use of the semivariogram in the study of stochastic processes is not widespread outside geostatistics.

The Semivariogram of the Random Function

Let us recall Assumption 4.4. Given two locations $\boldsymbol{x}$ and $\boldsymbol{x}+\boldsymbol{h}$ inside the field of a random function Z, if $\mathrm{E}\left[Z\left(\boldsymbol{x}\right)\right]$ is a constant, then the semivariogram is:

$$\gamma\left(\boldsymbol{h}\right)=\frac{1}{2}\mathrm{Var}\left[Z\left(\boldsymbol{x}\right)-Z\left(\boldsymbol{x}+\boldsymbol{h}\right)\right],$$

which is at the same time an assumption and a definition of the second order stationarity of the increments. The definition involves the assumption that although the random function varies from place to place, the semivariogram is independent of location and depends only on the distance of the pair of variates being considered, which is a necessity for the purpose of estimating the semivariogram.

One may note that unlike the other moments, such as the mean, the semivariogram is not a single number but a continuous function of a variable $\boldsymbol{h}$, called the *lag*. The lag is a vector, thus involving not only the magnitude of the separation but the azimuth of the line through the pairs.

Theorem 5.1

Given an intrinsic random function, the semivariogram is half the estimation variance produced by assigning $Z\left(\boldsymbol{x}+\boldsymbol{h}\right)$ to $Z\left(\boldsymbol{x}\right)$,

$$\gamma\left(\boldsymbol{x}+\boldsymbol{h},\,\boldsymbol{x}\right)=\frac{1}{2}\sigma_{OKI}^{2}\left(\boldsymbol{x}\right).$$

Proof

If one considers a sample of size 1 at $\boldsymbol{x}_1=\boldsymbol{x}+\boldsymbol{h}$, then Theorem 4.2 reduces to

$$\left.\begin{array}{l}\mu=-\gamma\left(\boldsymbol{x}+\boldsymbol{h},\,\boldsymbol{x}\right)\\ \lambda_1=1\end{array}\right\}.$$

From Theorem 4.3,

$$\sigma_{OKI}^{2}\left(\boldsymbol{x}\right)=\gamma\left(\boldsymbol{x}+\boldsymbol{h},\,\boldsymbol{x}\right)-\left(-\gamma\left(\boldsymbol{x}+\boldsymbol{h},\,\boldsymbol{x}\right)\right)$$

and the addition of the two terms proves the theorem. □

Both the semivariogram and the covariance refer to the rate of spatial variation of the random function, but while the semivariogram increases, the covariance decreases—the covariance measures similarity and the semivariogram dissimilarity.

Corollary 5.1

Let Z be a second order stationary random function. If Cov $(\boldsymbol{h}) = 0$, for $\boldsymbol{h} > \boldsymbol{a}$, then

$$\gamma(\boldsymbol{h}) = \text{Cov}\ (0)\,, \text{ for } \boldsymbol{h} > \boldsymbol{a}\,.$$

Proof

From Lemma 4.3,

$$\gamma(\boldsymbol{h}) = \text{Cov}\ (0) - \text{Cov}\,(\boldsymbol{h})\ ;$$

but, if Cov $(\boldsymbol{h}) = 0$, for $\boldsymbol{h} > \boldsymbol{a}$, then

$$\gamma(\boldsymbol{h}) = \text{Cov}\ (0)\,, \text{ for } \boldsymbol{h} > \boldsymbol{a}\,. \quad \square$$

Figure 5.1 illustrates Corollary 5.1. A semivariogram that increases in dissimilarity with distance over short lags and then levels off is called a transitive semivariogram. The lag $\boldsymbol{a}$ at which the semivariogram reaches a constant value is called the range. The value of the semivariogram beyond the range is referred to as the sill. Considering that by Theorem 5.1 two variates separated by a distance larger than the range are stochastically independent, the range is equivalent to the more intuitive notion of zone of influence of an observation.

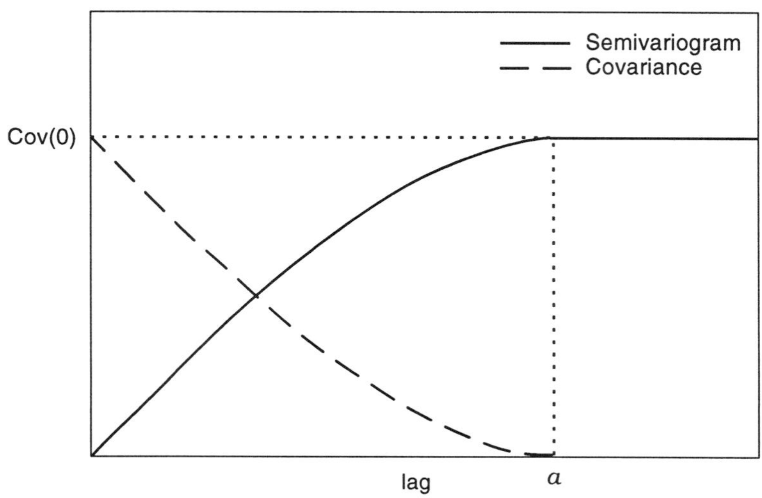

Figure 5.1. Idealized examples of covariance and semivariogram showing the location $\boldsymbol{a}$ of the range.

Corollary 5.2

Let Z be a second order stationary random function with sill C. Then

$$\text{Cov}\,(\boldsymbol{h}) = C - \gamma\,(\boldsymbol{h})\ .$$

Proof

From Lemma 4.3

$$\text{Cov}\,(\boldsymbol{h}) = \text{Cov}\,(0) - \gamma\,(\boldsymbol{h})$$

and the proof follows from Corollary 5.1. □

The equivalence allows the conversion of semivariograms into covariances.

The Experimental Semivariogram

The most commonly used estimator is based on Lemma 4.2 that proves that the random function semivariogram is equivalent to

$$\gamma\,(\boldsymbol{h}) = \frac{1}{2}\text{E}\left[\{Z\,(\boldsymbol{x}) - Z\,(\boldsymbol{x}+\boldsymbol{h})\}^2\right]\ .$$

Definition 5.1

Let Z be an intrinsic random function and let $Z\,(\boldsymbol{x}_i)$, for $i = 1,\,2,\,\ldots,\,N$, be a sampling of size N. Then the following expression is an unbiased estimator for the semivariogram of the random function:

$$\hat{\gamma}\,(\boldsymbol{h}) = \frac{1}{2n\,(\boldsymbol{h})}\sum_{i=1}^{n(\boldsymbol{h})}\left[Z\,(\boldsymbol{x}_i+\boldsymbol{h}) - Z\,(\boldsymbol{x}_i)\right]^2,$$

where $n\,(\boldsymbol{h})$ is the number of pairs of variables at distance $\boldsymbol{h}$ apart. □

An estimate of the semivariogram is also called an *experimental semivariogram*. The estimator in Definition 5.1 is not robust with respect to outliers, which has prompted the formulation of other estimators in a probabilistic framework (Cressie, 1991, p. 74–83) that is infrequently employed and will not be covered here.

Estimation of the semivariogram implies an ergodic assumption. Remember that according to Assumption 2.1 and Assumption 4.1 data are one partial realization of a random function, which is not, indeed, sufficient to infer the parent random function. Basically what an ergodic assumption says is that one realization is sufficient to determine properties of the ensemble of all possible realizations.

The expression for calculating the experimental semivariogram involves terms that depend on specific locations, namely $\boldsymbol{x}_i$ and $\boldsymbol{x}_i + \boldsymbol{h}$. The mean square average of the differences, however, is supposed to depend solely on the distance $\boldsymbol{h}$. This is a property imposed upon the random function model rather than a fact of life. Geostatistics does not have a test to verify that an attribute meets this assumption.

The fact that Definition 5.1 requires that the random function be intrinsically stationary implies that the mean must be a constant, a requirement that users tend to overlook. A gentle and systematic variation in the mean, such as the increase in temperature with depth, is called a *drift*. Proper semivariogram estimation practice requires the removal of this drift. Otherwise Lemma 4.2 does not hold; the mean square difference is no longer equal to the variance of the difference, the actual definition of a semivariogram.

Other than practical rules of thumb to guide the estimation, there are no statistical tests to assess how close an experimental semivariogram approximates the semivariogram of the exhaustive sampling or the semivariogram of the random function model used in the normal equations. Practical guidelines are:

1. Because the accuracy of the estimates is proportional to the number of pairs, the greater the lag, the fewer the pairs $n(\boldsymbol{h})$ for a given distance, thus the less reliable the estimates. Fortunately, in the applications of the semivariogram to kriging and simulation, it is the portion near the origin that requires the most accurate estimation because of its higher influence on the results derived from its use. A common yet unsubstantiated practice is to limit estimation to lags with a minimum of 30 pairs (Journel and Huijbregts, 1978, p. 194).
2. If it is necessary to estimate the semivariogram for a large lag similar in magnitude to the diameter of the sampling area, the only pairs of observations that are far enough apart are those comprising sampling sites located at opposite extremes of the sampling area; thus, the central points are excluded from the analysis. Such a consideration is the justification for a second practical rule that advises us to limit the lag of the experimental semivariogram to half the extreme distance in the sampling domain for the direction to be analyzed (Journel and Huijbregts, 1978, p. 194; Clark, 1979, p. 14).

Although the semivariogram is, in general, a continuous vectorial function of $\boldsymbol{h}$, discontinuous sampling precludes estimation of the experimental semivariogram for every possible value of the lag. In practice the estimation may be limited to a few points and directions after sorting of the pairs of sampling sites not only by distance but by spatial orientation as well.

In terms of calculation, a favorable situation occurs when sampling is performed at the nodes of a square grid such as that shown in Figure 5.2. Then estimates can be made at fixed multiples in directions other than that of sampling, which typically consists of one-directional transects—the columns, for instance. The symmetry of the configuration permits investigation in other directions by the taking of successive readings from different rows and columns. For example, in a two-dimensional square grid oriented north–south and east–west, directions

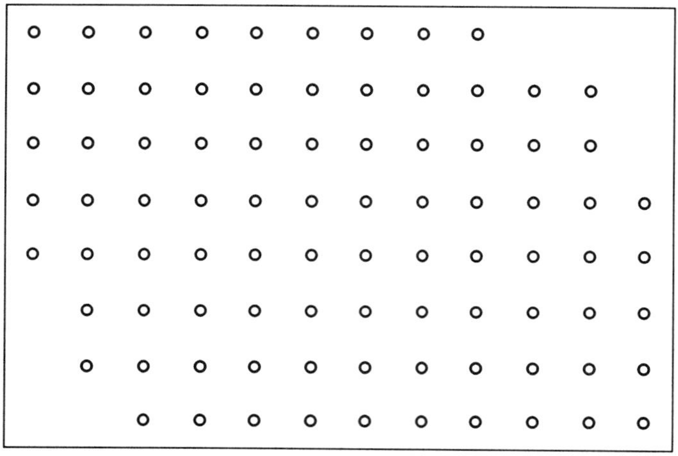

Figure 5.2. A regular square grid.

that may be analyzed in addition to, say, the N–S direction of sampling include E–W, NW–SE and NE–SW. Notice that the minimum spacing possible in the NW–SE or NE–SW directions is $\sqrt{2}$ times the spacing along the N–S or E–W directions.

Somewhat less convenient than grids in terms of semivariogram estimation are linear traverses sampled at regular intervals. Calculations along a traverse work as well as calculations along a row or column in a grid, but the regularity of the locations along directions other than the direction of sampling is lost.

The worst configuration in terms of convenience when estimating a semivariogram is also the most common: data at irregular locations. The practice in this case calls for the grouping of distances and directions into classes (Figure 5.3). A safe practice is to take a lag interval equal to the average distance to the nearest neighbor, lateral bounds equal to twice the lag interval, and an angular tolerance of 22.5 degrees.

Algorithm 5.1

In order to obtain an experimental semivariogram for an attribute honoring the intrinsic hypothesis (Assumptions 4.3 and 4.4), perform the following steps:

1. Collect a sampling, preferably one that exactly follows or closely approximates a regular pattern. Figure 2.5 shows the regular patterns in two dimensions.

2. Inspect the data and correct any mistakes that the recording of the sampling may have contributed either to the coordinates of the sampling sites or the measurements of the attribute.

3. Set the lag increment and its tolerance in the case of an irregular sampling. Limit the maximum lag to no more than one half of the maximum dimension of the sampling area.

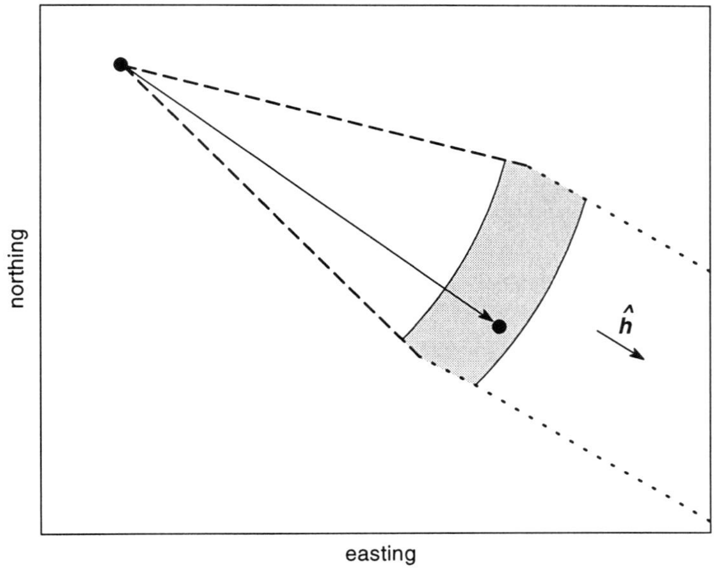

Figure 5.3. A distance class for semivariogram estimation along direction $\hat{\boldsymbol{h}}$. All points in the shaded area belong to the same distance class relative to the point in the upper left corner. The dotted line denotes the lateral bounds and the segmented line the angular tolerance. The lag interval is the distance between the arc segments bounding the shaded area.

4. Either set the directional search to some preferential direction or else ignore direction by running an omnidirectional search that focuses solely on the distance regardless of orientation.

5. Use the estimator in Definition 5.1 to generate the experimental semivariogram. Discard values for which $n(\boldsymbol{h}) < 30$. □

An estimation is good only if the experimental semivariogram is in agreement with the semivariogram that would be expected from an exhaustive sampling. If this necessary condition has not been met, inspect and correct the data; or change the distance class parameters, run the algorithm again, draw conclusions, and make a decision. Keep in mind that the final judgment may be that the sampling is unsuitable for calculation of a reliable experimental semivariogram, particularly in the case of small samplings with outliers.

EXERCISE 5.1

Given the sampling

x	$Z(x)$	x	$Z(x)$
7.0	3.2	10.0	7.5
7.5	4.3	10.5	7.3
8.0	5.0	11.0	6.7
8.5	6.5	11.5	5.8
9.0	7.9		
9.5	8.1		

calculate the experimental semivariogram along the traverse for lags from 0 to 4.5 at increments of 0.5.

Solution

Before starting any calculations, one should inspect the sequence for systematic variations in the mean. For this traverse the sample arithmetic mean is 6.23. There are six values above the mean and four below, with an initial increasing sequence of five values followed by a decreasing sequence of four values. Then one can safely assume that the sequence does not have a systematic variation in the mean, and calculation of the experimental semivariogram can proceed. Table 5.1 is a convenient arrangement displaying all necessary calculations in the unlikely event of manual computation.

Table 5.1. Estimation of the experimental semivariogram in Exercise 5.1.

x_i	$z(x_i)$	Δ_0^2	$\Delta_{0.5}^2$	$\Delta_{1.0}^2$	$\Delta_{1.5}^2$	$\Delta_{2.0}^2$	$\Delta_{2.5}^2$	$\Delta_{3.0}^2$	$\Delta_{3.5}^2$	$\Delta_{4.0}^2$	$\Delta_{4.5}^2$
7.0	3.2	0.0	1.21	3.24	10.89	22.09	24.01	18.49	16.81	12.25	6.76
7.5	4.3	0.0	0.49	4.84	12.96	14.44	10.24	9.00	5.76	2.25	
8.0	5.0	0.0	2.25	8.41	9.61	6.25	5.29	2.89	0.64		
8.5	6.5	0.0	1.96	2.56	1.00	0.64	0.04	0.49			
9.0	7.9	0.0	0.04	0.16	0.36	1.44	4.41				
9.5	8.1	0.0	0.36	0.64	1.96	5.29					
10.0	7.5	0.0	0.04	0.64	2.89						
10.5	7.3	0.0	0.36	2.25							
11.0	6.7	0.0	0.81								
11.5	5.8	0.0									
	$\sum\Delta_h^2$	0.0	7.52	22.74	38.97	50.15	43.99	30.87	23.21	14.50	6.76
	$n(\boldsymbol{h})$	10	9	8	7	6	5	4	3	2	1
	$\gamma(\boldsymbol{h})$	0.0	0.42	1.42	2.78	4.18	4.40	3.85	3.86	3.62	3.38
					Mean:	6.23					
					Variance:	2.39					

For clarity in the mechanics of the manual calculation, the sequence is purposely shorter than it should be; it does not honor the 30-pair rule even for the shortest lag. By Step 5 in Algorithm 5.1 the sequence is too short for any of the values to be significant. □

It is interesting to note in this example the operation of a rule that applies to any semivariogram estimation: The number of pairs tends to decrease with increases in the lag up to a certain limit beyond which there are no more pairs of data.

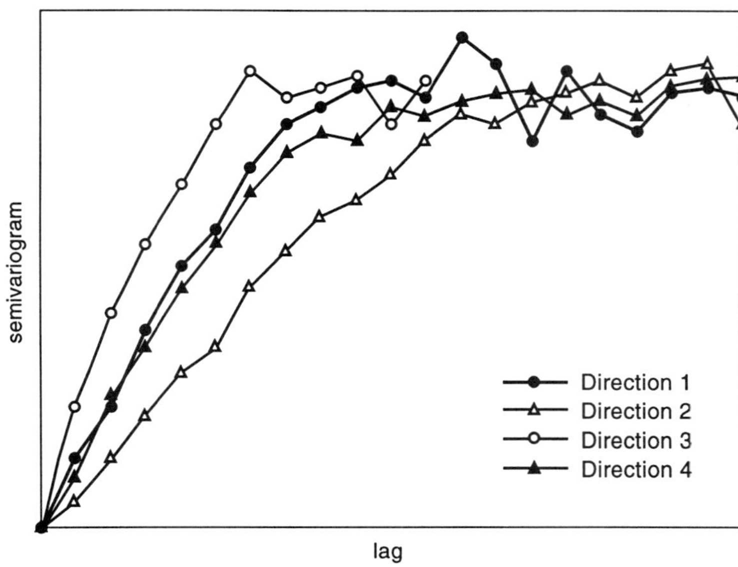

Figure 5.4. An example semivariogram with geometric anisotropy.

Anisotropy and Drift

Experimental semivariograms must be prepared for as many different directions as are feasible given the configuration of the data, with four equally spaced directions as a bare minimum. The semivariogram is considered to be isotropic when variations in the azimuth do not produce significant changes in the experimental semivariogram and otherwise is anisotropic. An anisotropic semivariogram is a true function of $\boldsymbol{h}$ rather than just of h. Directional analysis has the inconvenience of reducing the number of pairs available for the estimation, reduction that may completely preclude a directional analysis for small data sets.

In the presence of isotropy or in the absence of a large sample size, pairs can be pooled regardless of the azimuth. The resulting experimental semivariogram is said to be omnidirectional. Figure 5.4 illustrates the most common and simplest case of anisotropy—geometric anisotropy. In this situation, one model can be used to fit all experimental semivariograms by making the range a function of the azimuth or, equivalently, by rescaling the multidimensional sampling space to force it to be isotropic. The common choice is to model an elliptical variation for the range, in which case the user must find both the size and the orientation of the minor and major axes.

A drift also causes the experimental semivariogram to vary with the azimuth. In the presence of linear drift, as shown in Figure 5.5, the experimental semivariogram in the direction of maximum dip has a slow increase near the origin approximating a parabola, whereas in a perpendicular direction the experimental semivariogram is drift-free. For more details, see Exercise 6.1.

Using the experimental semivariogram in the drift-free direction as a surrogate semivariogram approximating the one that would be obtained after actually removing the drift is the simplest way to eliminate the effect of a drift on the semivariogram. The main inconvenience of this approach is the impossibility of directly detecting anisotropy in the semivariogram of the residuals.

Semivariogram Models

As seen in Definition 4.2 and the derivation of Theorem 4.2, the direct use of the experimental semivariogram or derived interpolated values can lead to singular matrices, multiple solutions, or negative mean square errors in the kriging system of equations because a tabulation of points plus its derived interpolated values may not be negative definite. The solution is to replace the tabulation obtained for the experimental semivariogram by negative definite functions, also called permissible functions (Christakos, 1984), thus assuring a unique solution to the normal equations and a nonnegative mean square error.

Although the number of permissible functions is infinite, in practice the options reduce to a handful. In approximately decreasing frequency of use, for a distance h in a given direction, the analytical expressions for the most common models shown in Figure 5.6 are given in Definitions 5.2–5.8.

Definition 5.2

Let h be the lag. Then the spherical semivariogram is the model

$$\gamma(h) = \begin{cases} C\left(\frac{3}{2}\frac{h}{a} - \frac{1}{2}\left(\frac{h}{a}\right)^3\right) & , 0 \leq h < a \\ C & , a \leq h , \end{cases}$$

where C is the sill and a is the range. □

Near the origin this semivariogram is close to a line in shape. The model is said to be transitive because it reaches a finite sill at a finite range. The tangent to this semivariogram at the origin intersects the sill at two-thirds of the range. The model is not permissible for a space with a dimension larger than 3.

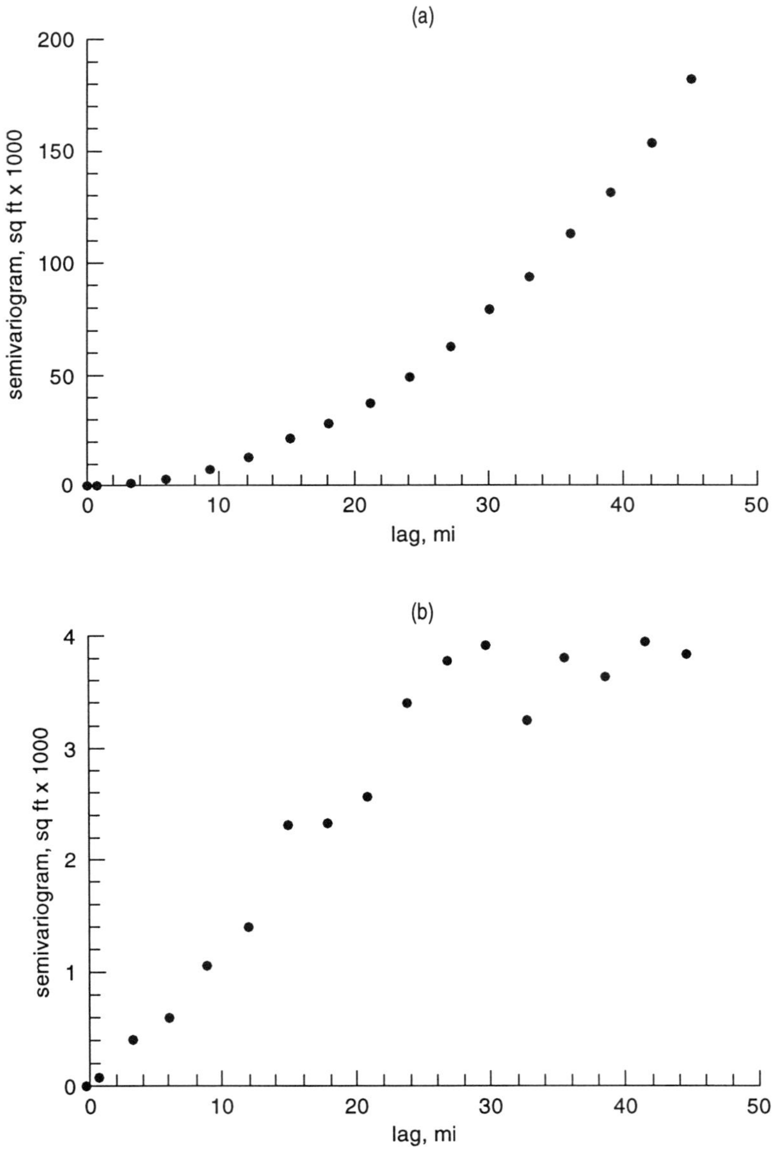

Figure 5.5. Semivariograms for the water table elevation in the High Plains aquifer of Kansas during the winter 1980–81. (a) Increasing semivariogram along direction of maximum dip, N70E, for the linear trend. (b) Transitive semivariogram along N20W. Notice that the vertical scale in (a) is 50 times that in (b).

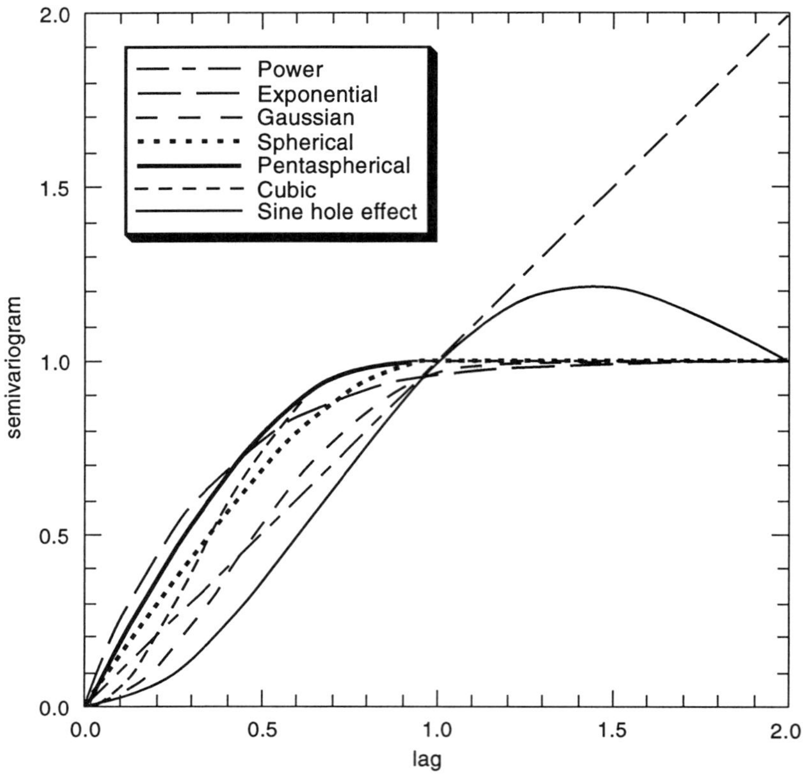

Figure 5.6. Seven simple semivariogram models when the two model parameters in Definitions 5.2 to 5.8 are both equal to 1.

Definition 5.3

Let h be the lag. Then the exponential semivariogram is the model

$$\gamma(h) = C\left(1 - e^{-\frac{3h}{a}}\right). \quad \square$$

The model approaches the sill asymptotically. A practical definition of the range is the distance a where the semivariogram is $0.95C$. Geometrically, a tangent at the origin intersects the asymptote C at lag $a/3$.

Definition 5.4

Let h be the lag. Then the Gaussian semivariogram is the model

$$\gamma(h) = C\left(1 - e^{-3\left(\frac{h}{a}\right)^2}\right). \quad \square$$

The sill is approached asymptotically. A practical rule is to consider the range to be the distance a, for which the semivariogram is $0.95C$. A graph of the model has a parabolic form near the origin. Despite its theoretical permissibility, normal equations involving Gaussian models may be numerically singular (Davis and Morris, 1997).

Definition 5.5

Let h be the lag. Then the power semivariogram is the model

$$\gamma(h) = \alpha\, h^{\beta}\,,\ 0 < \beta < 2\,. \quad \square$$

This is a non-transitive model. The linear semivariogram is a special case for $\alpha = 1$. The main advantage of the power model is its simplicity. The power semivariogram may be used to model an experimental semivariogram that is not transitive or to model transitive semivariograms if the lag in the normal equations will always be smaller than the range, making the form of the semivariogram beyond the range immaterial.

Definition 5.6

Let h be the lag. Then the cubic semivariogram is the model

$$\gamma(h) = \begin{cases} C\left(7\left(\frac{h}{a}\right)^2 - \frac{35}{4}\left(\frac{h}{a}\right)^3 + \frac{7}{2}\left(\frac{h}{a}\right)^5 - \frac{3}{4}\left(\frac{h}{a}\right)^7\right), & 0 \le h < a \\ C & , a \le h\,. \end{cases} \quad \square$$

Definition 5.7

Let h be the lag. Then the pentaspherical semivariogram is the model

$$\gamma(h) = \begin{cases} C\left(\frac{15}{8}\frac{h}{a} - \frac{5}{4}\left(\frac{h}{a}\right)^3 + \frac{3}{8}\left(\frac{h}{a}\right)^5\right) & , 0 \le h < a \\ C & , a \le h\,. \end{cases} \quad \square$$

This model is permissible up to a dimension of 5.

Definition 5.8

Let h be the lag. Then the sine hole effect semivariogram is the model

$$\gamma(h) = C\left(1 - \frac{\sin\left(\pi\frac{h}{a}\right)}{\pi\frac{h}{a}}\right)\,. \quad \square$$

The model is not monotonic. It reaches a first maximum of 1.217 at 1.43—the global maximum—and then it continues a damped oscillation around the sill with period a.

For all models, both the range a and the sill C must be larger than zero. They are factors that change the scale, but not the shape, of these models, generically called transitive models. The power model is the only one in which radical variations in shape occur in response to changes in the exponent β.

The range of possibilities for increasing monotonically from zero to a constant value are well covered by the seven models; this is what experimental semivariograms are supposed to do. Because of their simplicity and availability in geostatistical software packages, the first four models are more widely used than the last three.

As seen in Figure 5.6, among the more common models the exponential model is the steepest and the Gaussian is the flattest near the origin. The exponential and Gaussian semivariograms cross at a lag equal to the practical range where, by definition, both are equal to 0.95 times the value of the sill; the linear and spherical semivariograms cross for the same lag, where they have a value of 1. The Gaussian semivariogram is the only one out of the four that has an inflection point.

Among the last three models the cubic and the pentaspherical models have one inflection point each while the sine hole effect model has an infinite number of these points, as it approximates the sill as a damped sinusoid that extends beyond the area displayed by the figure.

A pure nugget effect model is a special, degenerate case of a transitive semivariogram with an infinitesimal range—the semivariogram surges directly from zero to a constant value.

Definition 5.9

Let h be the lag. Then the pure nugget effect semivariogram is the model

$$\gamma(h) = C\big(1 - H(0)\big)$$

where C is the nugget effect and $H(0)$ is the Heaviside function, which is 1 at lag 0 and zero otherwise. □

The experimental semivariogram of a unidimensional sequence of random numbers follows a pure nugget effect model.

Additivity

Adding to the complexity of the modeling is the additivity property of permissible semivariogram models. Any linear combination of permissible models with positive coefficients is also permissible and is called a nested semivariogram. For instance, the sum of a pure nugget effect semivariogram model plus two spherical semivariograms with different parameters is an example of a nested semivariogram. By convention, however, a pure nugget effect model plus another model is not regarded as a nested model. Instead it is referred to as a model with a nugget effect, such as an exponential semivariogram with a nugget effect (Olea, 1991, p. 52).

Single models are used for the modeling of experimental semivariograms that are close in shape to any of the permissible models or for the approximate fitting of complex structural functions. Nested models are used for better fitting of complex structural functions.

Parsimony should prevail in order to avoid overfitting with nested models. The number of combinations of models can be staggering even after limiting both the number of simple models and the degree of nesting. On the other hand, the choice of distance class parameters for irregularly spaced data or the insertion or deletion of a single observation in a small- to medium-sized data set often makes more difference in the experimental semivariogram to be fit than the analytical difference between alternative fitting models.

Augmenting an additive model with a nugget value has beneficial effects in the numerical stability of normal equations, particularly in the case of Gaussian models (Davis and Morris, 1997). At the same time, however, this makes honoring of the data by a grid of values impossible if the data do not coincide exactly with the nodes, since the point estimation is discontinuous at the sampling locations because of the discontinuity at the origin in the semivariogram model (Christensen, 1991, p. 276).

The apparent presence of a nugget effect may result from sampling that is insufficient for resolution of the smallest features or may be the consequence of second order stationary measurement errors (Dubrule, 1994).

Parameter Estimation by Trial and Error

Once the user selects a semivariogram model—simple or nested—the problem remains to estimate its parameters, either automatically or by trial and error.

Traditionally the predominant practice has been first to calculate an experimental semivariogram and then to fit a model by trial and error with the help of computer software (Deutsch and Journel, 1992; Englund and Sparks, 1988). The appeal of this approach comes from the opportunity to visually compare the model to a few estimated points.

The following hints are useful for speeding up the manual fitting:

1. The range is often close to the average diameter of anomalies in the spatial fluctuation of the attribute.
2. The sample variance is a good approximation for the magnitude of the sill when the observations are evenly distributed over the area of interest and the dimensions of the area are at least three times the range (Barnes, 1991).
3. The simplest model that best approximates the fluctuations in the experimental semivariogram should be selected.
4. If the value of the experimental semivariogram for the first lag is several times the increment between the second and the third, then the model may need a nugget effect or some short range nesting.

In the absence of a reliable experimental semivariogram, rather than forcing a fit to an ill-behaved set of experimental values it may be better to postulate a model directly based on hints 1 and 2 or to adopt a model from an area judged similar to the area of interest. Examples of ill-behaved experimental semivariograms are those that have:

i) the highest value near the origin and then decrease monotonically as the lag increases, and
ii) an overall monotonic increase with lag, interrupted by a few values that fluctuate by more than 50% of the value of the sill. The abnormal fluctuations are particularly disruptive if they occur for short lags.

An experimental semivariogram with erratic, small amplitude fluctuations around a constant value—although an undesirable result—is not an example of an ill-behaved experimental semivariogram. In this situation the attribute follows a pure nugget effect model because the attribute lacks spatial continuity, at least at the scale of the minimum sampling distance. Notice, however, that lack of spatial correlation may not be real, but only the result of too sparse sampling. As seen in Corollaries 2.9 and 4.9, in the absence of spatial correlation, geostatistics corresponds to classical statistics in its predictions.

Procedure 5.1

For the manual fitting of a model to an experimental semivariogram, do the following:

1. Display the experimental semivariogram and, before proceeding to any model fitting, do your best to verify that the experimental semivariogram does not openly contradict the semivariogram that one would expect with an exhaustive sampling.

2. Select a parsimonious model based on the appearance of the experimental semivariogram.

3. Guess the parameters of the model based on the pattern of fluctuations in the experimental semivariogram and information that may be available about the attribute of interest.

4. Graphically display an overlay of the experimental semivariogram and the model.

5. If this is not the first trial compare the model with previous fittings. If any of the models at hand are deemed adequate, the best parameters are those of the most satisfactory model. Stop. Otherwise keep a copy of the display, change one parameter or the model, and go back to Step 4. □

Defendants of this approach argue that automatic fitting cannot take into account increasing estimation variance with lag for the semivariogram estimator, the disproportionate influence of anomalous estimates, the nonlinearity of most models with respect to the parameters, and the correlation among errors that depart from normality.

Automatic Parameter Fitting

Conceivably, semivariogram modeling could be made completely transparent to the user. Consider the case of kriging, for example. Kriging requires the same data needed for semivariogram modeling, which in theory could be made an integral part of the kriging program as a preprocessing step; the user would not even notice that there are semivariogram terms behind the calculations. This temptation has been avoided because of the numerous problems that such a practice might leave undetected. Visual inspection during modeling remains a requirement in geostatistical analysis and estimation. This has nothing to do with attempts to try to automate the search for models.

All analytical methods for the automatic modeling of semivariograms fall into two categories, maximum likelihood methods and least squares methods:

1. Maximum likelihood methods calculate parameters directly from the raw data provided by the attribute sampling (Kitanidis, 1997, chapter 4). The problems with these methods are that they rely heavily on the Gaussian distributional assumption and the estimates are biased (Cressie, 1991). Besides, according to McBratney and Webster (1986), the methods are unacceptably slow for samplings of sizes larger than 150, an opinion that is corroborated by Kitanidis (1997, p. 98).
2. Least squares methods require a preliminary estimation of the experimental semivariogram at a few lags. This is then followed by the estimation of the parameters in the model by automatically minimizing the mean square differences between the experimental points and those values of the semivariogram predicted by the model.

Differences in weighting schemes lead to three least squares methods: ordinary, weighted, and general. Let $\hat{\gamma}(\boldsymbol{h})$ be the vectorial form of a discrete experimental semivariogram containing k estimates of the semivariogram for increasing values of the lag

$$\hat{\gamma}(\boldsymbol{h}) = \left[\hat{\gamma}(\boldsymbol{h}_1)\ \hat{\gamma}(\boldsymbol{h}_2)\ \ldots\ \hat{\gamma}(\boldsymbol{h}_k)\right]$$

and let $\gamma(\boldsymbol{h};\theta)$ be a vector with values for the semivariogram model of interest with unknown parameters $\Theta = [\theta_1,\ \theta_2, \ldots, \theta_p]$

$$\gamma(\boldsymbol{h};\theta) = \left[\gamma(\boldsymbol{h}_1,\theta)\ \gamma(\boldsymbol{h}_2,\theta)\ \ldots\ \gamma(\boldsymbol{h}_k,\theta)\right] .$$

The best set of parameters is the one that minimizes the sum of the squared differences, R, between the experimental values and those predicted by the model,

$$R = \left[\hat{\gamma}(\boldsymbol{h}) - \gamma(\boldsymbol{h};\theta)\right]' \mathbf{V}^{-1} \left[\hat{\gamma}(\boldsymbol{h}) - \gamma(\boldsymbol{h};\theta)\right] .$$

In the simplest case—ordinary least squares—$\mathbf{V}$ is equal to the identity matrix and the minimization of R is straightforward. The method, however, makes the unrealistic assumption that the differences are normally distributed, are independent of one another, and that all the estimated values have the same variance. On these grounds the method has been rightly criticized to be unsuited for the fitting of semivariogram models.

Most of the objections are overcome by making $\mathbf{V}$ equal to the covariance matrix of the experimental semivariogram, a procedure known as generalized least squares. Because $\mathbf{V}$ is unknown, one has to assume it and the solution for the best θ turns into an iterative procedure that does not always converge (Gotway, 1991) and sometimes requires extensive processing time due to the inversion of large matrices.

The third approach is a heuristic solution that considers only the variance terms in the main diagonal of the covariance matrix (McBratney and Webster, 1986; Cressie, 1985 and 1991). This approach is a special case of weighted least squares and is the one favored in the implementation of automatic parameter estimation. The solution also involves iteration, but goes much faster than generalized least squares because all elements in the matrix are known and its inversion is trivial: each difference is weighted directly by the inverse of the variance of the experimental semivariogram. The method is capable of producing completely satisfactory results when the experimental semivariogram is well behaved, proving that the objections to automatic fitting by the proponent of manual fitting are exaggerated.

The only public domain computer program for the weighted least squares fitting of a semivariogram (Jian *et al.*, 1996) is a strict application of weighted least squares using a modified version from Press *et al.* (1992) of the Levenberg–Marquardt method (Levenberg, 1944; Marquardt, 1963), which Bard (1970) found to be among the best solutions for the parameter estimation of nonlinear functional models.

For the weight at lag $\boldsymbol{h}$ this program uses the variance of the $n(\boldsymbol{h})$ square differences in Definition 5.1 as the estimate for the variance of the experimental semivariogram. Consequently, modeling semivariograms by weighted least squares requires three pieces of information per record: lag, semivariogram estimate, and estimate of standard deviation.

The goodness of fit is reported employing both the sum of the square of the weighted differences, R_m, and the Akaike information criterion, AIC. AIC takes into account not only the goodness of fit but the parsimony of the model as well. The estimate of this criterion is:

$$\widehat{AIC} = n \ln\left(\frac{R_m}{n}\right) + 2p\,,$$

where n is the number of points in the experimental semivariogram and p is the number of parameters in the model. The advantage of the Akaike information criterion over the sum of the square of the weighted differences is that the criterion is sensitive to sample size and the number of parameters. The latter is particularly useful for decisions about the parsimony of nested modeling. Its disadvantage is that it requires independence of the squared differences, which is acceptable, as one intuitively expects that on a good fitting the experimental values should randomly alternate on different sides of the fitted model.

In the case of fitting more than one model, the one with the smallest $\widehat{AIC}$ is best. Notice that in comparing models with the same number of points n and parameters p, the ranking by weighted differences R_m or Akaike information criterion $\widehat{AIC}$ is the same, making selection of the best model independent of the criterion.

Algorithm 5.2

For the automatic fitting of a semivariogram model by weighted least squares, do the following:

1. Display the experimental semivariogram and before proceeding to any model fitting, make sure that it is a reasonable approximation to the semivariogram such as would be obtained with an exhaustive sampling.
2. Fit all models in Figure 5.6.
3. Select the simple model with the lowest $\widehat{AIC}$.
4. If the best-fitting simple model would likely be improved by nesting, try some nested models and keep the model with the lowest $\widehat{AIC}$, simple or nested. □

Despite the fact that the search is iterative, the process is transparent to the user and, for all practical purposes, the fitting is instantaneous. In this sense the method is "friendlier" than manual fitting by trial and error, which may be time-consuming even for the experienced eye and is certainly confusing for the novice.

EXERCISE 5.2

Given the following omnidirectional experimental semivariogram and its corresponding standard deviation for the $n(h)$ squared differences:

	Lag	Experimental semivariogram	Standard deviation
1	1039.02941	185.55762	309.53909
2	1671.37619	222.71608	429.62136
3	2411.96731	283.78858	546.77823
4	3238.55621	350.41225	624.94125
5	3986.34431	409.28920	709.26128
6	4798.25527	425.10046	700.69357
7	5622.82245	479.82902	781.57852
8	6390.44643	515.66721	793.59968
9	7219.57553	512.68859	754.06766
10	7991.70332	557.21386	784.46418
11	8818.26894	553.13395	753.51006
12	9594.10504	608.88869	793.58481
13	10398.11004	588.75490	760.50014
14	11206.04916	603.36585	768.43620
15	11990.76099	624.89485	794.51177
16	12809.31622	607.05642	781.33360
17	13593.52361	622.04279	809.45208
18	14406.22013	619.60402	801.96384
19	15195.65893	599.43991	811.63138
20	15992.80252	589.11388	808.16050
21	16799.44020	601.43822	818.22600
22	17603.98278	582.36187	826.22142

1. Find the best simple model in a weighted least squares sense.
2. Is a double nested spherical model better than the best simple model?

Solution

1. From Figure 5.7, the model with the lowest $\widehat{AIC}$ is the spherical model with a value of -164.3. This is the best simple model.

2. The $\widehat{AIC}$ for a model comprising two spherical semivariograms at the bottom of Figure 5.8 is -168.7. Although the improvement is marginal, this nested model fits the experimental semivariogram even better than the best simple model and the reduction in the mean squared errors offsets the model's loss in simplicity. □

As illustrated by the exercise, besides the advantages of speed and a systematic approach to the search for the best parameters, the method converges on the same solution aspired to in a manual fitting and provides the Akaike information criterion, which gives a quantitative measure for selecting the best model among several candidate models.

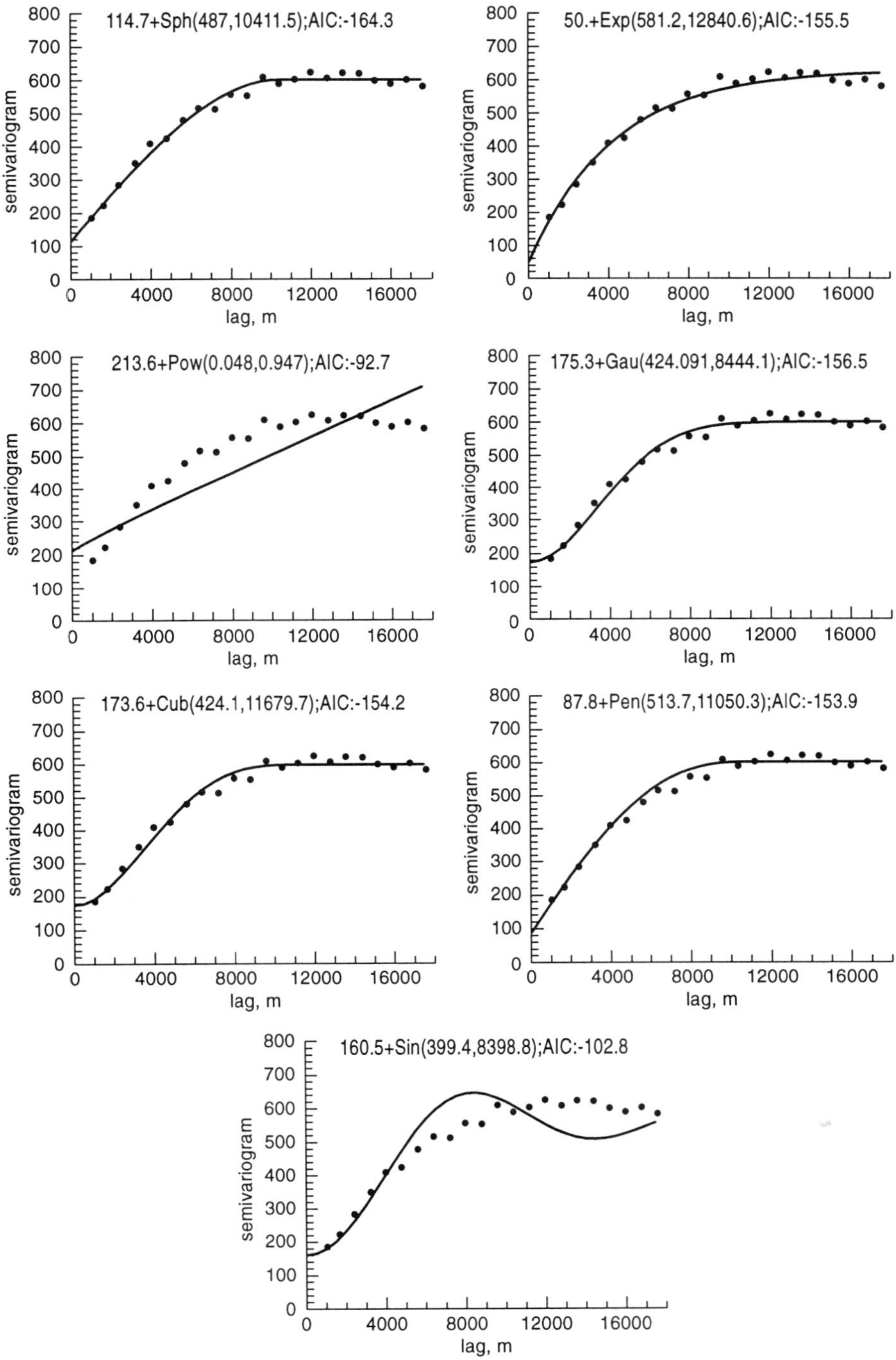

Figure 5.7. Best fitting of seven simple models to the data in Exercise 5.2.

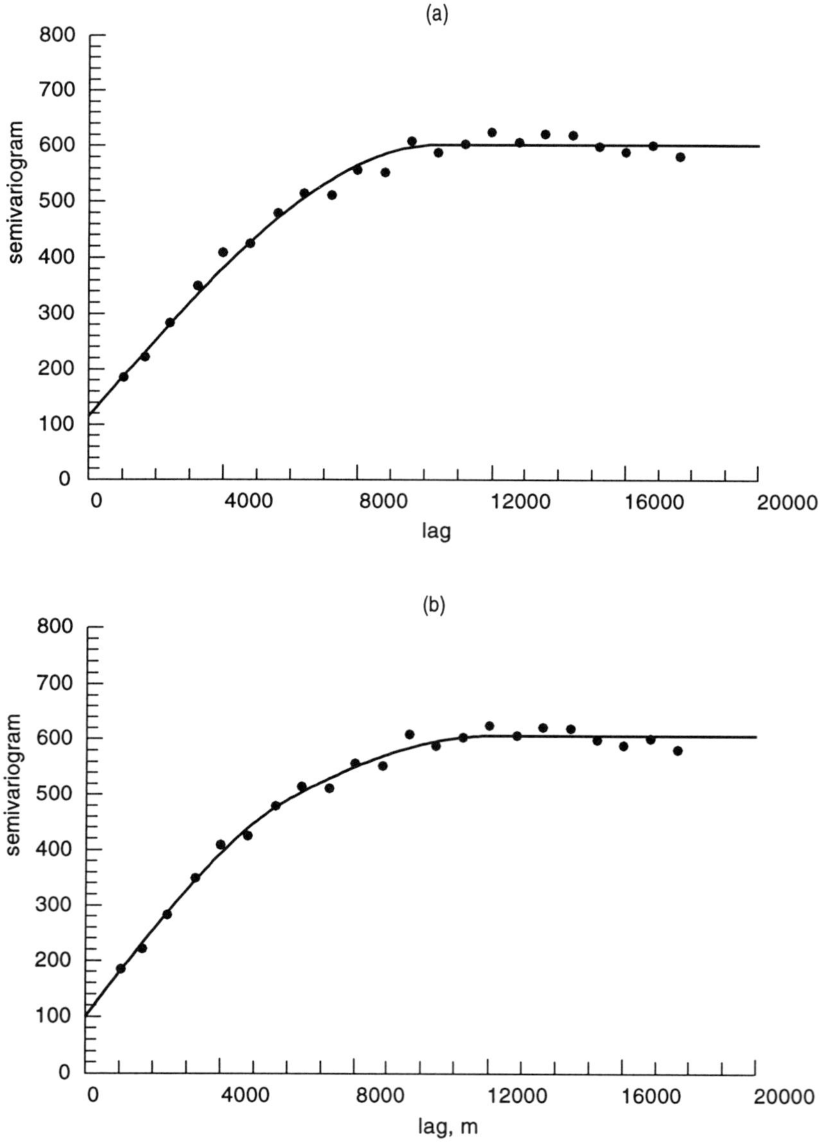

Figure 5.8. Simple and nested modeling of the same experimental semivariogram. (a) Best simple model in Figure 5.7. The model is spherical with a nugget effect of 114.73, (sill – nugget) equal to 486.98, and range of 10411 meters: 114.73 + sph(486.98, 10411). The sum of the weighted square differences (R_m) is 0.010 and the Akaike information criterion ($\widehat{AIC}$) is -164.3. (b) A model comprising two spherical semivariograms: 99.7 + sph(338.41, 12339) + sph(167.85, 6420), with $R_m = 0.007$ and $\widehat{AIC} = -168.7$.

Support

Some of the geologic attributes of interest in geostatistics, such as formation thickness, are point measurements in the sense that there is a location associated with each measurement, but not a volume. Other attributes, such as permeability, are spatial averages and, indeed, have a volume associated with each measurement. In this case, the location usually refers to the coordinates of the center of gravity of the analyzed volume. The geostatistical name for this type of volume is *support*, and its complete specification includes geometrical shape, size, and spatial orientation.

A change in any of the characteristics of a support defines a new regionalized variable. At a given location and elevation, the permeability of a cylindrical core 3 inches long may be completely different from the permeability for the central 1 inch of the same core. Thus an additional caution in semivariogram analysis is to make sure that data for the estimation of any semivariogram relate to the same support. In general, the effect of a larger support is to reduce variability, resulting in semivariograms with smaller sills and larger ranges.

A different but related problem has to do with the heterogeneity of equipment, operating conditions, and experimental procedures used in obtaining the measurements. It is not uncommon to discover that variations in measurements are not related to actual spatial variation but rather to changes in other confounding factors, such as techniques employed by the contractor who collected the data or the laboratory that made the analyses.

There are analytical procedures to account for the variations in the semivariogram due to changes in support (Clark, 1979, chapter 3; Journel and Huijbregts, 1978, chapter 2). When the sampling is comprised of mixtures of supports, the different supports should be processed separately as different regionalized variables. Change of support transformations should be used to consolidate the results.

As shown in Chapter 12, kriging may require semivariogram models for supports other than points. Today most computer programs only rely on point semivariograms. Necessary semivariogram transformations are computed internally by the software and are transparent to the user.

Direct Applications

Although rare, calculation of the experimental semivariogram, in itself, can be the final objective of a study intended to make use of the capability of the semivariogram to assess the average deterioration in similarity between two observations as their mutual distance increases. Such analyses may be of particular interest for detecting anisotropy.

Reliable experimental semivariograms can be used for comparative purposes and for sampling design. In a general way, similar semivariograms for the same type of attribute indicate a common origin and significantly different semivariograms denote a distinct genesis. Longer ranges and smaller sills and nugget effects relate to more continuous processes having smoother and less frequent fluctuations.

Proper sampling of a regionalized variable calls for an initial survey involving a few traverses sampled at a short interval to allow for a good semivariogram estimation. Further densification of the sampling should then be done, preferably following a regular pattern. At a bare minimum the sampling density should be sufficient to assure that no location in the field is farther away from the closest observation than a distance equal to the range. Otherwise important structures could escape detection and the advantage of geostatistical estimations based on spatial correlation would be lost. As demonstrated by Corollaries 2.9 and 4.9, when the mean is constant, geostatistics, like classical statistics, predicts that the most likely value at those isolated locations is the sample mean.

In the presence of anisotropy, the sampling interval should be directional, vary proportionally to fluctuations in the range, and be inversely proportional to variations in the sill.

Chapter 6
UNIVERSAL KRIGING

In Chapter 4 we saw a first attempt to generalize simple kriging by removing the requirement that the constant mean of the random function be known and an effort to lift the constant mean requirement by moving from a global constant mean to a locally constant mean. There are several instances of attributes—water depth near the shore, temperature in the upper part of the earth's crust, water table elevation in the High Plains aquifer of Kansas—that have a clear systematic variation. Models that presume constancy of the mean are inadequate for the characterization of such attributes.

Universal kriging is a generalization that takes simple kriging one step beyond the ordinary kriging expansion. In addition to removing the requirement of simple kriging that the mean be known, universal kriging fully removes from ordinary kriging the need to assume a constant mean. A.G. Journel and G. Matheron coined the term and formulated the method to address problems in the mapping of submarine topography while the former was on military duty at the French Navy Cartographic Center (Journel, 1969). Currently Journel favors the more descriptive label of *kriging with a trend model* (Deutsch and Journel, 1992, p. 64).

The Estimator

At the core of the generalization that characterizes universal kriging is the same idea as that in Corollary 2.11 of automatically filtering the mean.

Definition 6.1

Let $Z(\boldsymbol{x}_i)$ be k random variables inside a neighborhood $d \subset D$ of the spatial domain. The universal kriging estimator $\hat{Z}(\boldsymbol{x}_0)$ for the random function $Z(\boldsymbol{x})$ is the linear combination

$$\hat{Z}(\boldsymbol{x}_0) = \sum_{i=1}^{k} \lambda_i Z(\boldsymbol{x}_i) . \quad \square$$

Use of a neighborhood is not a fundamental theoretical requirement. It is employed mostly to better fulfill the assumptions and to reduce the size of the system of normal equations. Although in practice the use of simple kriging and stationary ordinary kriging involves using only a few of the closest observations, theoretically they do not implicitly restrict the estimation to a neighborhood.

Definition 6.2

Let $Z(\boldsymbol{x})$ be a random function. Then the drift $m(\boldsymbol{x})$ is

$$m(\boldsymbol{x}) = \mathrm{E}[Z(\boldsymbol{x})] \; ,$$

where $\mathrm{E}[\cdot]$ denotes expected value. □

As shown in Figure 6.1, the drift is a mathematical characterization of the intuitive notion of trend, a gentle underlying fluctuation in the regionalized variable. A trend is a somewhat arbitrary but convenient notion that is impossible to sample and whose perception depends on the scale of the study. The same feature may be regarded as part of a trend at the local level, as an anomaly at the regional level, and as an insignificant detail at the continental scale.

Assumptions

In its most general form, universal kriging makes the following assumptions that are needed to make the derivations and estimate the semivariogram.

Assumption 6.1

A sampling is a partial realization of a random function. □

Assumption 6.2

The residuals $Y(\boldsymbol{x})$ in Definition 2.2 honor the intrinsic hypothesis:

$$\left.\begin{array}{l}\mathrm{E}[Y(\boldsymbol{x})] \text{ is constant} \\ \mathrm{Var}[Y(\boldsymbol{x}) - Y(\boldsymbol{x}+\boldsymbol{h})] = 2\gamma_Y(\boldsymbol{h})\end{array}\right\} \text{ for all } \boldsymbol{x}, \; \boldsymbol{x}+\boldsymbol{h} \text{ in } d \; ,$$

where $\boldsymbol{h}$ is a Euclidean distance and $\gamma_Y(\boldsymbol{h})$ is the semivariogram of the residuals. □

The null expectation for the residuals is not exactly an assumption, but a fact proved by Lemma 2.5.

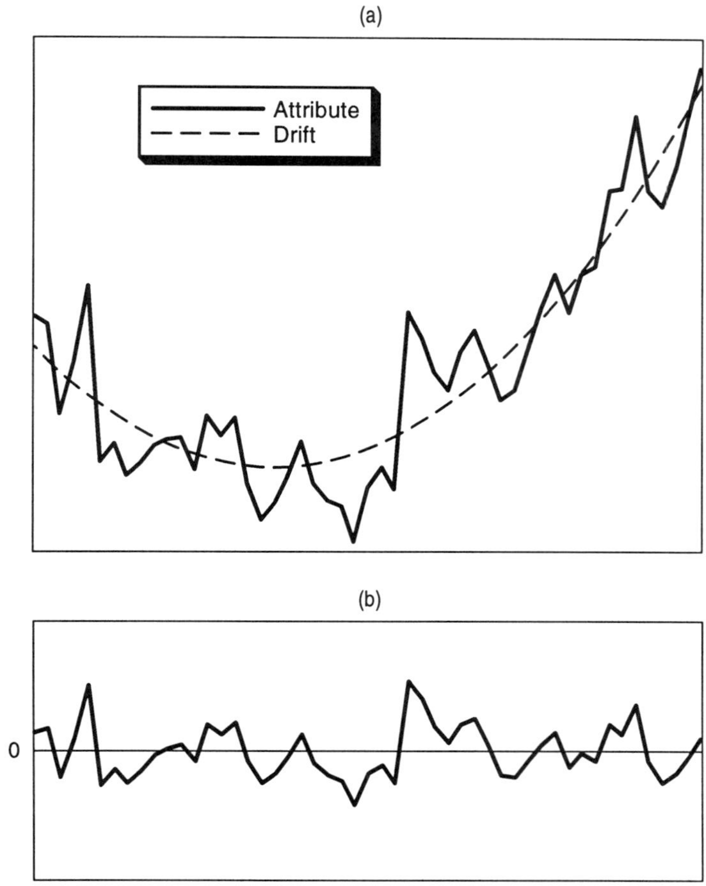

Figure 6.1. Decomposition of a regionalized variable: (a) attribute and its drift, (b) residual after subtraction of the drift from the attribute.

Assumption 6.3

The deterministic and unknown drift can be modeled as a linear combination of analytical functions. □

Among numerous possibilities, here I will restrict attention to the most common alternative—polynomials.

Definition 6.3

Let $m(\boldsymbol{x})$ be the drift in Definition 6.2. A polynomial drift model $f_l(\boldsymbol{x})$ is the summation

$$m(\boldsymbol{x}) = \sum_{l=0}^{n} a_l f_l(\boldsymbol{x}), \text{ with } f_0(\boldsymbol{x}) = 1$$

that is valid inside the same neighborhood $d \subset D$ in Definition 6.1. □

The most common form of $f_l(\boldsymbol{x})$ is that of a monomial in the geographical coordinates, in which case the expression for the drift takes, for example, the form $m(\boldsymbol{x}) = a_0 + a_1 x + a_2 y + a_3 x^2 + a_4 xy + a_5 y^2$, where x denotes easting and y denotes northing in a two-dimensional space.

Handling of the mean is done automatically in universal kriging—it is transparent to the user. Quasi-stationary ordinary kriging is a special case of universal kriging when $m(\boldsymbol{x}) = a_0 = m$.

The derivation of the universal kriging weight estimation follows steps similar to those of ordinary kriging:

1. Find the requirement to have an unbiased estimator.
2. Calculate an analytical expression for the estimation error.
3. Solve the constraint minimization problem by the Lagrange method of multipliers.
4. Evaluate the estimation error for the optimal parameters.

Unbiasedness

Finding the necessary and sufficient conditions to have an unbiased universal kriging estimator is more straightforward if one first proves the following lemma.

Lemma 6.1

Let $\hat{Z}(\boldsymbol{x}_0)$ be the estimator in Definition 6.1 and let $m(\boldsymbol{x})$ be the drift in Definition 6.3. Then

$$\mathrm{E}\left[\hat{Z}(\boldsymbol{x}_0)\right] = \sum_{l=0}^{n}\sum_{i=1}^{k} a_l \lambda_i f_l(\boldsymbol{x}_i)\,.$$

Proof

From Definition 6.1,

$$\mathrm{E}\left[\hat{Z}(\boldsymbol{x}_0)\right] = \mathrm{E}\left[\sum_{i=1}^{k} \lambda_i Z(\boldsymbol{x}_i)\right].$$

One can interchange expectation and summation and remove the parameter λ_i from the expectation:

$$\mathrm{E}\left[\hat{Z}(\boldsymbol{x}_0)\right] = \sum_{i=1}^{k} \lambda_i \mathrm{E}\left[Z(\boldsymbol{x}_i)\right].$$

But the expected value is the drift at $\boldsymbol{x}_i$. Hence, by Definitions 6.2 and 6.3,

$$\mathrm{E}\left[\hat{Z}(\boldsymbol{x}_0)\right] = \sum_{i=1}^{k} \lambda_i \sum_{l=0}^{n} a_l f_l(\boldsymbol{x}_i)\,.$$

The term λ_i can be placed in the second summation because λ_i is independent of l. Therefore, interchanging summation and reversing the product,

$$\mathrm{E}\left[\hat{Z}(\boldsymbol{x}_0)\right] = \sum_{l=0}^{n}\sum_{i=1}^{k} a_l \lambda_i f_l(\boldsymbol{x}_i) \,. \quad \square$$

Theorem 6.1

Let $\hat{Z}(\boldsymbol{x}_0)$ be the estimator in Definition 6.1 and $m(\boldsymbol{x})$ be the drift in Definition 6.3. Then the estimator is unbiased if and only if

$$\sum_{i=1}^{k} \lambda_i f_l(\boldsymbol{x}_i) = f_l(\boldsymbol{x}_0) \ , \ \text{for } l = 0,\, 1,\, \ldots,\, n \,.$$

Proof

The relationship $\mathrm{E}\left[\hat{Z}(\boldsymbol{x}_0) - Z(\boldsymbol{x}_0)\right] = 0$ is the mathematical way of saying that the estimator is unbiased.

(a) Let us suppose that the estimator is unbiased. Then

$$\mathrm{E}\left[\hat{Z}(\boldsymbol{x}_0)\right] = \mathrm{E}\left[Z(\boldsymbol{x}_0)\right] .$$

But by Lemma 6.1 the left side can be changed to

$$\sum_{l=0}^{n}\sum_{i=1}^{k} a_l \lambda_i f_l(\boldsymbol{x}_i) = \mathrm{E}\left[Z(\boldsymbol{x}_0)\right] ,$$

and by Definitions 6.2 and 6.3 the right side can be

$$\sum_{l=0}^{n}\sum_{i=1}^{k} a_l \lambda_i f_l(\boldsymbol{x}_i) = \sum_{l=0}^{n} a_l f_l(\boldsymbol{x}_0)$$

or

$$\sum_{l=0}^{n} a_l \sum_{i=1}^{k} \lambda_i f_l(\boldsymbol{x}_i) = \sum_{l=0}^{n} a_l f_l(\boldsymbol{x}_0) \,,$$

which is true only if

$$\sum_{i=1}^{k} \lambda_i f_l(\boldsymbol{x}_i) = f_l(\boldsymbol{x}_0) \ , \ \text{for } l = 0,\, 1,\, 2,\, \ldots,\, n \,.$$

(b) Now let

$$\sum_{i=1}^{k} \lambda_i f_l(\boldsymbol{x}_i) = f_l(\boldsymbol{x}_0) \ , \ \text{for } l = 0,\, 1,\, 2,\, \ldots,\, n \,.$$

The equality persists if every lth equality is multiplied by a_l. The summation of the $n+1$ equations yields

$$\sum_{l=0}^{n} a_l \sum_{i=1}^{k} \lambda_i f_l(\boldsymbol{x}_i) = \sum_{l=0}^{n} a_l f_l(\boldsymbol{x}_0) .$$

On the left side of the equation, after moving a_l to the second summation, according to Lemma 6.1 the double summation is equal to the expected value of $\hat{Z}(\boldsymbol{x}_0)$. On the right side, by Definitions 6.2 and 6.3, the summation is equal to the expected value of $Z(\boldsymbol{x}_0)$. Therefore,

$$\mathrm{E}\left[\hat{Z}(\boldsymbol{x}_0) - Z(\boldsymbol{x}_0)\right] = 0 .$$

Proofs (a) and (b) prove the theorem. □

Notice that the constraint $\sum_{i=1}^{k} \lambda_i = 1$ is included in the set of unbiasedness constraints for $l = 0$. For ordinary kriging, the constraint $\sum_{i=1}^{k} \lambda_i = 1$ simultaneously takes care of filtering the mean and causing the estimator to be unbiased. In universal kriging, making the weights sum to 1 is only one of the necessary conditions for an unbiased estimator.

Estimation Variance

The mean square error is used to measure the efficiency of the estimator, which will be established in terms of the unknown weights after proving a lemma and a theorem.

Lemma 6.2

Let $\hat{Z}(\boldsymbol{x}_0)$ be the linear unbiased estimator for $Z(\boldsymbol{x})$ in Definition 6.1 and let λ_i be the weights in the same definition. Then if $Y(\boldsymbol{x})$ is the residual in Definition 2.2,

$$\hat{Z}(\boldsymbol{x}_0) - Z(\boldsymbol{x}_0) = \sum_{i=1}^{k} \lambda_i Y(\boldsymbol{x}_i) - Y(\boldsymbol{x}_0) .$$

Proof

Replacing $\hat{Z}(\boldsymbol{x}_0)$ by its Definition 6.1 and splitting the random variable into the drift and its residual in Definition 2.2,

$$\hat{Z}(\boldsymbol{x}_0) - Z(\boldsymbol{x}_0) = \sum_{i=1}^{k} \lambda_i \{Y(\boldsymbol{x}_i) + \mathrm{E}[Z(\boldsymbol{x}_i)]\} - \{Y(\boldsymbol{x}_0) + \mathrm{E}[Z(\boldsymbol{x}_0)]\} .$$

After expanding the summation, interchanging summation and expectation, and rearranging terms, the relationship changes to

$$\hat{Z}(\boldsymbol{x}_0) - Z(\boldsymbol{x}_0) = \sum_{i=1}^{k} \lambda_i Y(\boldsymbol{x}_i) - Y(\boldsymbol{x}_0) + \left\{ \mathrm{E}\left[\sum_{i=1}^{k} \lambda_i Z(\boldsymbol{x}_i)\right] - \mathrm{E}\Big[Z(\boldsymbol{x}_0)\Big] \right\} .$$

But by Definition 6.1, the last summation is equal to $\hat{Z}(\boldsymbol{x}_0)$, so

$$\hat{Z}(\boldsymbol{x}_0) - Z(\boldsymbol{x}_0) = \sum_{i=1}^{k} \lambda_i Y(\boldsymbol{x}_i) - Y(\boldsymbol{x}_0) + \mathrm{E}\left[\hat{Z}(\boldsymbol{x}_0) - Z(\boldsymbol{x}_0)\right] .$$

The expectation is zero because, by Theorem 6.1, the estimator is unbiased. Hence,

$$\hat{Z}(\boldsymbol{x}_0) - Z(\boldsymbol{x}_0) = \sum_{i=1}^{k} \lambda_i Y(\boldsymbol{x}_i) - Y(\boldsymbol{x}_0) \ . \quad \square$$

Theorem 6.2

Let $\hat{Z}(\boldsymbol{x}_0)$ be the linear unbiased estimator for $Z(\boldsymbol{x})$ in Definition 6.1 and let $\gamma_Y(\boldsymbol{h})$ be the semivariogram of the residuals in Assumption 6.2. Then the estimation variance $\sigma^2(\boldsymbol{x}_0)$ is equal to

$$2\sum_{i=1}^{k} \lambda_i \gamma_Y(\boldsymbol{x}_0, \boldsymbol{x}_i) - \sum_{i=1}^{k}\sum_{j=1}^{k} \lambda_i \lambda_j \gamma_Y(\boldsymbol{x}_i, \boldsymbol{x}_j) \ .$$

Proof

By definition,

$$\sigma^2(\boldsymbol{x}_0) = \mathrm{Var}\left[\hat{Z}(\boldsymbol{x}_0) - Z(\boldsymbol{x}_0)\right] .$$

By Lemma 6.1, one can change the argument to

$$\sigma^2(\boldsymbol{x}_0) = \mathrm{Var}\left[\sum_{i=1}^{k} \lambda_i Y(\boldsymbol{x}_i) - Y(\boldsymbol{x}_0)\right]$$

Considering that by Theorem 6.1, $\sum_{i=1}^{k} \lambda_i = 1$, then

$$\sigma^2(\boldsymbol{x}_0) = \mathrm{Var}\left[\sum_{i=1}^{k} \lambda_i \big[Y(\boldsymbol{x}_i) - Y(\boldsymbol{x}_0)\big]\right]$$

and, by Lemma 2.4,

$$\sigma^2(\boldsymbol{x}_0) = \sum_{i=1}^{k}\sum_{j=1}^{k} \lambda_i \lambda_j \mathrm{Cov}\big(Y(\boldsymbol{x}_i) - Y(\boldsymbol{x}_0), Y(\boldsymbol{x}_j) - Y(\boldsymbol{x}_0)\big),$$

which by Lemma 4.4 is equivalent to

$$\sigma^2(\boldsymbol{x}_0) = \sum_{i=1}^{k}\sum_{j=1}^{k} \lambda_i \lambda_j \{\gamma_Y(\boldsymbol{x}_0 - \boldsymbol{x}_i) + \gamma_Y(\boldsymbol{x}_0 - \boldsymbol{x}_j) - \gamma_Y(\boldsymbol{x}_i - \boldsymbol{x}_j)\}.$$

Expanding, because of the constraint $\sum_{i=1}^{k} \lambda_i = 1$,

$$\sigma^2(\boldsymbol{x}_0) = \sum_{i=1}^{k} \lambda_i \gamma_Y(\boldsymbol{x}_0 - \boldsymbol{x}_i) + \sum_{j=1}^{k} \lambda_j \gamma_Y(\boldsymbol{x}_0 - \boldsymbol{x}_j) - \sum_{i=1}^{k}\sum_{j=1}^{k} \lambda_i \lambda_j \gamma_Y(\boldsymbol{x}_i - \boldsymbol{x}_j).$$

The first two terms on the right-hand side of the equation are the same, which proves the theorem. □

Optimization

The purpose of universal kriging is to calculate the weights in such a way that the estimator in Definition 6.1 is unbiased and has minimum mean square error. Finding the optimum weights for this estimator can be done employing the Lagrange method of multipliers (Hillier and Lieberman, 1995, p. 961) for minimizing the estimation variance in Theorem 6.2, subject to the constraint in Theorem 6.1.

Definition 6.4

Let $\sigma^2(\boldsymbol{x}_0)$ be the estimation variance in Theorem 6.2 for an estimation at $\boldsymbol{x}_0$, let λ_i be the weights in Definition 6.1, and let μ_l be Lagrange multipliers. The Lagrangian function for universal kriging is

$$L(\lambda_i, \mu_l) = \sigma^2(\boldsymbol{x}_0) + 2\mu_0 \left(\sum_{i=1}^{k} \lambda_i - 1\right) + 2\sum_{l=1}^{n} \mu_l \left(\sum_{i=1}^{k} \lambda_i f_l(\boldsymbol{x}_i) - f_l(\boldsymbol{x}_0)\right). \quad \square$$

Theorem 6.3

Let λ_i be the weights in Definition 6.1, let μ_l be the Lagrange multiplier in Definition 6.4, and let $\gamma_Y(\boldsymbol{h})$ be the semivariogram for the residuals in Assumption 6.2. If the semivariogram is negative definite (Definition 4.2), then

(a) the weights that produce the unique minimum estimation variance are the solution to

$$\left.\begin{array}{l}
\sum_{i=1}^{k} \lambda_i \gamma_Y(\boldsymbol{x}_i, \boldsymbol{x}_1) - \mu_0 - \sum_{l=1}^{n} \mu_l f_l(\boldsymbol{x}_1) = \gamma_Y(\boldsymbol{x}_0, \boldsymbol{x}_1) \\
\sum_{i=1}^{k} \lambda_i \gamma_Y(\boldsymbol{x}_i, \boldsymbol{x}_2) - \mu_0 - \sum_{l=1}^{n} \mu_l f_l(\boldsymbol{x}_2) = \gamma_Y(\boldsymbol{x}_0, \boldsymbol{x}_2) \\
\cdots\cdots\cdots\cdots\cdots\cdots\cdots\cdots\cdots\cdots \\
\sum_{i=1}^{k} \lambda_i \gamma_Y(\boldsymbol{x}_i, \boldsymbol{x}_k) - \mu_0 - \sum_{l=1}^{n} \mu_l f_l(\boldsymbol{x}_k) = \gamma_Y(\boldsymbol{x}_0, \boldsymbol{x}_k) \\
\sum_{i=1}^{k} \lambda_i = 1 \\
\sum_{i=1}^{k} \lambda_i f_1(\boldsymbol{x}_i) = f_1(\boldsymbol{x}_0) \\
\sum_{i=1}^{k} \lambda_i f_2(\boldsymbol{x}_i) = f_2(\boldsymbol{x}_0) \\
\cdots\cdots\cdots\cdots\cdots \\
\sum_{i=1}^{k} \lambda_i f_n(\boldsymbol{x}_i) = f_n(\boldsymbol{x}_0)
\end{array}\right\}$$

and (b) the kriging variance is never negative.

Proof

(a) Replacing the value of $\sigma^2(\boldsymbol{x}_0)$ in Definition 6.4 by that given by Theorem 6.2,

$$L(\lambda_i, \mu_l) = 2\sum_{i=i}^{k} \lambda_i \gamma_Y(\boldsymbol{x}_0, \boldsymbol{x}_i) - \sum_{i=1}^{k}\sum_{j=1}^{k} \lambda_i \lambda_j \gamma_Y(\boldsymbol{x}_i, \boldsymbol{x}_j) + 2\mu_0 \left(\sum_{i=1}^{k} \lambda_i - 1\right)$$
$$+ 2\sum_{l=1}^{n} \mu_l \left(\sum_{i=1}^{k} \lambda_i f_l(\boldsymbol{x}_i) - f_l(\boldsymbol{x}_0)\right),$$

one can see that the Lagrangian function is a quadratic expression of the unknown weights and the constraint on the weights is linear. In such a case, the necessary and sufficient condition for having a unique global minimum is that the term $-\sum_{i=1}^{k}\sum_{j=1}^{k} \lambda_i \lambda_j \gamma_Y(\boldsymbol{x}_i, \boldsymbol{x}_j)$ must be larger than or equal to zero (Cooper, 1974, p. 65), which by Definition 4.2 is mathematically equivalent to saying that the semivariogram of the residuals must be negative definite.

The minimum mean square error is given by those weights that make zero all first derivatives of the Lagrangian function with respect to the unknowns:

$$\frac{\partial L}{\partial \lambda_i} = 2\gamma_Y(\boldsymbol{x}_0, \boldsymbol{x}_i) - 2\sum_{j=1}^{k} \lambda_j \gamma_Y(\boldsymbol{x}_i, \boldsymbol{x}_j) + 2\mu_0 + 2\sum_{l=1}^{n} \mu_l f_l(\boldsymbol{x}_i), \quad \text{for } i = 1, 2, \ldots, k$$

$$\frac{\partial L}{\partial \mu_0} = 2\left(\sum_{i=1}^{k} \lambda_i - 1\right)$$

$$\frac{\partial L}{\partial \mu_l} = 2\left(\sum_{i=1}^{k} \lambda_i f_l(\boldsymbol{x}_i) - f_l(\boldsymbol{x}_0)\right), \text{ for } l = 1, 2, \ldots, n\,.$$

The solution of the system of equations comes from setting these partial derivatives equal to zero, dividing both sides by -2, and adding $\gamma_Y(\boldsymbol{x}_0, \boldsymbol{x}_i)$ to both sides of the first k equations.

(b) One can rewrite Theorem 6.2 as

$$\sigma^2(\boldsymbol{x}_0) = -\sum_{i=0}^{k}\sum_{j=0}^{k} \lambda_i \lambda_j \gamma_Y(\boldsymbol{x}_i, \boldsymbol{x}_j)\,,$$

with $\lambda_0 = -1$. The proof follows directly from the negative definite condition. □

Minimum Mean Square Error

The expression for the estimation variance given by Theorem 6.2 is valid for any weights λ_i and multipliers μ_l. The next theorem provides the value of $\sigma^2(\boldsymbol{x}_0)$ for the special case of the optimal parameters.

Lemma 6.3

Let λ_i be the optimal weights for the universal kriging estimator in Definition 6.1 and let μ_l be the Lagrange multiplier in Definition 6.4. If $\gamma_Y(\boldsymbol{h})$ is the semivariogram for the residuals in Assumption 6.2, then

$$\sum_{i=1}^{k}\sum_{j=1}^{k} \lambda_i \lambda_j \gamma_Y(\boldsymbol{x}_i, \boldsymbol{x}_j) = \sum_{i=1}^{k} \lambda_i \gamma_Y(\boldsymbol{x}_0, \boldsymbol{x}_i) - \mu_0 - \sum_{l=1}^{n} \mu_l f_l(\boldsymbol{x}_0)\,.$$

Proof

From Theorem 6.3,

$$\sum_{j=1}^{k} \lambda_j \gamma_Y(\boldsymbol{x}_i, \boldsymbol{x}_j) = \gamma_Y(\boldsymbol{x}_0, \boldsymbol{x}_i) - \mu_0 - \sum_{l=1}^{n} \mu_l f_l(\boldsymbol{x}_i)\,, \text{ for } i = 1, 2, \ldots, k\,.$$

Each of the k expressions is still true if one multiplies both sides of the ith equation by λ_i. Adding them all together,

$$\sum_{i=1}^{k} \lambda_i \sum_{j=1}^{k} \lambda_j \gamma_Y (\boldsymbol{x}_i, \boldsymbol{x}_j) = \sum_{i=1}^{k} \lambda_i \gamma_Y (\boldsymbol{x}_0, \boldsymbol{x}_i) - \sum_{i=1}^{k} \lambda_i \left\{ \mu_0 + \sum_{l=1}^{n} \mu_l f_l (\boldsymbol{x}_i) \right\}.$$

The proof follows by introducing λ_i under the other two summations and by Theorem 6.1. □

Theorem 6.4

Let λ_i be the optimal weights and μ_l be the optimal Lagrange multipliers for the universal kriging estimator in Definition 6.1. If $\gamma_Y (\boldsymbol{h})$ is the semivariogram for the residuals in Assumption 6.2 and the $f_l (\boldsymbol{x}_0)$ are the terms in Definition 6.3 evaluated at the estimation location, then the minimum mean square error of universal kriging, $\sigma^2_{UKI} (\boldsymbol{x}_0)$, is

$$\sigma^2_{UKI} (\boldsymbol{x}_0) = \sum_{i=i}^{k} \lambda_i \gamma_Y (\boldsymbol{x}_0, \boldsymbol{x}_i) - \mu_0 - \sum_{l=1}^{n} \mu_l f_l (\boldsymbol{x}_0) .$$

Proof

From Theorem 6.2,

$$\sigma^2_{UKI} (\boldsymbol{x}_0) = 2 \sum_{i=1}^{k} \lambda_i \gamma_Y (\boldsymbol{x}_0, \boldsymbol{x}_i) - \sum_{i=1}^{k} \sum_{j=1}^{k} \lambda_i \lambda_j \gamma_Y (\boldsymbol{x}_i, \boldsymbol{x}_j)$$

and by Lemma 6.3

$$\sigma^2_{UKI} (\boldsymbol{x}_0) = 2 \sum_{i=1}^{k} \lambda_i \gamma_Y (\boldsymbol{x}_0, \boldsymbol{x}_i) - \sum_{i=1}^{k} \lambda_i \gamma_Y (\boldsymbol{x}_0, \boldsymbol{x}_i) + \mu_0 + \sum_{l=1}^{n} \mu_l f_l (\boldsymbol{x}_0) ,$$

which, by addition of the terms in $\sum_{i=1}^{k} \lambda_i \gamma_Y (\boldsymbol{x}_0, \boldsymbol{x}_i)$, reduces to

$$\sigma^2_{UKI} (\boldsymbol{x}_0) = \sum_{i=1}^{k} \lambda_i \gamma_Y (\boldsymbol{x}_0, \boldsymbol{x}_i) + \mu_0 + \sum_{l=1}^{n} \mu_l f_l (\boldsymbol{x}_0) . \quad \square$$

Algorithm for Intrinsically Stationary Residuals

For more compact display of results, let us define some matrices.

Definition 6.5

Let $\mathbf{A}$ be the matrix

$$\mathbf{A} = \begin{bmatrix} \gamma_Y(\boldsymbol{x}_1,\boldsymbol{x}_1) & \gamma_Y(\boldsymbol{x}_2,\boldsymbol{x}_1) & \dots & \gamma_Y(\boldsymbol{x}_k,\boldsymbol{x}_1) & 1 & f_1(\boldsymbol{x}_1) & f_2(\boldsymbol{x}_1) & \dots & f_n(\boldsymbol{x}_1) \\ \gamma_Y(\boldsymbol{x}_1,\boldsymbol{x}_2) & \gamma_Y(\boldsymbol{x}_2,\boldsymbol{x}_2) & \dots & \gamma_Y(\boldsymbol{x}_k,\boldsymbol{x}_2) & 1 & f_1(\boldsymbol{x}_2) & f_2(\boldsymbol{x}_2) & \dots & f_n(\boldsymbol{x}_2) \\ \dots & \dots & \dots & \dots & \dots & \dots & \dots & \dots & \dots \\ \gamma_Y(\boldsymbol{x}_1,\boldsymbol{x}_k) & \gamma_Y(\boldsymbol{x}_2,\boldsymbol{x}_k) & \dots & \gamma_Y(\boldsymbol{x}_k,\boldsymbol{x}_k) & 1 & f_1(\boldsymbol{x}_k) & f_2(\boldsymbol{x}_k) & \dots & f_n(\boldsymbol{x}_k) \\ 1 & 1 & \dots & 1 & 0 & 0 & 0 & \dots & 0 \\ f_1(\boldsymbol{x}_1) & f_1(\boldsymbol{x}_2) & \dots & f_1(\boldsymbol{x}_k) & 0 & 0 & 0 & \dots & 0 \\ f_2(\boldsymbol{x}_1) & f_2(\boldsymbol{x}_2) & \dots & f_2(\boldsymbol{x}_k) & 0 & 0 & 0 & \dots & 0 \\ \dots & \dots & \dots & \dots & \dots & \dots & \dots & \dots & \dots \\ f_n(\boldsymbol{x}_1) & f_n(\boldsymbol{x}_2) & \dots & f_n(\boldsymbol{x}_k) & 0 & 0 & 0 & \dots & 0 \end{bmatrix},$$

where $\boldsymbol{x}_0$ is the estimation location, the $\boldsymbol{x}_i$'s are the sampling sites inside the neighborhood in Definition 6.1, the $\gamma_Y(\boldsymbol{x}_i, \boldsymbol{x}_j)$'s are residual semivariograms in Assumption 6.2, and the $f_i(\boldsymbol{x})$'s are the functions in Definition 6.3. □

Definition 6.6

Let $\mathbf{X}$ be the matrix

$$\mathbf{X} = [\,\lambda_1 \quad \lambda_2 \quad \dots \quad \lambda_k \quad \mu_0 \quad \mu_1 \quad \mu_2 \quad \dots \quad \mu_n\,]'\ ,$$

where the λ_i's are the optimal weights for the estimator in Definition 6.1 and the μ_l's are Lagrange multipliers in Definition 6.4. □

Definition 6.7

Let $\mathbf{a}$ be the following matrix:

$$\mathbf{a} = [\,\gamma_Y(\boldsymbol{x}_0,\boldsymbol{x}_1)\ \gamma_Y(\boldsymbol{x}_0,\boldsymbol{x}_2)\ \dots\ \gamma_Y(\boldsymbol{x}_0,\boldsymbol{x}_k)\ 1\ f_1(\boldsymbol{x}_0)\ f_2(\boldsymbol{x}_0)\ \dots\ f_n(\boldsymbol{x}_0)\,]'\ ,$$

where $\boldsymbol{x}_0$ is the estimation location, the $\boldsymbol{x}_i$'s are sampling sites inside the neighborhood in Definition 6.1, the $\gamma_Y(\boldsymbol{x}_0, \boldsymbol{x}_i)$'s are residual semivariograms, and the $f_i(\boldsymbol{x}_0)$'s are the functions in Definition 6.3. □

Algorithm 6.1

This is an algorithm for the estimation of a random function honoring Assumptions 6.1, 6.2, and 6.3.

1. Calculate each term in matrix $\mathbf{A}$ in Definition 6.5.
2. Calculate each term in matrix $\mathbf{a}$ in Definition 6.7.

3. Solve the system of equations $\mathbf{AX} = \mathbf{a}$, where $\mathbf{X}$ is the matrix in Definition 6.6.

4. Using the weights in $\mathbf{X}$, compute the estimate

$$\hat{Z}_{UKI}(\boldsymbol{x}_0) = \sum_{i=1}^{k} \lambda_i Z(\boldsymbol{x}_i) .$$

5. Calculate the universal kriging estimation variance

$$\sigma^2_{UKI}(\boldsymbol{x}_0) = \mathbf{X}'\mathbf{a} . \quad \square$$

As long as the assumptions are met, the algorithm can be employed to characterize any spatial attribute regardless of its physical nature.

Second Order Stationary Universal Kriging

Using Corollary 4.1, findings obtained assuming the intrinsic hypothesis can be translated into equivalent results assuming second order stationarity instead of the intrinsic hypothesis of Assumption 6.2.

Assumption 6.4

The residuals $Y(\boldsymbol{x})$ in Definition 2.2 are second order stationary if

$$\left.\begin{array}{l} \mathrm{E}[Y(\boldsymbol{x})] = 0 \\ \mathrm{E}[Y(\boldsymbol{x})\,Y(\boldsymbol{x}+\boldsymbol{h})] = \mathrm{Cov}_Y(\boldsymbol{h}) \end{array}\right\} \text{ for all } \boldsymbol{x}, \boldsymbol{x}+\boldsymbol{h} \in d ,$$

where $\mathrm{E}[\cdot]$ denotes expected value, $\boldsymbol{h}$ is a distance in the spatial domain of interest, and $\mathrm{Cov}_Y(\boldsymbol{h})$ is the covariance of the residuals. □

By Lemma 2.5, the constant value of $\mathrm{E}[Y(\boldsymbol{x})]$ is zero. Assumptions 6.1 and 6.3 remain in effect and unchanged from the intrinsically stationary case. I will limit the conversion to the final and most relevant results only.

Theorem 6.5

Let λ_i be the weights in Definition 6.1, let μ_l be the Lagrange multiplier in Definition 6.4, and let $\mathrm{Cov}_Y(\boldsymbol{h})$ be the covariance for second order stationary residuals. Then the weights that produce the minimum estimation variance are the solution to

$$\left.\begin{array}{l}
\sum\limits_{i=1}^{k} \lambda_i \mathrm{Cov}_Y(\boldsymbol{x}_i, \boldsymbol{x}_1) + \mu_0 + \sum\limits_{l=1}^{n} \mu_l f_l(\boldsymbol{x}_1) = \mathrm{Cov}_Y(\boldsymbol{x}_0, \boldsymbol{x}_1) \\
\sum\limits_{i=1}^{k} \lambda_i \mathrm{Cov}_Y(\boldsymbol{x}_i, \boldsymbol{x}_2) + \mu_0 + \sum\limits_{l=1}^{n} \mu_l f_l(\boldsymbol{x}_2) = \mathrm{Cov}_Y(\boldsymbol{x}_0, \boldsymbol{x}_1) \\
\cdots\cdots\cdots\cdots\cdots\cdots\cdots\cdots\cdots\cdots\cdots\cdots \\
\sum\limits_{i=1}^{k} \lambda_i \mathrm{Cov}_Y(\boldsymbol{x}_i, \boldsymbol{x}_k) + \mu_0 + \sum\limits_{l=1}^{n} \mu_l f_l(\boldsymbol{x}_k) = \mathrm{Cov}_Y(\boldsymbol{x}_0, \boldsymbol{x}_k) \\
\sum\limits_{i=1}^{k} \lambda_i = 1 \\
\sum\limits_{i=1}^{k} \lambda_i f_1(\boldsymbol{x}_i) = f_1(\boldsymbol{x}_0) \\
\sum\limits_{i=1}^{k} \lambda_i f_2(\boldsymbol{x}_i) = f_2(\boldsymbol{x}_0) \\
\cdots\cdots\cdots\cdots\cdots \\
\sum\limits_{i=1}^{k} \lambda_i f_n(\boldsymbol{x}_i) = f_n(\boldsymbol{x}_0)
\end{array}\right\} .$$

Proof

From Theorem 6.3,

$$\left.\begin{array}{l}
\sum\limits_{j=1}^{k} \lambda_j \gamma_Y(\boldsymbol{x}_i, \boldsymbol{x}_j) - \mu_0 - \sum\limits_{l=1}^{n} \mu_l f_l(\boldsymbol{x}_i) = 2\gamma_Y(\boldsymbol{x}_0, \boldsymbol{x}_i), \text{ for } i = 1,\ 2,\ \ldots,\ k \\
\sum\limits_{i=1}^{k} \lambda_i = 1 \\
\sum\limits_{i=1}^{k} \lambda_i f_l(\boldsymbol{x}_i) = f_l(\boldsymbol{x}_0), \text{ for } l = 1,\ 2,\ \ldots,\ n
\end{array}\right\} .$$

The last $n+1$ equations are the same. Only the first k equations have to be converted. From Corollary 4.1,

$$\sum_{j=1}^{k} \lambda_j \{\mathrm{Cov}_Y(0) - \mathrm{Cov}_Y(\boldsymbol{x}_i, \boldsymbol{x}_j)\} - \mu_0 - \sum_{l=1}^{n} \mu_l f_l(\boldsymbol{x}_i) = \mathrm{Cov}_Y(0) - \mathrm{Cov}_Y(\boldsymbol{x}_0, \boldsymbol{x}_i), \text{for } i = 1,\ 2,\ \ldots,\ k .$$

Expanding the summation,

$$\sum_{j=1}^{k} \lambda_j \mathrm{Cov}_Y(0) - \sum_{j=1}^{k} \lambda_j \mathrm{Cov}_Y(\boldsymbol{x}_i, \boldsymbol{x}_j) - \mu_0 - \sum_{l=1}^{n} \mu_l f_l(\boldsymbol{x}_i) = \mathrm{Cov}_Y(0) - \mathrm{Cov}_Y(\boldsymbol{x}_0, \boldsymbol{x}_i), \text{for } i = 1,\ 2,\ \ldots,\ k .$$

$\mathrm{Cov}_Y(0)$ is a constant; and, because according to the $(k+1)$th equation, $\sum\limits_{i=1}^{k} \lambda_i = 1$, the terms in $\mathrm{Cov}_Y(0)$ cancel out. The proof follows after multiplying both sides of the first k equations by -1. □

$\text{Cov}_Y(\cdot)$ must be positive definite (Definition 2.3) for the system to have all the favorable characteristics of a quadratic system.

Theorem 6.6

Let λ_i be the optimal weights for the simple kriging estimator in Definition 6.1 and let μ_l be the Lagrange multiplier in Definition 6.4. If $\text{Cov}_Y(\boldsymbol{h})$ is the covariance for the residuals, then the minimum mean square error for universal kriging, $\sigma^2_{UK}(\boldsymbol{x}_0)$, is

$$\sigma^2_{UK}(\boldsymbol{x}_0) = \text{Cov}_Y(0) - \sum_{i=i}^{k} \lambda_i \text{Cov}_Y(\boldsymbol{x}_0, \boldsymbol{x}_i) - \mu_0 - \sum_{l=1}^{n} \mu_l f_l(\boldsymbol{x}_0) .$$

Proof

From Theorem 6.4,

$$\sigma^2_{UK}(\boldsymbol{x}_0) = \sum_{i=i}^{k} \lambda_i \gamma_Y(\boldsymbol{x}_0, \boldsymbol{x}_i) - \mu_0 - \sum_{l=1}^{n} \mu_l f_l(\boldsymbol{x}_0) ,$$

but, by Corollary 4.1,

$$\sigma^2_{UK}(\boldsymbol{x}_0) = \sum_{i=i}^{k} \lambda_i \left\{\text{Cov}_Y(0) - \text{Cov}_Y(\boldsymbol{x}_0, \boldsymbol{x}_i)\right\} - \mu_0 - \sum_{l=1}^{n} \mu_l f_l(\boldsymbol{x}_0) .$$

Expanding the summation,

$$\sigma^2_{UK}(\boldsymbol{x}_0) = \sum_{i=i}^{k} \lambda_i \text{Cov}_Y(0) - \sum_{i=i}^{k} \lambda_i \text{Cov}_Y(\boldsymbol{x}_0, \boldsymbol{x}_i) - \mu_0 - \sum_{l=1}^{n} \mu_l f_l(\boldsymbol{x}_0) ;$$

the proof follows by the constraint on the weights in Theorem 6.1 and the fact that $\text{Cov}_Y(0)$ is a constant. □

If four more matrices are defined, a second algorithm can be formulated taking advantage of compact matrix notation.

Definition 6.8

Let $\boldsymbol{x}_i$ be the sampling sites inside the neighborhood in Definition 6.1 and let $\text{Cov}_Y(\boldsymbol{x}_i, \boldsymbol{x}_j)$ be the semivariogram. Then $\mathbf{P}$ is the matrix

$$\mathbf{P} = \begin{bmatrix} \text{Cov}_Y(\boldsymbol{x}_1, \boldsymbol{x}_1) & \dots & \text{Cov}_Y(\boldsymbol{x}_k, \boldsymbol{x}_1) & 1 & f_1(\boldsymbol{x}_1) & f_2(\boldsymbol{x}_1) & \dots & f_n(\boldsymbol{x}_1) \\ \dots & \dots & \dots & \dots & \dots & \dots & \dots & \dots \\ \text{Cov}_Y(\boldsymbol{x}_1, \boldsymbol{x}_k) & \dots & \text{Cov}_Y(\boldsymbol{x}_k, \boldsymbol{x}_k) & 1 & f_1(\boldsymbol{x}_k) & f_2(\boldsymbol{x}_k) & \dots & f_n(\boldsymbol{x}_k) \\ 1 & \dots & 1 & 0 & 0 & 0 & \dots & 0 \\ f_1(\boldsymbol{x}_1) & \dots & f_1(\boldsymbol{x}_k) & 0 & 0 & 0 & \dots & 0 \\ f_2(\boldsymbol{x}_1) & \dots & f_2(\boldsymbol{x}_k) & 0 & 0 & 0 & \dots & 0 \\ \dots & \dots & \dots & \dots & \dots & \dots & \dots & \dots \\ f_n(\boldsymbol{x}_1) & \dots & f_n(\boldsymbol{x}_k) & 0 & 0 & 0 & \dots & 0 \end{bmatrix} . \square$$

Definition 6.9

Let λ_i's be the optimal weights for the estimator in Definition 2.1. Then $\mathbf{Q}$ is the matrix

$$\mathbf{Q} = [\,\lambda_1 \quad \lambda_2 \quad \cdots \quad \lambda_k \quad -\mu_0 \quad -\mu_1 \quad -\mu_2 \quad \cdots \quad -\mu_n\,]' . \quad \square$$

Definition 6.10

Let $\text{Cov}_Y\,(\boldsymbol{x}_i, \boldsymbol{x}_j)$ be the covariance of a second order stationary random function, $\boldsymbol{x}_0$ be the estimation location, $\boldsymbol{x}_i$ be sampling sites inside the neighborhood in Definition 6.1, and $f_2(\boldsymbol{x}_0)$ be the functions in Definition 6.3. Then $\mathbf{p}$ is the matrix

$$\mathbf{p} = [\text{Cov}_Y(\boldsymbol{x}_0, \boldsymbol{x}_1) \quad \cdots \quad \text{Cov}_Y(\boldsymbol{x}_0, \boldsymbol{x}_k) \;\; 1 \;\; f_1(\boldsymbol{x}_0) \;\; f_2(\boldsymbol{x}_0) \quad \cdots \quad f_n(\boldsymbol{x}_0)\,]' . \quad \square$$

Definition 6.11

Let $Z\,(\boldsymbol{x}_i)$ be random variables of a random function sampled at sites $\boldsymbol{x}_i$ inside the neighborhood in Definition 6.1. Then the matrix $\mathbf{Z}$ is the vectorial matrix

$$\mathbf{Z} = [\,Z\,(\boldsymbol{x}_1) \quad Z\,(\boldsymbol{x}_2) \quad \cdots \quad Z\,(\boldsymbol{x}_k) \quad 0 \quad 0 \quad \cdots \quad 0\,]'$$

containing $n+1$ zeros, where n is the number of monomial terms in the drift model in Definition 6.3. $\square$

Algorithm 6.2

This is a universal kriging algorithm for the estimation of attributes honoring Assumptions 6.1, 6.3, and 6.4.

1. Calculate each term in matrix $\mathbf{P}$ in Definition 6.8.
2. Calculate each term in matrix $\mathbf{p}$ in Definition 6.10.
3. Solve the system of equations

$$\mathbf{PQ} = \mathbf{p}\,,$$

where $\mathbf{Q}$ is the matrix in Definition 6.9.

4. Compute the estimate

$$\hat{Z}_{UK}\,(\boldsymbol{x}_0) = \mathbf{Z}'\mathbf{Q} = \mathbf{Z}'\mathbf{P}^{-1}\mathbf{p}\,,$$

where $\mathbf{Z}$ is the matrix in Definition 6.11.

5. Calculate the universal kriging estimation variance

$$\sigma^2_{UK}\,(\boldsymbol{x}_0) = \text{Cov}_Y\,(0) - \mathbf{p}'\mathbf{Q} = \text{Cov}_Y\,(0) - \mathbf{p}'\mathbf{P}^{-1}\mathbf{p}\,. \quad \square$$

Despite the fact that Algorithm 6.1 is more general than Algorithm 6.2, in practice the latter is implemented more widely than the former. There are two reasons for this:

i) Attributes with infinite variance are a curiosity rather than a common occurrence.

ii) The system of equations in Algorithm 6.2 is better behaved than the algorithm for intrinsically stationary random functions. In the first k rows of matrix $\mathbf{P}$ in Algorithm 6.2, the largest elements are along the main diagonal, while all the diagonal elements in the equivalent matrix $\mathbf{A}$ in Algorithm 6.1 are zero.

Practice

Universal kriging shares with ordinary kriging all its properties: it is a minimum mean square error, unbiased, exact interpolator that automatically corrects for clustering in the sampling, and the observations take weights under a screen effect. The method is fairly robust to parameter misspecifications and inadequate assumptions. Notice that although the drift model must be specified, the drift is handled automatically in the systems of equations without the necessity of resorting to the estimation of the coefficients a_l for the model in Definition 6.3.

Perhaps the most serious drawback of universal kriging is the need to specify the semivariogram. The method requires the semivariogram of the residuals, not the semivariogram of the original regionalized variable. Although there are methods for estimating the residuals (Chapter 8), these presume knowledge of the semivariogram of the residuals, which leads to a circular problem. As mentioned in Chapter 5, the easiest and most common way to estimate the semivariogram of the residuals is to do the modeling along trend-free directions in the original data. For example, for an attribute such as the water table elevation in the High Plains aquifer of Kansas, the trend is a plane dipping approximately east; hence, fluctuations approximately along the north–south direction of the strike are free of the effect of the trend. A semivariogram estimated along that direction is the same as would be obtained removing the trend. The main limitation of the approach is that it precludes using the experimental semivariogram to explore the presence of anisotropies, which may be accounted for by the trend if they are all large in scale.

Next in difficulty is the problem of drift selection. Once more there is no direct statistical test to guide a search. The decision is up to the user and is more relevant in extrapolations beyond the convex hull of the sampling space (Figure 2.1). In unequivocal cases like the water table elevation in the High Plains aquifer of Kansas, the decision—although personal—is clear. But in other circumstances, even an experienced geostatistician will have problems determining whether or not there is a drift. The ambiguity can be resolved by heeding the following advice:

(a) We have seen that along directions free of the effect of trend the semivariogram will have the normal appearance of the semivariogram of an attribute honoring stationarity—steep increase for short lags and variation around a constant value beyond a range. Along directions affected by the trend, pseudo-anisotropy will occur—the semivariogram will have different properties along different directions. For a linear trend, for instance, the semivariogram will remain small for short lags and then will start to increase quadratically without bound. Such pseudo-anisotropy, illustrated in Figure 5.5, is a clear indication of the influence of a trend.

(b) Borderline alternatives such as those which may arise when deciding whether to select kriging with or without a drift are cases to be solved through the crossvalidation procedure in Chapter 7. Provided that all other parameters have been set properly, a clear advantage in favor of one case should be a strong hint to select the alternative producing the best crossvalidation. In the face of marginal advantages, for the sake of a parsimonious estimation, the consideration of a drift should be dropped.

Similarly, if a drift is warranted, prior knowledge about the attribute and crossvalidation are the best criteria to guide the selection of the number of terms in the drift expression in Definition 6.3.

EXERCISE 6.1

Appendix B contains measurements of water table elevation in northwestern Kansas. The maximum deviation from normality is 0.086. Knowing that the water table closely follows a gentle and steady decline in elevation from west to east, then:

1. Make a decision about the multivariate normality of the data.
2. Find the experimental semivariogram for the residuals.
3. Model the semivariogram for the residuals.
4. Use the six closest observations to estimate the weights and Lagrange multipliers at (60, 193).
5. Find the estimate and its kriging variance.
6. Produce a map for the water table elevation.

Solution

1. Considering that the maximum discrepancy of the univariate cumulative distribution differs by less than 0.1, it can be assumed that the sampling approximates a realization of a multivariate normal random function. The precision of the estimates will not be improved significantly by performing a normal score transformation.

2. The suggestion that, in general, the water table elevation decreases steadily eastwards is a warning for the existence of a trend. A first pass at directional estimation of semivariograms shows that the direction of minimal trend influence is in the NNW octant. The preliminary search is started from the north, investigating directions at 22.5° increments clockwise until reaching south. There is no need to investigate the other half of the directions from south continuing

clockwise until reaching the north direction because by the symmetry of the semivariogram $\gamma(\boldsymbol{h}) = \gamma(-\boldsymbol{h})$; that is, the semivariogram in the east direction, for example, is equal to the semivariogram in the west direction. A small tolerance is especially recommended in the presence of a trend. I used an angular tolerance of 5° and a bandwidth of 3 miles.

A finer investigation of the NNW octant at 1° increments with the same tolerance parameters reveals that the sill stabilizes for the smallest value in a N20W direction. According to Chapter 5, this is the best surrogate for the semivariogram for the residuals. Figure 6.2 is a rendition of the third round of results, again using direction increments of 22.5° and the same tolerance. Notice that the largest values in the experimental semivariogram along N20W are one to two orders of magnitude smaller than those for the other directions dominated by the trend influence. Estimation of an omnidirectional semivariogram is not of interest for the same reason.

3. Figure 6.3 shows the best simple model for the residuals. Considering the fitting of the simple model and the noise level in the experimental semivariogram, fitting a nested model is not recommended.

4. Given the hint that a trend has been confirmed by the semivariogram analysis, the model of choice is universal kriging. Considering the additional suggestion that the water table elevation closely follows the topography and that the topography approximates an eastward dipping plane, the clear choice for trend model is a first degree polynomial

$$m(\boldsymbol{x}) = a_0 + a_1 \cdot \text{easting} + a_2 \cdot \text{northing}\ .$$

The six closest observations are those from wells 993, 1002, 1003, 1502, 1504, and 1505. The matrix of sampling distances is

$$\begin{bmatrix} 0 & & & & & \\ 3.34 & 0 & & & & \\ 7.27 & 7.37 & 0 & & & \\ 8.77 & 5.44 & 10.03 & 0 & & \\ 9.91 & 9.17 & 3.09 & 10.20 & 0 & \\ 8.72 & 6.57 & 5.34 & 5.85 & 4.47 & 0 \end{bmatrix}$$

and the vector of distances to the estimation location is

$$[5.09 \quad 3.45 \quad 4.36 \quad 5.92 \quad 5.73 \quad 3.64]'\ .$$

By Definition 6.8,

$$\mathbf{P} = \begin{bmatrix} 3710.5 & & & & & & & & \\ 3489.4 & 3710.5 & & & & & & & \\ 2970.1 & 2954.3 & 3710.5 & & & & & & \\ 2707.3 & 3250.3 & 2471.3 & 3710.5 & & & & & \\ 2494.0 & 2633.9 & 3511.0 & 2438.4 & 3710.5 & & & & \\ 2716.0 & 3084.4 & 3262.7 & 3192.0 & 3372.7 & 3710.5 & & & \\ 1 & 1 & 1 & 1 & 1 & 1 & 0 & & \\ 61.56 & 62.94 & 55.68 & 64.96 & 54.80 & 59.12 & 0 & 0 & \\ 197.85 & 194.81 & 193.56 & 189.77 & 190.60 & 189.47 & 0 & 0 & 0 \end{bmatrix}$$

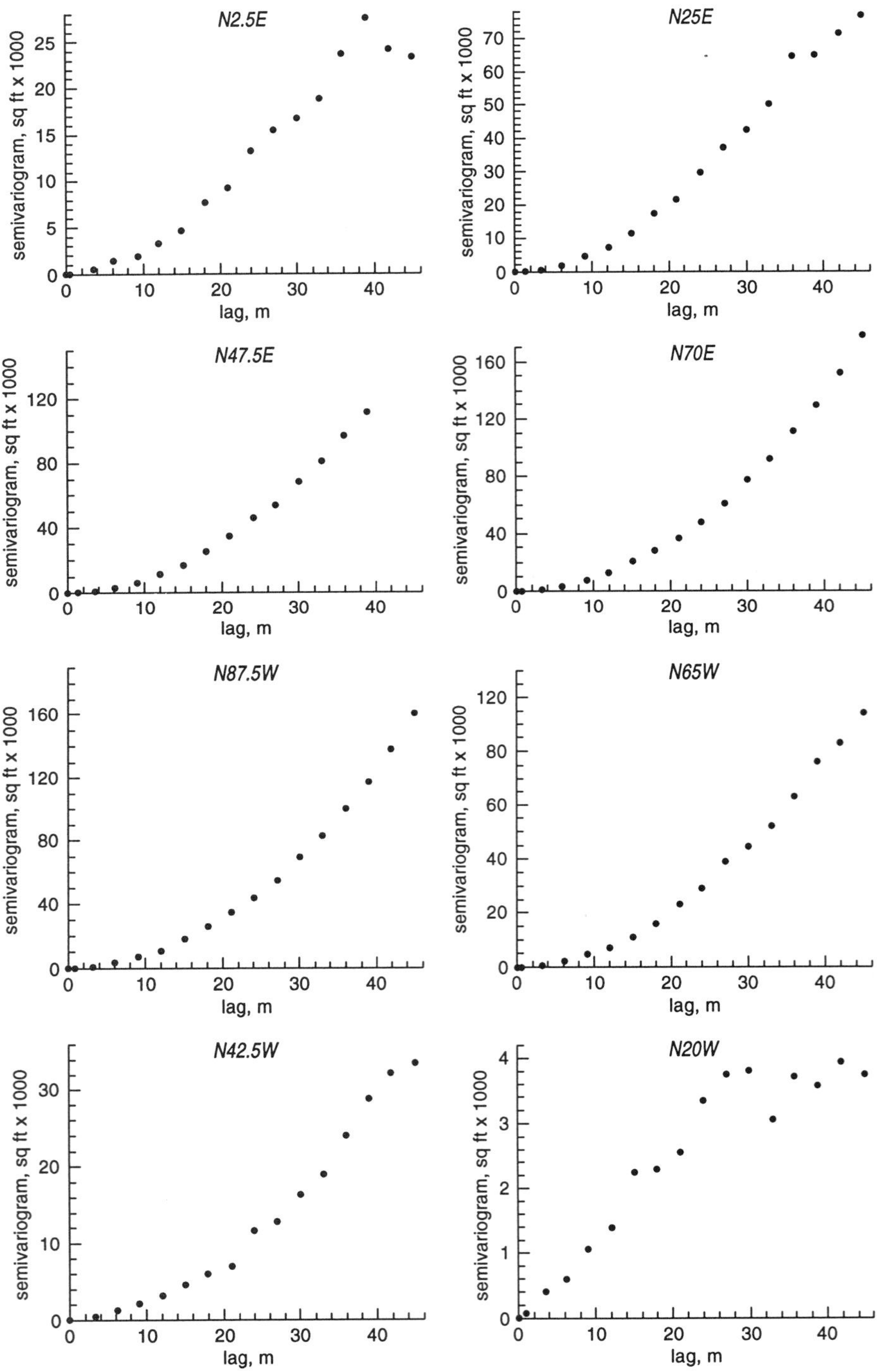
N2.5E
N25E
N47.5E
N70E
N87.5W
N65W
N42.5W
N20W
semivariogram, sq ft x 1000
lag, m

[Figure 6.2]

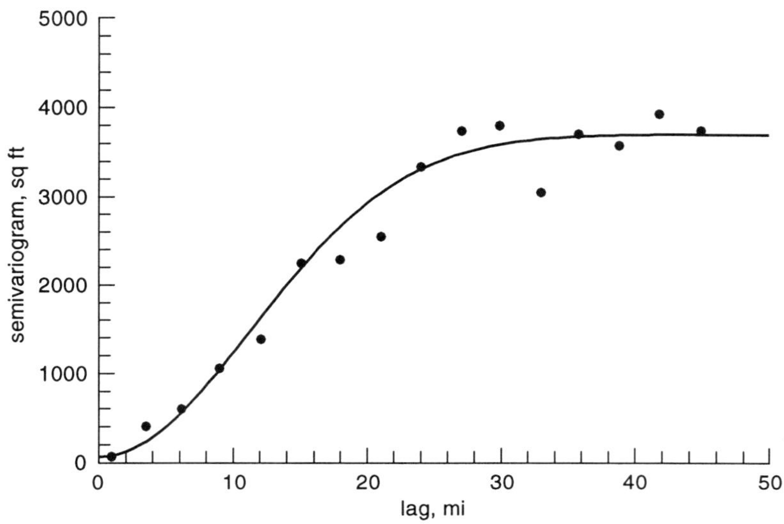

Figure 6.3. Semivariogram model for the residuals. The model is Gaussian with a nugget of 67.9 ft^2, (sill − nugget) of 3642.6 ft^2, a range of 27.9 miles, and an $\widehat{AIC}$ of −69.2.

and by Definition 6.10,

$$\mathbf{p} = [\,3296.1 \;\; 3478.8 \;\; 3385.5 \;\; 3181.3 \;\; 3209.4 \;\; 3461.4 \;\; 1 \;\; 60.0 \;\; 193.0\,]' \; .$$

By Algorithm 6.2, the universal kriging weights and Lagrange multipliers are:

$$\mathbf{P}^{-1}\mathbf{p} = [\,0.073 \;\; 0.348 \;\; 0.274 \;\; 0.012 \;\; -0.056 \;\; 0.350 \;\; 33.95 \;\; 0.27 \;\; 0.53\,]' \; .$$

5. By Algorithm 6.2, the estimate is

$$Z_{UK}(60, 193) = [\,3065.0 \;\; 3099.4 \;\; 3200.0 \;\; 3114.9 \;\; 3217.1 \;\; 3189.7 \;\; 0 \;\; 0 \;\; 0\,] \begin{bmatrix} 0.073 \\ 0.348 \\ 0.274 \\ 0.012 \\ -0.056 \\ 0.350 \\ 33.95 \\ 0.27 \\ 0.53 \end{bmatrix}$$

$$= 3149.6$$

[Opposite page, left.]

Figure 6.2. Experimental semivariograms along eight directions for the water table elevation (in feet) in the High Plains aquifer, northwestern Kansas, winter 1980–81. The average lag increment is 3 miles and the tolerance parameters are 3 miles for the bandwidth and 5° for the angular tolerance.

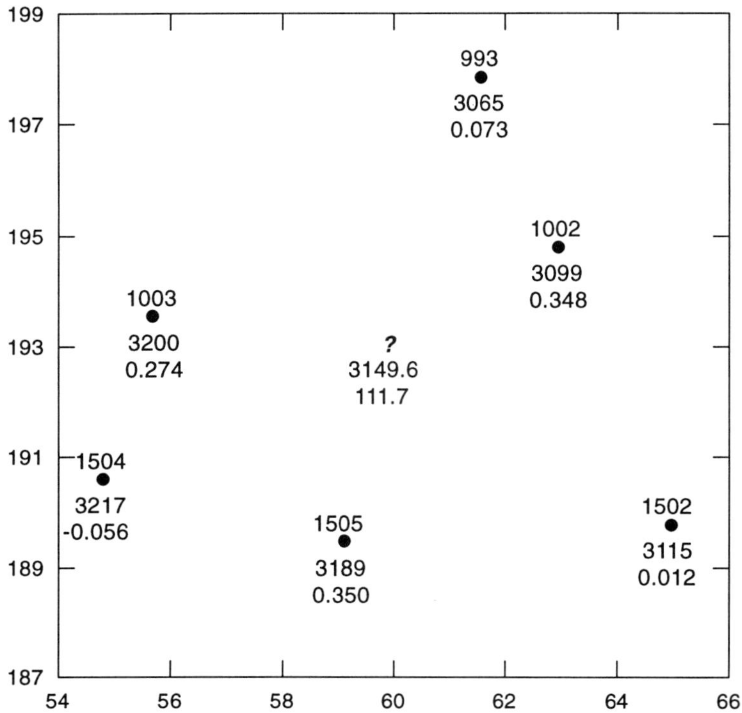

Figure 6.4. Wells and estimation location in Exercise 6.1. A solid dot denotes a well location and the labels are, from top to bottom, the identification number, water table elevation (in feet), and the universal kriging weight. The question mark shows the estimation location and the labels are the estimate (top; ft) and its kriging variance (bottom; sq. ft).

and the kriging variance is

$$\sigma^2_{UK}(60, 193) =$$

$$3710.5 - [3296.1 \;\; 3478.8 \;\; 3385.5 \;\; 3181.3 \;\; 3209.4 \;\; 3461.4 \;\; 1 \;\; 60.0 \;\; 193.0] \begin{bmatrix} 0.073 \\ 0.348 \\ 0.274 \\ 0.012 \\ -0.056 \\ 0.350 \\ 33.95 \\ 0.27 \\ 0.53 \end{bmatrix}$$

$$= 111.71 \; .$$

Figure 6.4 summarizes the results for the last two steps. Notice that the triangle formed by the closest wells (1003, 1002, and 1505) is screening the other three wells.

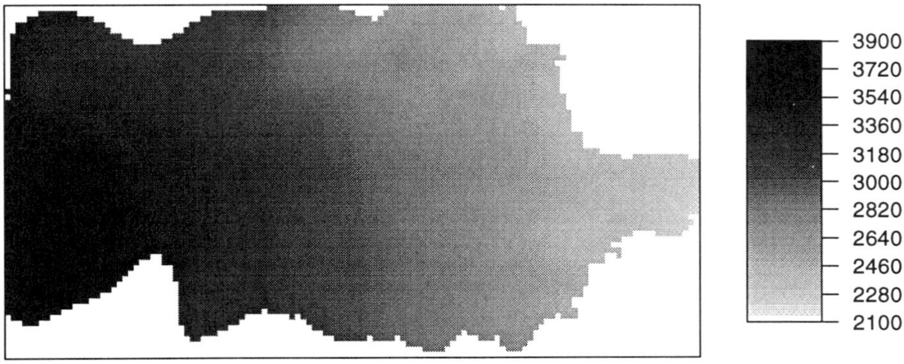

Figure 6.5. Water table elevation (in feet) in the High Plains aquifer, northwestern Kansas, winter 1980–81. For a color rendition of this illustration, visit Internet site http://www.kgs.ukans.edu/Mathgeo/Books/Geostat/index.html.

6. Figure 6.5 is a map for the water table elevation. The map confirms that the water table dips gently in the N70E direction. □

Chapter 7
CROSSVALIDATION

Upon selecting any estimator and its parameters and producing estimations for a given spatial attribute, there is curiosity and the need to know more about the quality of the job done.

The ultimate approach would be to run an exhaustive sampling and then assess the quality of the estimation by comparing the new data and the corresponding estimates produced with the original data. Unquestionably, the best estimator would be the one producing the best statistic on the basis of a previously selected criterion, such as the mean square error.

The implementation of such a conclusive approach, however, if not impossible, would be ruinous at best. With most of the spatial attributes having a continuous variation over large volumes or regions, the amount of time and financial resources required to accomplish a truly exhaustive sampling would be colossal, not to mention the fact that by the time the evaluation was finished, the exhaustive sampling would have thwarted any interest in using the estimates.

Alternative Evaluation Method

Crossvalidation is an ingenious alternative evaluation scheme for solving these inconveniences by hiding information—one observation at a time—instead of requiring additional data. Given a sampling of size n, each observation is dropped once with replacement and for each discarded observation, an estimate is computed at the location of the discarded observation by using at most the remaining $(n-1)$ measurements. By pretending that an observation was never taken, one can actually attempt to produce an estimate. And bringing the observation back provides an instant and free measurement for the evaluation of the estimation.

In principle, for parametric methods, every parameter (such as the covariance function) should be reevaluated if the discarded point was used in the parameter estimation. In practice such parameters are presumed to be robust enough to be insensitive to the discarding of a single observation and one does not recalculate the parameters.

The final product of the comparison is a file in which each record contains the original information plus the corresponding estimation variance, if the estimator happens to be kriging.

Algorithm 7.1

Execution of this algorithm will result in the generation of crossvalidation errors.

1. Select an estimator.

2. Set all parameters that the estimator may require.

3. Drop $z(\boldsymbol{x}_i)$ from the sampling, where $\boldsymbol{x}_i$ is a sampling site not yet visited.

4. Estimate $z(\boldsymbol{x}_i)$, employing the estimator selected in Steps 1 and 2. If the estimator is kriging, calculate the estimation variance.

5. Calculate the estimation error $\hat{z}(\boldsymbol{x}_i) - z(\boldsymbol{x}_i)$.

6. Replace $z(\boldsymbol{x}_i)$ into the sampling.

7. If there are more sites to visit, select a site not yet visited and go back to Step 3. Otherwise, stop. □

Notice that the algorithm is not limited to geostatistics—a crossvalidation can be run for any spatial estimator. In the case of simple or ordinary kriging, aside from the selective dropping of observations, crossvalidation is an extension of Algorithm 2.1, 4.1, or 4.2, in which, instead of generating one estimate, one must generate as many estimates as there are measurements.

Crossvalidation is the only way to indirectly test anything about the kriging or the data. In geostatistics there are no null hypotheses or levels of significance for acceptance or rejection. Crossvalidation does not indicate whether an observation, estimate, parameter, or assumption is correct. All crossvalidation does is generate errors associated with different selections; the user must draw conclusions by comparing the errors.

As imperfect as crossvalidation may be, drawing conclusions by crossvalidation is in many instances better than the alternative—making arbitrary decisions or assumptions. One must use crossvalidation with caution, however. Complex interdependencies among the errors preclude finding the error distribution, thus hindering a rigorous analysis or testing of the results. Yet in most circumstances geostatisticians employ the results as a guide in the analysis and estimation of attributes. Special circumstances deserve extra caution:

i) The sampling pattern may lead to errors that are not representative of the errors that would arise from an exhaustive sampling. A typical situation is the sampling with clusters, wherein every sampling site will be close to at least one other sampling site. The crossvalidation errors will be deceptively small, because for every site $\boldsymbol{x}_i$ there will always be a close site among the retained $(n-1)$ sites in the sampling. Crossvalidation of such patterns does not include the large errors that the estimation will produce for those critical sites away from control points.

ii) In the case of kriging, by Corollaries 2.4 and 4.2, the estimate is insensitive to scaling of the semivariogram, thus making crossvalidation insensitive to changes in the sill.

Yet crossvalidation remains a useful tool, not necessarily to prove anything, but primarily to preclude blunders and draw honest conclusions by comparison.

EXERCISE 7.1

Given the sampling in Exercises 3.1 and 3.2, use crossvalidation to explore the convenience—in a mean square crossvalidation error sense—of using ordinary kriging instead of simple kriging for the estimation of the normal scores. At most, use only the eight closest observations no farther away than 0.4 miles.

Solution

The results of the crossvalidation are given in Table 7.1.

Table 7.1. Crossvalidation for Exercise 7.1.

		Simple kriging			Ordinary kriging		
Well	True	Est.	K.Var.	Error	Est.	K.Var.	Error
94	−1.089	−0.419	0.323	0.670	−0.455	0.324	0.633
92	−0.425	−1.339	0.229	−0.915	−1.283	0.230	−0.858
95	−2.059	−1.999	0.342	0.060	−2.106	0.345	−0.047
99	−1.840	−1.487	0.417	0.353	−1.746	0.433	0.094
93	−0.828	−0.935	0.221	−0.107	−0.904	0.222	−0.077
97	−1.458	−1.345	0.301	0.113	−1.356	0.301	0.102
96	−1.684	−1.525	0.241	0.159	−1.456	0.242	0.228
91	−2.480	−1.148	0.287	1.331	−1.156	0.287	1.323
80	2.480	−0.219	0.422	−2.698	−0.259	0.438	−2.739
82	−0.083	−0.319	0.231	−0.236	−0.303	0.232	−0.221
84	−0.216	−0.316	0.265	−0.099	−0.301	0.266	−0.085
89	0.389	−0.755	0.280	−1.144	−0.760	0.280	−1.149
83	−0.536	−0.300	0.241	0.236	−0.295	0.242	0.241
88	−1.031	−0.414	0.233	0.617	−0.395	0.235	0.635
81	−0.284	0.344	0.234	0.628	0.325	0.235	0.609
87	−0.498	−0.637	0.236	−0.139	−0.589	0.237	−0.090
86	−0.654	−0.692	0.260	−0.038	−0.696	0.261	−0.042
74	0.149	0.076	0.240	−0.073	0.089	0.241	−0.060
75	0.116	−0.224	0.251	−0.340	−0.224	0.253	−0.340
68	0.250	−0.251	0.226	−0.501	−0.224	0.227	−0.474

(cont.)

Table 7.1. continued

Well	True	Simple kriging Est.	Simple kriging K.Var.	Simple kriging Error	Ordinary kriging Est.	Ordinary kriging K.Var.	Ordinary kriging Error
73	0.083	0.229	0.263	0.147	0.211	0.263	0.129
76	−0.389	−0.317	0.240	0.071	−0.288	0.241	0.101
65	−0.574	−0.151	0.275	0.424	−0.153	0.275	0.422
66	0.284	−0.317	0.251	−0.601	−0.299	0.252	−0.583
67	−0.354	−0.073	0.224	0.281	−0.078	0.225	0.276
70	−1.217	−1.267	0.295	−0.051	−1.276	0.295	−0.059
77	−1.289	−0.745	0.312	0.544	−0.753	0.312	0.536
69	−1.150	−0.810	0.233	0.340	−0.788	0.235	0.363
40	−0.049	−0.002	0.140	0.048	−0.006	0.141	0.043
50	−0.149	0.021	0.234	0.170	0.039	0.236	0.188
52	−0.250	0.115	0.243	0.364	0.150	0.245	0.400
60	1.150	−0.458	0.232	−1.608	−0.434	0.234	−1.585
58	−0.461	−0.393	0.234	0.068	−0.386	0.235	0.075
49	−0.116	−0.051	0.269	0.065	−0.051	0.269	0.065
59	−0.924	0.076	0.226	1.000	0.073	0.227	0.997
37	1.458	0.551	0.240	−0.907	0.535	0.241	−0.923
45	0.461	0.334	0.230	−0.127	0.324	0.232	−0.137
36	0.498	0.754	0.287	0.256	0.771	0.287	0.273
46	−0.738	0.271	0.241	1.009	0.259	0.243	0.997
57	−0.695	−0.376	0.263	0.319	−0.376	0.263	0.319
47	0.049	−0.415	0.234	−0.464	−0.414	0.236	−0.463
39	−0.016	−0.003	0.151	0.013	−0.001	0.151	0.016
38	0.216	0.421	0.240	0.205	0.406	0.241	0.190
44	0.654	0.296	0.257	−0.358	0.294	0.257	−0.359
51	0.319	−0.481	0.239	−0.800	−0.492	0.240	−0.811
61	−0.782	−0.510	0.241	0.272	−0.480	0.242	0.302
41	−0.182	−0.044	0.259	0.138	−0.044	0.259	0.138
53	−0.875	−0.834	0.242	0.041	−0.819	0.243	0.056
62	−1.561	−1.140	0.246	0.421	−1.095	0.247	0.466
48	−0.614	−0.430	0.252	0.183	−0.416	0.253	0.198
63	−0.976	−1.212	0.420	−0.236	−1.412	0.437	−0.435
54	−1.369	−1.046	0.309	0.323	−1.063	0.310	0.306
25	0.738	0.705	0.233	−0.033	0.669	0.234	−0.069
29	0.875	0.773	0.150	−0.102	0.765	0.150	−0.110
31	1.031	0.908	0.187	−0.123	0.887	0.187	−0.144
30	0.425	0.767	0.266	0.342	0.789	0.266	0.364
33	0.016	0.512	0.226	0.495	0.487	0.227	0.470
32	0.924	0.672	0.240	−0.253	0.649	0.242	−0.276
14	2.059	1.427	0.271	−0.632	1.441	0.271	−0.618
24	1.369	0.826	0.239	−0.542	0.795	0.241	−0.574

(*cont.*)

Table 7.1. concluded

		Simple kriging			Ordinary kriging		
Well	True	Est.	K.Var.	Error	Est.	K.Var.	Error
15	1.684	1.411	0.239	−0.273	1.358	0.241	−0.326
23	0.695	0.906	0.203	0.211	0.921	0.203	0.226
19	0.976	1.125	0.236	0.149	1.078	0.237	0.101
18	0.828	1.438	0.265	0.611	1.408	0.265	0.580
16	0.536	1.174	0.271	0.638	1.144	0.272	0.609
26	0.354	0.856	0.254	0.502	0.833	0.255	0.480
34	0.574	0.059	0.274	−0.515	0.062	0.274	−0.512
21	1.289	0.604	0.432	−0.685	0.683	0.450	−0.606
20	1.089	0.830	0.239	−0.258	0.788	0.240	−0.301
10	1.840	1.617	0.320	−0.223	1.694	0.321	−0.146
11	1.561	1.380	0.229	−0.181	1.341	0.230	−0.220
5	1.217	1.071	0.302	−0.146	1.082	0.302	−0.135
6	0.182	0.131	0.144	−0.052	0.115	0.145	−0.067
7	−0.319	0.286	0.159	0.604	0.319	0.160	0.638
12	0.782	0.528	0.233	−0.254	0.516	0.233	−0.266
3	0.614	0.367	0.422	−0.246	0.454	0.452	−0.159

For simple kriging the mean square error is 0.576 and for ordinary kriging 0.575. Hence, in this example it is marginally better to ignore the known mean of zero and use ordinary kriging for the estimation despite the slight increase in the kriging variance from a mean value of 0.257 for simple kriging to 0.259 for ordinary kriging. □

Diagnostic Statistics

Comparison or remedial action based on crossvalidation is more straightforward and conclusive if the collection of errors is reduced to some key statistics. These statistics and relevant information include:

1. The 5th percentile, the median and the 95th percentile for the data, the estimates, the errors, and the standard scores. The *standard score* is the ratio of the error over the positive square root of the estimation variance. It is also known as the z-score.

Ideally the data percentiles should be the same as the corresponding percentiles for the estimates. In such a case, it is also true that if the estimates are subdivided into classes based on their magnitude, the average of the estimate will coincide with the average of the true values, a property called conditional unbiasedness. Identical distributions for true values and estimates imply conditional unbiasedness.

2. The mean and standard deviation for the same variables plus the square root of the mean square error. Again, conformity between the parameters for the data and the estimates denotes a thorough estimation procedure.

Zero is the ideal value for the error moments. A zero mean denotes global unbiasedness in the estimation procedure used in the generation of the errors. The standard deviation and the mean square error can only be zero if every single estimate is identical to the corresponding observation.

3. The symmetry coefficient, S, given by:

$$S = 100\frac{\hat{M} - \hat{\mu}}{p_{95} - p_5} ,$$

where $\hat{M}$ is the mode for the sample, $\hat{\mu}$ is the sample mean, p_5 is the 5th percentile, and p_{95} is the 95th percentile. The coefficient is a measurement of the discrepancy between the mean and the mode normalized by the spread of the variable given by the percentiles. The coefficient is a measure of symmetry that is less sensitive to outliers than the coefficient of skewness. A distribution with a tail extending to the left has a negative coefficient, symmetric distributions such as the normal distribution have a coefficient of zero, and distributions with positive coefficients have a tail extending to the right in a fashion similar to the lognormal distribution.

4. The maximum discrepancy between two distributions is a useful measure of their similarity, as seen in Chapter 3. In the crossvalidation analysis one is interested in two pairs of distributions: the distributions of data and estimates and the standard score and standard normal distributions.

The similarity between the distribution of the data and the distribution of the estimates is a sign of good performance by the estimation method.

If the standard scores are normally distributed, one cannot discard the assumption that the errors may be multinormally distributed, in which case—as shown in Chapter 10—the kriging variance can be used to assess the reliability of the estimate. If the standard scores are normally distributed, they show the properties of a normal standard distribution. For example, p_5 should be equal to -1.65, p_{95} should be equal to 1.65, the standard deviation should be close to 1, and the symmetry coefficient S and the mean should both be close to 0.

5. If the errors are multinormally distributed, then at any estimation location the distribution of the error is a normal distribution with mean equal to the kriging estimate and variance equal to the kriging variance. For the particular case of the estimation errors, the actual errors should conform globally with the confidence limits that can be derived from the normality of the errors. For example, an average 68.3% of the errors should be within ± 1 kriging standard deviation, which is equivalent to the statement that 68.3% of the absolute value of the errors should be less that the corresponding kriging standard deviation. Similarly, 95.5% of the absolute value of the errors should be less than 2 kriging standard deviations.

6. If estimates systematically deviate from true values, they are said to be conditionally biased. Conditional bias is a common drawback of spatial estimators and is indicated by a regression line between the true and estimated

values with a slope other than 1 and an intercept other than zero. Chapter 9 expands more on the subject.

7. The correlation coefficient between the true and the estimated values is a measure of the overall quality of an estimation capturing the dispersion around the regression line. The ideal value is 1.

8. Ensuring that the data are free of measurement and recording mistakes, even before attempting any serious estimation, can be a difficult task, especially for samplings that include thousands of observations. Listing the largest estimation crossvalidation errors is a convenient way to spot improper data preparation, which often relates to the geographic coordinates instead of the measurements of the spatial attribute.

9. Another important criterion includes posting of the errors for the purpose of discarding any spatial correlation, which can be demonstrated by proving that the best model for the semivariogram of the errors is a pure nugget effect.

An enhanced version of program **xvan** (Olea, 1996) can be used for the preparation of statistics 1 through 8. The experimental semivariogram of the errors can be obtained by using the standard software available for the calculation of experimental semivariograms.

There is a good deal of redundancy in the statistics. In some instances just one statistic will suffice to summarize the quality of the data preparation and the estimation procedure. Redundancy, however, allows the incorporation of personal preferences in an area in which standardized criteria simply do not exist. In addition, not all statistics point in the same direction all the time.

Large discrepancies, in general, move all the statistics in the same direction. However, for the usual subtle differences in the estimation errors that characterize sensitivity analysis (as in Exercise 7.1), it is not uncommon to observe one statistic show a slight improvement, the mean square error for example, whereas another, such as the correlation coefficient, may become slightly worse. In such cases, selection of the best estimation procedure depends on the selection of the statistics.

EXERCISE 7.2

Figure 7.1 provides the cumulative distribution and the semivariogram model for the sampling in Appendix C. Using ordinary kriging for the estimation and no less than three observations within $8,000$ m of the estimation site:

1. Find the optimal maximum number of observations for the estimation, in a mean square error sense.

2. Geologists suspect that the structures are elongated in a north–south direction. The small number of pairs precludes a more conclusive directional semivariogram testing. Does crossvalidation support such a presumption of anisotropy?

Using the minimum mean square error parameters that follow from 1 and 2 above, answer the following additional questions:

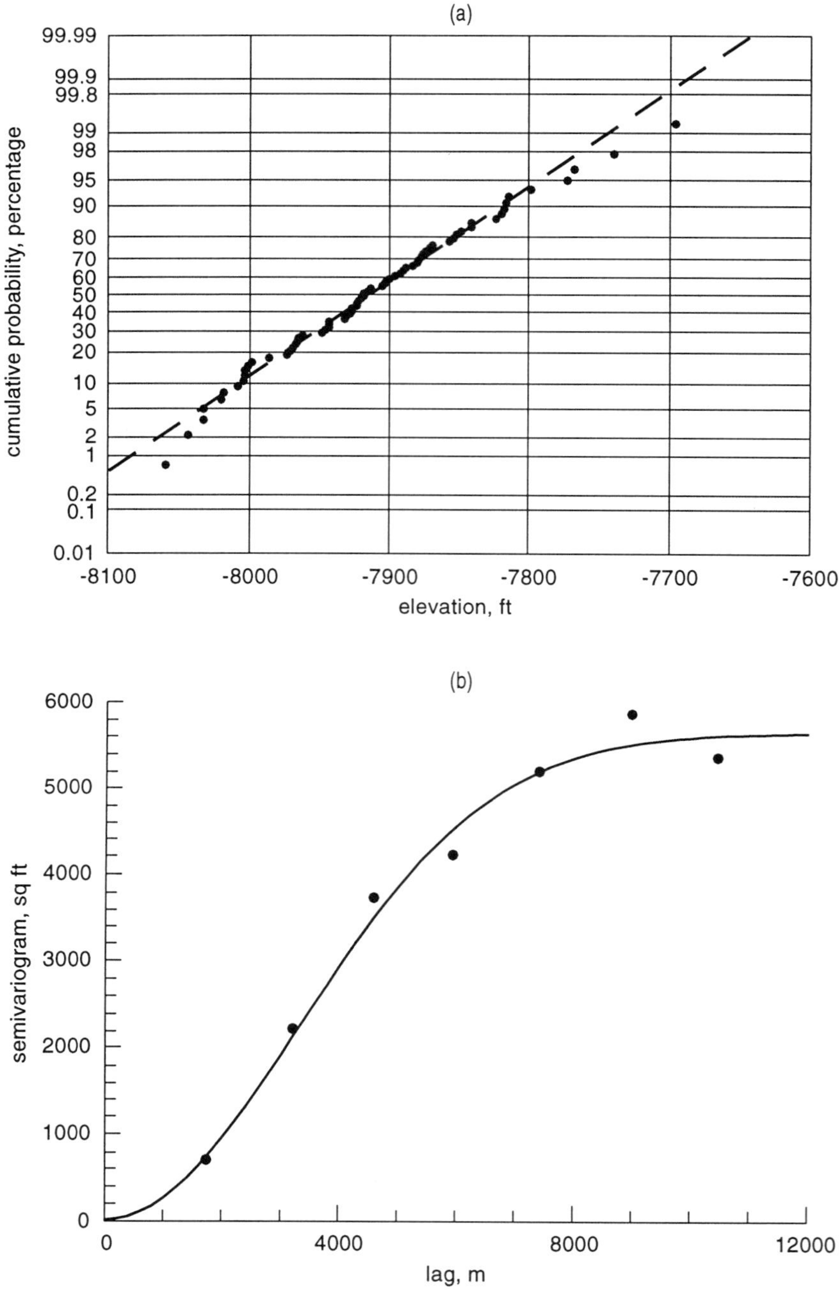

Figure 7.1. Depth to UNCF: (a) Dots denote cumulative distribution for sampling, segmented line is cumulative probabilities of normal distribution with same mean (−7912.2 ft) and standard deviation (74.5 sq. ft) as sampling; maximum discrepancy, 0.05. (b) Best semivariogram model in minimum weighted least squares sense: Gaussian type; nugget, 10; sill, 5646 sq. ft; range, 8107 m.

3. Are there some crossvalidation errors larger than 1 sampling standard deviation?

4. How close is the distribution for the data to the one for the estimates?

5. Is it safe to assume that the errors are multinormally distributed?

Solution

1. Table 7.2 shows that the mean square error is minimum when the maximum number of observations used in the estimation is 11.

Table 7.2. Sensitivity analysis to maximum number of estimation points.

Number of observations	Mean square error, sq. ft	Number of observations	Mean Square error, sq. ft
3	589.6	9	539.5
4	632.1	10	451.0
5	658.7	11	361.2
6	657.4	12	376.0
7	593.9	13	385.1
8	545.2		

2. According to Table 7.3, crossvalidation supports anisotropy. To be consistent with the idea that the omnidirectional semivariogram in Figure 7.1 represents an average value for spatial continuity, the major and minor axes were forced to deviate in the same amount from the range of the omnidirectional semivariogram—8,107 m. The analysis also demonstrates that the anisotropy is not exactly in the north–south direction but is rotated 10° to the east.

Table 7.3. Sensitivity to anisotropy.

Axes	Major axis orientation	Mean square error, sq. ft
8107/8107	N	361.2
8500/7714	N	302.8
9000/7214	N	257.7
9300/6914	N	247.0
9400/6814	N	246.0
9500/6714	N	246.2
9600/6614	N	247.5
9400/6814	N20E	253.8
9400/6814	N10E	243.6

Relative to the isotropic case, there is a 32.5% drop in mean square error by considering a maximum range of 9,400 m in the N10E direction and is 6,814 m along the perpendicular direction N80W.

3. From the analysis it follows that the optimal estimation parameters are:

Minimum number of observations:	3
Maximum number of observations:	11
Neighborhood radius:	8,000 m

and the best semivariogram model is:

Type:	Gaussian
Nugget:	10
Sill:	5,646
Range in the principal N10E direction:	9,400 m
Minor range:	6,814 m

Table 7.4 contains the complete crossvalidation analysis for the optimal neighborhood and semivariogram parameters. All errors are below 1 data standard deviation, thus it is safe to assume that sampling is free of blunders.

Table 7.4. Statistical analysis of crossvalidation errors.

Observations actually crossvalidated: 70

Observations excluded from the analysis: 0

Percentile and moments

	5th percentile	95th percentile	Median	Mean	Coeff. symm.	Std. dev.	Sq. rt. mn. sq.
True value	−8032.000	−7773.000	−7918.000	−7912.157	2.256	74.539	
Estimate	−8024.040	−7806.905	−7919.650	−7914.555	3.451	70.285	
Error	−33.973	18.889	−1.344	−2.398	−1.994	15.535	15.608
Std. score	−1.836	1.368	−0.229	−0.062	5.215	0.992	

Regression line

Correlation coefficient: 0.979

Intercept: −611.902

Slope: 0.923

Maximum distributional discrepancies

Between true values and estimates: 0.06

Between standard scores and a standard normal distribution: 0.09

Proportions

Proportion of errors under 1 kriging standard deviation: 70.0%

Proportion of errors under 2 kriging standard deviations: 95.7%

(cont.)

Table 7.4. concluded

Largest errors

	ID	Easting (m)	Northing (m)	Error (ft)
1	10	38024.000	76170.000	−69.188
2	2078	37100.000	62700.000	−40.556
3	4568	45500.000	64710.000	−36.151
4	12	46955.000	76620.000	−33.973
5	5036	49520.000	65940.000	−31.888
6	1	32000.000	87015.000	−25.283
7	5022	42740.000	75120.000	−20.040
8	5024	43730.000	74610.000	20.001
9	2068	29840.000	67560.000	22.421
10	6202	49400.000	84000.000	32.324

4. According to Table 7.4, the maximum discrepancy between the distribution for the data and that of the estimates is low. The average properties of the kriging surface using the optimal parameters should be the same as those of the sampling.

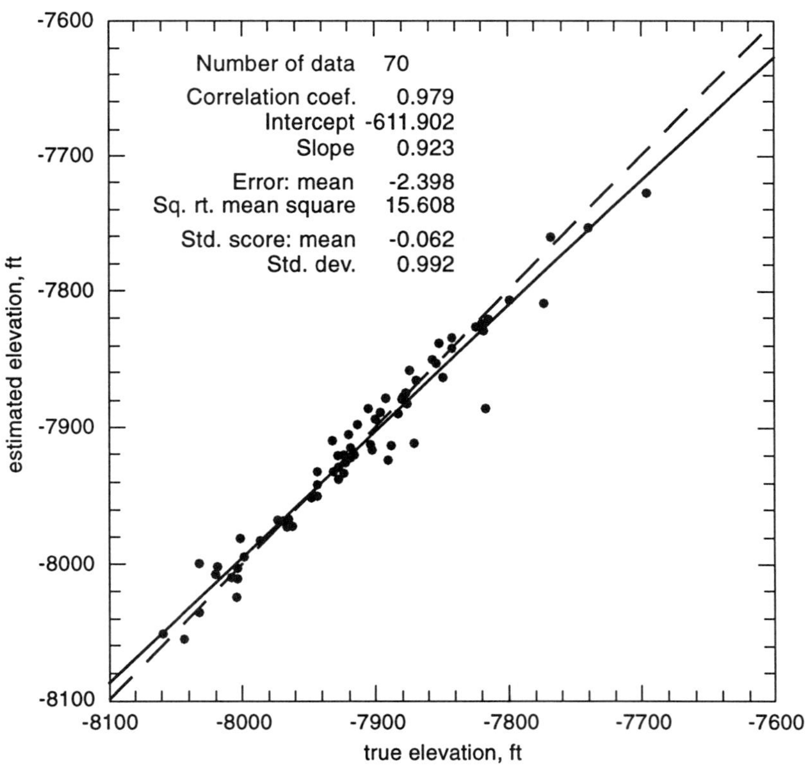

Figure 7.2. True values and estimates.

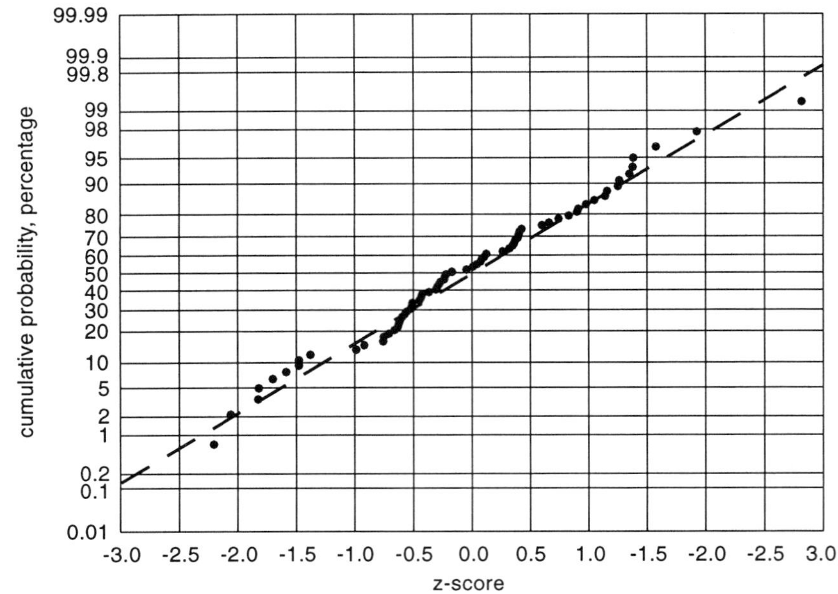

Figure 7.3. Cumulative distribution for standard scores. The maximum discrepancy with the cumulative distribution for the standard normal distribution is 0.09.

Figure 7.2 confirms the finding through the correlation coefficient between the true value and the estimates, which at 0.979 is almost identical to the ideal value of 1. Except for observation 10, all estimates are remarkably similar to the true values.

5. From Table 7.4 and Figure 7.3 the z-scores approximate well a standard normal distribution. Thus, it is acceptable to assume that the kriging errors are normally distributed.

Figure 7.4 confirms that the errors honor the 68.3/95.5% rule remarkably well. □

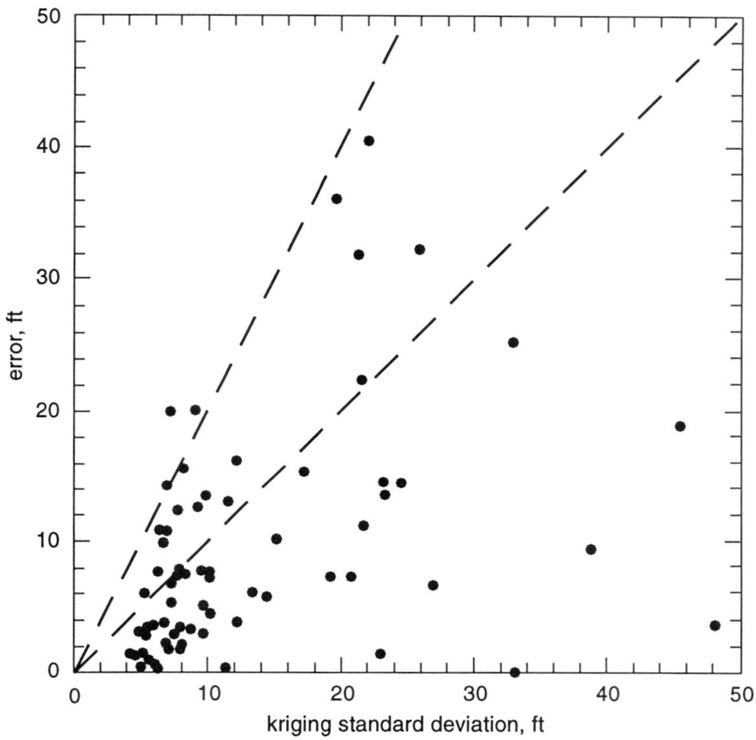

Figure 7.4. Kriging estimation variance and errors. The proportion of errors smaller than 1 kriging standard deviation is 70.0% and the proportion smaller than 2 standard deviations is 95.7%.

Chapter 8
DRIFT AND RESIDUALS

In Chapter 6, we have formulated universal kriging for the modeling of an attribute with a systematic trend. In some instances, however, the characteristic of interest is the trend itself rather than the attribute or, even more commonly, it is the residuals resulting from subtracting the drift from the kriging modeling of the attribute.

I will follow the convention that regards the trend as an intuitive concept denoting low frequency, continuous variation of a spatial attribute at a large scale. Drift is the mathematical expression modeling such gentle fluctuations as the expected value of the random function modeling the attribute (Olea, 1991).

Definition 8.1

Let $Z(\boldsymbol{x}_i)$ be k random variables inside a $d \subset D$ neighborhood of the spatial domain centered at $\boldsymbol{x}_0$. The drift estimator $\hat{m}(\boldsymbol{x}_0)$ of the variate $Z(\boldsymbol{x})$ is the linear combination

$$\hat{m}(\boldsymbol{x}_0) = \sum_{i=1}^{k} \lambda_i Z(\boldsymbol{x}_i) . \quad \square$$

Differing from other techniques attempting the same purpose, such as trend surface analysis, the drift is not given by a single analytical expression—it is approximated numerically as a collection of estimates at the center of overlapping neighborhoods.

Assumptions

The following assumptions are necessary for the formulation of the method to estimate the drift and infer the moments.

Assumption 8.1

A sampling is a partial realization of a random function. $\square$

Assumption 8.2

The residuals $Y(\boldsymbol{x})$ in Definition 2.2 are second order stationary,

$$\left.\begin{array}{l}\mathrm{E}\left[Y(\boldsymbol{x})\right]=0 \\ \mathrm{E}\left[Y(\boldsymbol{x})\,Y(\boldsymbol{x}+\boldsymbol{h})\right]=\mathrm{Cov}_Y(\boldsymbol{h})\end{array}\right\}\ \text{for all } \boldsymbol{x}, \boldsymbol{x}+\boldsymbol{h} \text{ in } d\ ,$$

where $\mathrm{E}[\cdot]$ denotes expected value, $\boldsymbol{h}$ is a distance in the spatial domain of interest, $\mathrm{Cov}_Y(\boldsymbol{h})$ is the covariance of the residuals, and d is the neighborhood in Definition 8.1. □

The null expectation for the residuals is not exactly an assumption in this case, but a fact proved by Lemma 2.5.

Assumption 8.3

One can model the drift as a linear combination of analytical functions. □

Among numerous possibilities, here I will restrict the attention to the most common alternative—polynomials, as those in Definition 6.3.

Work in the following sections leads to the calculation of weights λ_i such that the drift estimator in Definition 8.1 is unbiased with minimum mean square error.

Unbiasedness

Finding the necessary and sufficient conditions to have an unbiased drift estimator is more straightforward if one first proves the following lemma:

Lemma 8.1

Let $\hat{m}(\boldsymbol{x}_0)$ be the estimator in Definition 8.1 and $m(\boldsymbol{x})$ be the drift in Definition 6.3. Then

$$E\left[\hat{m}(x_0)\right]=\sum_{l=0}^{n}\sum_{i=1}^{k}a_l\lambda_i f_l(\boldsymbol{x}_i)\ .$$

Proof

From Definition 8.1,

$$\mathrm{E}\left[\hat{m}(\boldsymbol{x}_0)\right]=\mathrm{E}\left[\sum_{i=1}^{k}\lambda_i Z(\boldsymbol{x}_i)\right]\ .$$

One can interchange expectation and summation and remove the parameter λ_i from the expectation

$$\mathrm{E}\left[\hat{m}(\boldsymbol{x}_0)\right]=\sum_{i=1}^{k}\lambda_i \mathrm{E}\left[Z(\boldsymbol{x}_i)\right]\ .$$

But the expected value is the drift at $\boldsymbol{x}_i$. Hence, by Definitions 6.2 and 6.3,

$$\mathrm{E}\left[\hat{m}\left(\boldsymbol{x}_0\right)\right] = \sum_{i=1}^{k} \lambda_i \sum_{l=0}^{n} a_l f_l\left(\boldsymbol{x}_i\right)$$

One can place the term λ_i in the second summation because λ_i is independent of l. Therefore, interchanging summation and reversing the product,

$$\mathrm{E}\left[\hat{m}\left(\boldsymbol{x}_0\right)\right] = \sum_{l=0}^{n} \sum_{i=1}^{k} a_l \lambda_i f_l\left(\boldsymbol{x}_i\right) . \quad \square$$

Theorem 8.1

Let $\hat{m}\left(\boldsymbol{x}_0\right)$ be the estimator in Definition 8.1 and $m\left(\boldsymbol{x}\right)$ be the drift model in Definition 6.3. Then the estimator is unbiased if and only if

$$\sum_{i=1}^{k} \lambda_i f_l\left(\boldsymbol{x}_i\right) = f_l\left(\boldsymbol{x}_0\right) \quad \text{for } l = 0,\, 1,\, \ldots,\, n .$$

Proof (a)

The relationship $\mathrm{E}\left[\hat{m}\left(\boldsymbol{x}_0\right) - m\left(\boldsymbol{x}_0\right)\right] = 0$ is the mathematical way of saying that the estimator is unbiased. Let us suppose that the estimator is unbiased. Then

$$\mathrm{E}\left[\hat{m}\left(\boldsymbol{x}_0\right)\right] = m\left(\boldsymbol{x}_0\right) .$$

But by Lemma 8.1, the left side can change to

$$\sum_{l=0}^{n} \sum_{i=1}^{k} a_l \lambda_i f_l\left(\boldsymbol{x}_i\right) = m\left(\boldsymbol{x}_0\right)$$

and by Definition 6.3, the right side can be

$$\sum_{l=0}^{n} \sum_{i=1}^{k} a_l \lambda_i f_l\left(\boldsymbol{x}_i\right) = \sum_{l=0}^{n} a_l f_l\left(\boldsymbol{x}_0\right)$$

or

$$\sum_{l=0}^{n} a_l \sum_{i=1}^{k} \lambda_i f_l\left(\boldsymbol{x}_i\right) = \sum_{l=0}^{n} a_l f_l\left(\boldsymbol{x}_0\right) ,$$

which can be true only if

$$\sum_{i=1}^{k} \lambda_i f_l\left(\boldsymbol{x}_i\right) = f_l\left(\boldsymbol{x}_0\right) \quad \text{for } l = 0,\, 1,\, 2,\, \ldots,\, n .$$

Proof (b)

Now let $\sum_{i=1}^{k} \lambda_i f_l(\boldsymbol{x}_i) = f_l(\boldsymbol{x}_0)$ for $l = 0, 1, 2, \ldots, n$. The equality persists if one multiplies every lth equality by a_l. The summation of the $n+1$ equations yields

$$\sum_{l=0}^{n} a_l \sum_{i=1}^{k} \lambda_i f_l(\boldsymbol{x}_i) = \sum_{l=0}^{n} a_l f_l(\boldsymbol{x}_0) .$$

On the left side of the equation after moving a_l to the second summation, according to Lemma 8.1, the double sum is equal to the expected value of $\hat{m}(\boldsymbol{x}_0)$. On the right side, by Definitions 6.2 and 6.3, the summation yields the expected value of $m(\boldsymbol{x}_0)$. Therefore,

$$\mathrm{E}[\hat{m}(\boldsymbol{x}_0) - m(\boldsymbol{x}_0)] = 0 .$$

Proofs (a) and (b) prove the theorem. □

Notice that a constraint $\sum_{i=1}^{k} \lambda_i = 1$ is incorporated in the set of unbiasedness constraints for $l = 0$.

Estimation Variance

The mean square error is used to measure the efficiency of the estimator, which will be established after proving a lemma and a theorem.

Lemma 8.2

Let $\hat{m}(\boldsymbol{x}_0)$ be the linear unbiased estimator for $m(\boldsymbol{x})$ in Definition 8.1 and let λ_i be the weights in the same definition. Then if $Y(\boldsymbol{x})$ is the residual in Definition 2.2,

$$\hat{m}(\boldsymbol{x}_0) - m(\boldsymbol{x}_0) = \sum_{i=1}^{k} \lambda_i Y(\boldsymbol{x}_i) .$$

Proof

Replacing $\hat{m}(\boldsymbol{x}_0)$ by its Definition 8.1, and splitting the random variable into the drift and its residual in Definition 2.2,

$$\hat{m}(\boldsymbol{x}_0) - m(\boldsymbol{x}_0) = \sum_{i=1}^{k} \lambda_i \big(Y(\boldsymbol{x}_i) + \mathrm{E}[Z(\boldsymbol{x}_i)]\big) - m(\boldsymbol{x}_0) .$$

After expanding the summation, interchanging summation and expectation, and rearranging terms, the relationship changes to

$$\hat{m}(\boldsymbol{x}_0) - m(\boldsymbol{x}_0) = \sum_{i=1}^{k} \lambda_i Y(\boldsymbol{x}_i) + \mathrm{E}\left[\sum_{i=1}^{k} \lambda_i Z(\boldsymbol{x}_i)\right] - m(\boldsymbol{x}_0) .$$

But by Definition 8.1, the first expectation is equal to $\hat{m}(\boldsymbol{x}_0)$, so

$$\hat{m}(\boldsymbol{x}_0) - m(\boldsymbol{x}_0) = \sum_{i=1}^{k} \lambda_i Y(\boldsymbol{x}_i) + \mathrm{E}\left[\hat{m}(\boldsymbol{x}_0) - m(\boldsymbol{x}_0)\right] .$$

Because the estimator is unbiased, the expectation is zero according to Theorem 8.1. Hence,

$$\hat{m}(\boldsymbol{x}_0) - m(\boldsymbol{x}_0) = \sum_{i=1}^{k} \lambda_i Y(\boldsymbol{x}_i) . \quad \square$$

Theorem 8.2

Let $\hat{m}(\boldsymbol{x}_0)$ be the linear unbiased estimator for $m(\boldsymbol{x}_0)$ in Definition 8.1 and let $\mathrm{Cov}_Y(\boldsymbol{h})$ be the covariance of the residuals in Assumption 8.2. Then the estimation variance $\sigma^2(\boldsymbol{x}_0)$ is equal to

$$\sum_{i=1}^{k}\sum_{j=1}^{k} \lambda_i \lambda_j \,\mathrm{Cov}_Y(\boldsymbol{x}_i, \boldsymbol{x}_j) .$$

Proof

By definition,

$$\sigma^2(\boldsymbol{x}_0) = \mathrm{Var}\left[\hat{m}(\boldsymbol{x}_0) - m(\boldsymbol{x}_0)\right] ,$$

which by Lemma 8.2 is equivalent to

$$\sigma^2(\boldsymbol{x}_0) = \mathrm{Var}\left[\sum_{i=1}^{k} \lambda_i Y(\boldsymbol{x}_i)\right] .$$

Lemma 2.4 proves the theorem. $\square$

Optimal Estimator

We will produce a drift estimator that will be unbiased and will have minimum mean square error. Finding the optimum weights for such an estimator can be done by employing the Lagrange method of multipliers (Hillier and Lieberman, 1995, p. 961) to minimize the estimation variance in Theorem 8.2, subject to the constraints in Theorem 8.1.

Definition 8.2

Let $\sigma^2(\boldsymbol{x}_0)$ be the estimation variance in Theorem 8.2 for an estimation at $\boldsymbol{x}_0$, let λ_i be the weights in Definition 8.1, and let μ_l be Lagrange multipliers. The Lagrangian function for the unconstrained objective function is

$$L(\lambda_i, \mu_l) = \sigma^2(\boldsymbol{x}_0) + 2\mu_0\left(\sum_{i=1}^{k} \lambda_i - 1\right) + 2\sum_{l=1}^{n} \mu_l \left(\sum_{i=1}^{k} \lambda_i f_l(\boldsymbol{x}_i) - f_l(\boldsymbol{x}_0)\right) . \quad \square$$

Replacing the value of $\sigma^2(\boldsymbol{x}_0)$ in Definition 8.2 by that given by Theorem 8.2,

$$L(\lambda_i, \mu_l) = \sum_{i=1}^{k}\sum_{j=1}^{k} \lambda_i \lambda_j \operatorname{Cov}_Y(\boldsymbol{x}_i, \boldsymbol{x}_j) + 2\mu_0 \left(\sum_{i=1}^{k} \lambda_i - 1\right)$$
$$+ 2\sum_{l=1}^{n} \mu_l \left(\sum_{i=1}^{k} \lambda_i f_l(\boldsymbol{x}_i) - f_l(\boldsymbol{x}_0)\right).$$

The Lagrangian function is a quadratic expression in the unknown weights with linear constraints. If the covariance is positive definite as in Definition 2.3,

(a) the optimum is a global optimum,
(b) the solution is unique,
(c) the objective function for the optimum weights is a minimum, and
(d) the value of the objective function—the drift estimate variance—is zero or positive.

Theorem 8.3

Let λ_i be the weights in Definition 8.1, let μ_l be the Lagrange multiplier in Definition 8.2, and let $\operatorname{Cov}_Y(\boldsymbol{h})$ be the covariance for the residuals in Assumption 8.2. Then the weights that produce the minimum estimation variance are the solution to

$$\left.\begin{array}{l}
\sum_{i=1}^{k} \lambda_i \operatorname{Cov}_Y(\boldsymbol{x}_i, \boldsymbol{x}_1) + \mu_0 + \sum_{l=1}^{n} \mu_l f_l(\boldsymbol{x}_1) = 0 \\
\sum_{i=1}^{k} \lambda_i \operatorname{Cov}_Y(\boldsymbol{x}_i, \boldsymbol{x}_2) + \mu_0 + \sum_{l=1}^{n} \mu_l f_l(\boldsymbol{x}_2) = 0 \\
\cdots\cdots\cdots\cdots\cdots\cdots\cdots\cdots\cdots \\
\sum_{i=1}^{k} \lambda_i \operatorname{Cov}_Y(\boldsymbol{x}_i, \boldsymbol{x}_k) + \mu_0 + \sum_{l=1}^{n} \mu_l f_l(\boldsymbol{x}_k) = 0 \\
\sum_{i=1}^{k} \lambda_i = 1 \\
\sum_{i=1}^{k} \lambda_i f_1(\boldsymbol{x}_i) = f_1(\boldsymbol{x}_0) \\
\sum_{i=1}^{k} \lambda_i f_2(\boldsymbol{x}_i) = f_2(\boldsymbol{x}_0) \\
\cdots\cdots\cdots\cdots \\
\sum_{i=1}^{k} \lambda_i f_n(\boldsymbol{x}_i) = f_n(\boldsymbol{x}_0)
\end{array}\right\}.$$

Proof

Differentiating the Lagrangian function with respect to the unknowns, by Definition 8.2 and Theorem 8.2,

$$\frac{\partial L}{\partial \lambda_i} = 2\sum_{j=1}^{k} \lambda_j \mathrm{Cov}_Y(\boldsymbol{x}_i, \boldsymbol{x}_j) + 2\mu_0 + 2\sum_{l=1}^{n} \mu_l f_l(\boldsymbol{x}_i) \quad \text{for } i = 1,\ 2,\ \ldots,\ k$$

$$\frac{\partial L}{\partial \mu_0} = 2\left(\sum_{i=1}^{k} \lambda_i - 1\right)$$

$$\frac{\partial L}{\partial \mu_l} = 2\left(\sum_{i=1}^{k} \lambda_i f_l(\boldsymbol{x}_i) - f_l(\boldsymbol{x}_0)\right) \quad \text{for } l = 1,\ 2,\ \ldots,\ n\,.$$

The solution of the system of equations comes from setting these partial derivatives equal to zero, dividing both sides by 2, adding 1 to both sides of the $(k+1)$th equation, and adding $f_l(\boldsymbol{x}_0)$ to the last n equations. □

Minimum Estimation Variance

The expression for the estimation variance given by Theorem 8.2 is valid for any weights λ_i and multipliers μ_l. The next lemma and theorem provide the value of $\sigma^2(\boldsymbol{x}_0)$ for the special case of the optimal parameters.

Lemma 8.3

Let λ_i be the optimal weights for the drift estimator in Definition 8.1 and let μ_l be the Lagrange multiplier in Definition 8.2. If $\mathrm{Cov}_Y(\boldsymbol{h})$ is the covariance for the residuals in Assumption 8.2, then

$$\sum_{i=1}^{k}\sum_{j=1}^{k} \lambda_i \lambda_j \mathrm{Cov}_Y(\boldsymbol{x}_i, \boldsymbol{x}_j) = -\mu_0 - \sum_{l=1}^{n} \mu_l f_l(\boldsymbol{x}_0)\,.$$

Proof

From Theorem 8.3,

$$\sum_{j=1}^{k} \lambda_j \mathrm{Cov}_Y(\boldsymbol{x}_i, \boldsymbol{x}_j) = -\mu_0 - \sum_{l=1}^{n} \mu_l f_l(\boldsymbol{x}_i) \quad \text{for } i = 1,\ 2,\ \ldots,\ k\,.$$

Each of the k expressions is still true if one multiplies both sides of the ith equation by λ_i. Adding them all together,

$$\sum_{i=1}^{k} \lambda_i \sum_{j=1}^{k} \lambda_j \mathrm{Cov}_Y(\boldsymbol{x}_i, \boldsymbol{x}_j) = -\sum_{i=1}^{k} \lambda_i \left\{\mu_0 + \sum_{l=1}^{n} \mu_l f_l(\boldsymbol{x}_i)\right\}.$$

The proof follows by introducing λ_i under the other two summations and by Theorem 8.1. □

Theorem 8.4

Let λ_i be the optimal weights and μ_l be the optimal Lagrange multipliers for the drift estimator in Definition 8.1. If $\mathrm{Cov}_Y(\boldsymbol{h})$ is the covariance for the residuals in Assumption 8.2 and the $f_l(\boldsymbol{x}_0)$ are the terms in Definition 6.3 evaluated at the estimation location, then the minimum mean square error $\sigma_m^2(\boldsymbol{x}_0)$ for drift estimation is

$$\sigma_m^2(\boldsymbol{x}_0) = -\mu_0 - \sum_{l=1}^{n} \mu_l f_l(\boldsymbol{x}_0) .$$

Proof

From Theorem 8.2,

$$\sigma^2(\boldsymbol{x}_0) = \sum_{i=1}^{k}\sum_{j=1}^{k} \lambda_i \lambda_j \mathrm{Cov}_Y(\boldsymbol{x}_i, \boldsymbol{x}_j)$$

and the proof follows directly from Lemma 8.3. □

Algorithmic Summary

First let us define a matrix for a more compact display of the results.

Definition 8.3

Let $\mathbf{t}$ be the matrix

$$\mathbf{t} = [0 \quad 0 \quad \ldots \quad 0 \quad 1 \quad f_1(\boldsymbol{x}_0) \quad f_2(\boldsymbol{x}_0) \quad \ldots \quad f_n(\boldsymbol{x}_0)]' ,$$

where $f_i(\boldsymbol{x}_0)$'s are the functions in Definition 6.3. □

Algorithm 8.1

This is an algorithm for the estimation of drift honoring Assumptions 8.1, 8.2, and 8.3.

1. Calculate each term in matrix $\mathbf{P}$ in Definition 6.8.
2. Calculate each term in matrix $\mathbf{t}$ in Definition 8.3.
3. Solve the system of equations $\mathbf{PQ} = \mathbf{t}$, where $\mathbf{Q}$ is the matrix in Definition 6.9.
4. Using the weights in $\mathbf{Q}$, compute the estimate

$$\hat{m}(\boldsymbol{x}_0) = \sum_{i=1}^{k} \lambda_i Z(\boldsymbol{x}_i) .$$

5. Calculate the drift estimation variance

$$\sigma_m^2(\boldsymbol{x}_0) = -\mathbf{Q}'\mathbf{t} . \quad \square$$

Similar remarks made in Chapter 6 about the semivariograms for the residuals apply here to the estimation of the covariance of the residuals required by matrices $\mathbf{P}$ and $\mathbf{p}$.

Residuals

In many problems the interest is neither in the attribute nor in the drift; it is in the residual in Definition 2.2.

Definition 8.4

Let $\hat{m}(\boldsymbol{x}_0)$ be the drift estimator in Algorithm 8.1 for an attribute whose estimator $\hat{Z}_{UK}(\boldsymbol{x}_0)$ comes from Algorithm 6.2. Then the estimate of the residual is

$$\hat{Y}(\boldsymbol{x}_0) = \hat{Z}_{UK}(\boldsymbol{x}_0) - \hat{m}(\boldsymbol{x}_0) \ . \quad \square$$

A typical example comes from geophysics. If one considers the intensity of the gravity field at any place on the surface of the planet, the interest in prospecting is not in the absolute intensity but in the difference between observed intensity and the average theoretical value of the geoid for the area—the gravimetric residual.

EXERCISE 8.1

Use the data and results of Exercise 6.1 to:

1. Estimate the weights and Lagrange multipliers of a drift estimate at (60, 193) considering only the six closest observations.
2. Find the estimate, its variance, and its residual.
3. Produce a map of residuals.

Solution

1. The six closest observations are those from wells 993, 1002, 1003, 1502, 1504, and 1505. From Exercise 6.1, the square matrix in Definition 6.5 is

$$\mathbf{P} = \begin{bmatrix} 3710.5 & & & & & & & & \\ 3489.4 & 3710.5 & & & & & & & \\ 2970.1 & 2954.3 & 3710.5 & & & & & & \\ 2707.3 & 3250.3 & 2471.3 & 3710.5 & & & & & \\ 2494.0 & 2633.9 & 3511.0 & 2438.4 & 3710.5 & & & & \\ 2716.0 & 3084.4 & 3262.7 & 3192.0 & 3372.7 & 3710.5 & & & \\ 1 & 1 & 1 & 1 & 1 & 1 & 0 & & \\ 61.56 & 62.94 & 55.68 & 64.96 & 54.80 & 59.12 & 0 & 0 & \\ 197.85 & 194.81 & 193.56 & 189.77 & 190.60 & 189.47 & 0 & 0 & 0 \end{bmatrix}$$

and by Definition 8.3,

$$\mathbf{t} = [0 \quad 0 \quad 0 \quad 0 \quad 0 \quad 0 \quad 1 \quad 60.0 \quad 193.0]' .$$

By Algorithm 8.1, the drift estimate weights and Lagrange multipliers are

$$\mathbf{P}^{-1}\mathbf{t} = [0.954 \;\; -0.906 \;\; -0.267 \;\; 0.959 \;\; 1.043 \;\; -0.783 \;\; -1318.44 \;\; 3.56 \;\; -7.87]'.$$

2. By Algorithm 8.1, the estimate is

$$\begin{aligned}\hat{m}(60,193) &= \\ &[3065.0 \;\; 3099.4 \;\; 3200.0 \;\; 3114.9 \;\; 3217.1 \;\; 3189.7 \;\; 0 \;\; 0 \;\; 0] \begin{bmatrix} 0.954 \\ -0.906 \\ -0.267 \\ 0.959 \\ 1.043 \\ -0.783 \\ -1318.44 \\ 3.56 \\ -7.87 \end{bmatrix} \\ &= 3106.6\end{aligned}$$

and the drift variance is

$$\begin{aligned}\sigma^2_{UK}(60,193) &= -\,[0 \;\; 0 \;\; 0 \;\; 0 \;\; 0 \;\; 0 \;\; 1 \;\; 60.0 \;\; 193.0] \begin{bmatrix} 0.954 \\ -0.906 \\ -0.267 \\ 0.959 \\ 1.043 \\ -0.783 \\ -1318.44 \\ 3.56 \\ -7.87 \end{bmatrix} \\ &= 2623.8\,.\end{aligned}$$

Figure 8.1 summarizes the results for the last two steps. It is interesting to observe that the three closest wells, 1003, 1002, and 1505, have negative weights and the others have positive weights.

3. Figure 8.2 is a map for the water table drift and Figure 8.3 is the map of residuals. □

From Definition 8.4 and Exercise 6.1 (5.), the estimate of the residual is

$$\begin{aligned}\hat{Y}(60,193) &= 3149.6 - 3106.6 \\ &= 43.0\,.\end{aligned}$$

In the case of the residuals in Exercise 8.1, the anomalies are closely related to the drainage system for the area. The positive ridge running east–west in the middle of Figure 8.3 separates the Republican River watershed to the north from the ones of the Smoky Hill and Solomon rivers to the south.

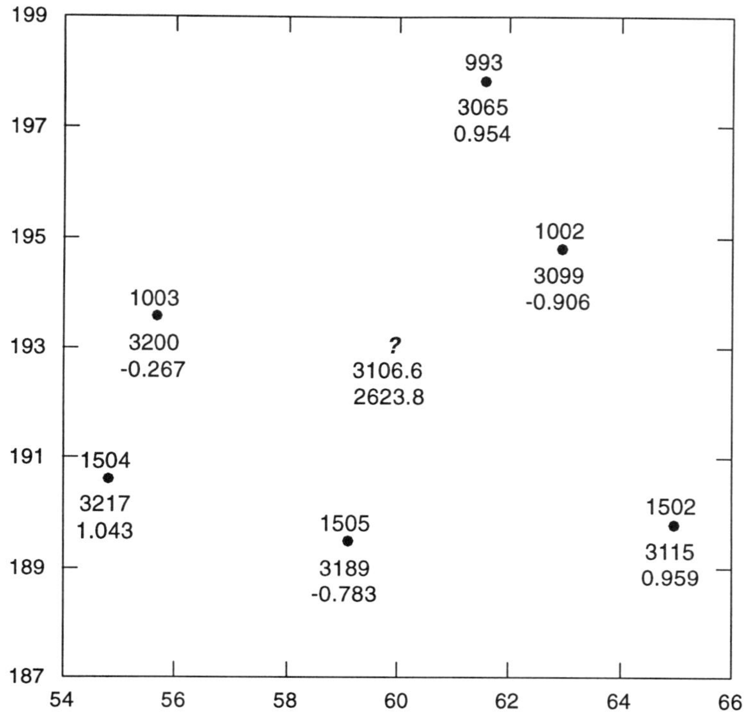

Figure 8.1. Wells and estimation location in Exercise 8.1. Solid dots denote well locations. Levels are, from top to bottom, identification number, water table elevation (ft), and drift estimate weight. Question mark indicates estimation location; annotations are the drift estimate (top; ft) and its drift variance (sq. ft).

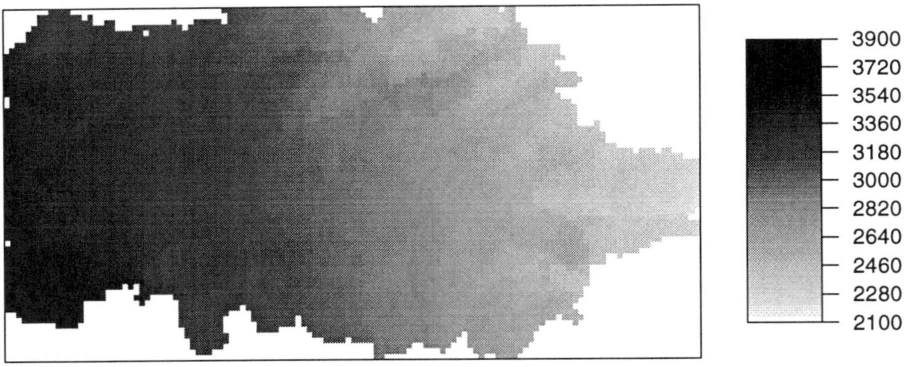

Figure 8.2. First order water table drift (in ft) in High Plains aquifer, northwestern Kansas, winter 1980-81. A color rendition of the map is available at Internet site http://www.kgs.ukans.edu/Mathgeo/Books/Geostat/index.html.

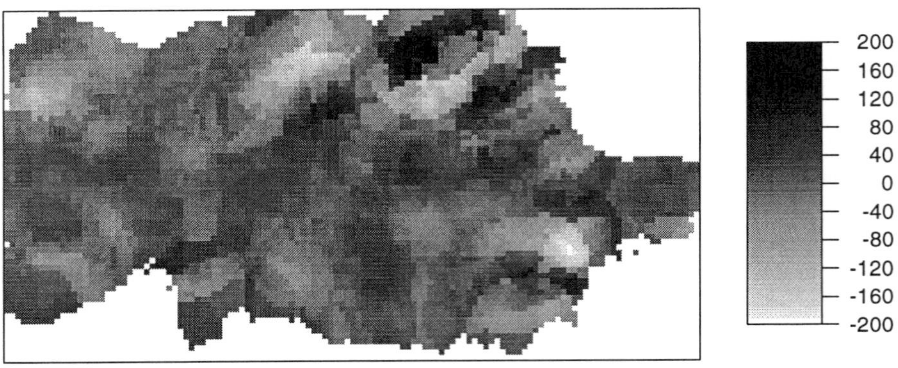

Figure 8.3. First order water table residuals (in ft) in High Plains aquifer, northwestern Kansas, winter 1980-81. For a color rendition, visit Internet site http://www.kgs.ukans.edu/Mathgeo/Books/Geostat/index.html.

Chapter 9

STOCHASTIC SIMULATION

If one analyzes a regular grid of kriging estimates for a spatial attribute such as that shown in Figure 9.1, one finds that there is an uneven smoothing in the grid of estimates—smoothing that is inversely proportional to the data density. Such distortion can be visualized in several ways:

(a) The experimental semivariogram of the estimates is different from the sampling experimental semivariogram. As illustrated in Figure 9.2, the experimental semivariogram for the grid has a smaller sill and a larger range than the experimental semivariogram for the sampling, denoting an exaggerated continuity in the estimated values.

(b) The histogram of the sampling is different than the histogram of the estimated values. Relative to the sample histogram, the histogram for the estimated values has fewer values in the tails and a larger proportion close to the mean. In Figure 9.3, the quartile deviation of the sampling is 36 ft, while that for the grid values is only 26.1 ft.

(c) Crossvalidation of the sampling reveals that there is a tendency of kriging to underestimate values above the sample mean and to overestimate those below the mean, which results in a regression line such as that shown in Figure 9.4. Its slope is less steep than the ideal slope of 1.0 for the main diagonal—a distortion called *conditional bias* in the estimation.

Geostatistics addresses smoothing in kriging by means of stochastic simulation. Unfortunately simulation is only a trade-off, not a perfect solution; stochastic realizations are not error-free renditions of reality. For any realization, the errors, on average, are even larger than those for a kriging estimation, which is the least appealing characteristic of stochastic simulation.

Lacking a perfect solution, the choice between kriging and simulation must be decided based upon what is more relevant for each specific application: minimum local estimation errors in a mean square sense or correct spatial continuity.

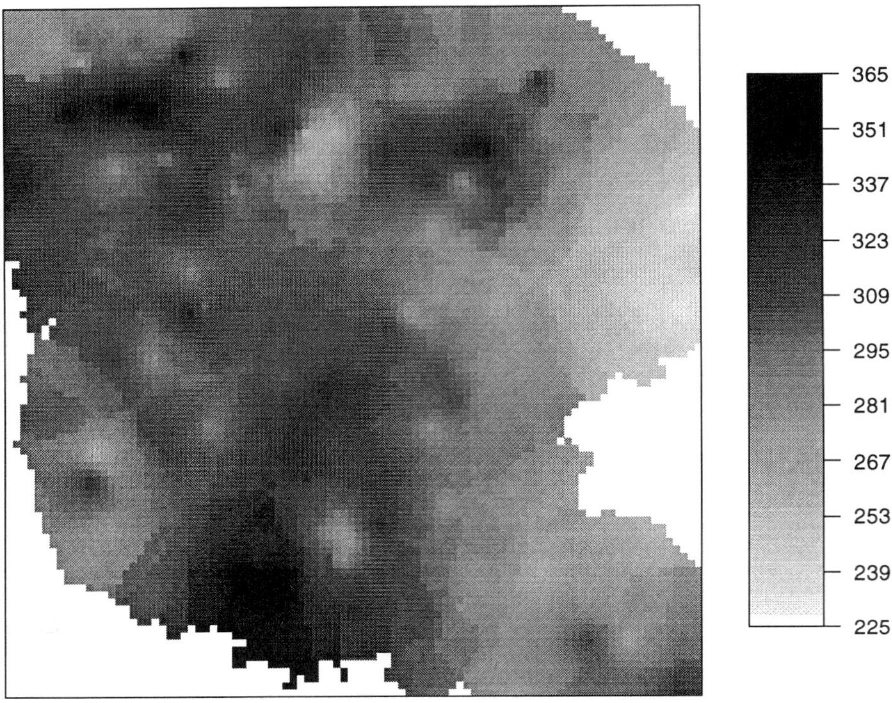

Figure 9.1. Ordinary kriging estimation of Dakota formation thickness (in ft), south-central Hodgeman County, Kansas. For a color rendition, visit Internet site http://www.kgs.ukans.edu/Mathgeo/Books/Geostat/index.html.

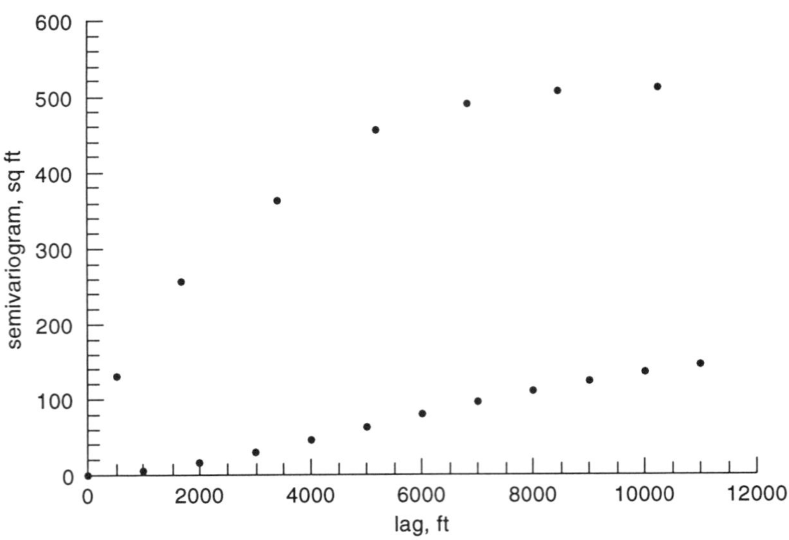

Figure 9.2. Semivariograms for Dakota thickness. Upper sequence of dots is experimental semivariogram for sampling, lower is ordinary kriging grid.

Among the most commonly used simulation methods on a node-by-node basis we have:

i) sequential Gaussian simulation,
ii) simulated annealing,
iii) the lower–upper triangular method, and
iv) the turning bands method.

Any of these techniques have the potential to generate an infinite number of equiprobable realizations. A great number of them are discarded if the solution is constrained to honor the data; this is referred to as *conditional simulation*. Other possibilities not covered here include object-oriented methods (Haldorsen and Damsleth, 1990; Damsleth *et al.*, 1992) and truncated plurigaussian methods (Galli *et al.*, 1994).

Sequential Gaussian Simulation

Sequential Gaussian simulation is a method for the generation of partial realizations using multivariate normal random functions. The cornerstone of the method is the following theorem demonstrating the equivalence between drawing from a multivariate distribution and drawing from a sequence of univariate distributions conditional to univariate realizations.

Theorem 9.1

Let $Z(\boldsymbol{x})$ be a subset of N variates of a random function and let $z(\boldsymbol{x})$ be a sampling of size n. Then the conditional cumulative frequency distribution function $F(x_1, x_2, \ldots, x_N;\ t_1, t_2, \ldots, t_N | n)$ is given by

$$\begin{aligned} &F(x_1, x_2, \ldots, x_N;\ t_1, t_2, \ldots, t_N | n) \\ &\quad = F(x_1;\ t_1 | n)\ F(x_2;\ t_2 | n+1) \cdots F(x_n;\ t_n | n+N-1)\ . \end{aligned}$$

Proof

Let us consider first the case of only two variates. By Bayes' theorem one can break the bivariate cumulative frequency distribution function into the product of two univariate cumulative frequency distribution functions:

$$F(x_1, x_2;\ t_1, t_2 | n) = F(x_1;\ t_1 | n)\ F(x_2;\ t_2 | n+1)\ ,$$

where $F(x_2;\ t_2 | n+1)$ is the probability that $\text{Prob}[Z(\boldsymbol{x}_2) \leq t_2]$ is conditional on the original sampling, plus the value $z(\boldsymbol{x}_1)$ drawn from the distribution $F(x_1;\ t_1 | n)$. Recursive application of Bayes' theorem proves the theorem. □

Implementation of the idea employing kriging requires the assumption of multinormality.

Theorem 9.2

If at location $\boldsymbol{x}_0$ the kriging error for estimate $\hat{Z}(\boldsymbol{x}_0)$ is normally distributed according to $N(0, \sigma^2(\boldsymbol{x}_0))$, then the probability distribution for the true value is $N(\hat{z}(\boldsymbol{x}_0), \sigma^2(\boldsymbol{x}_0))$.

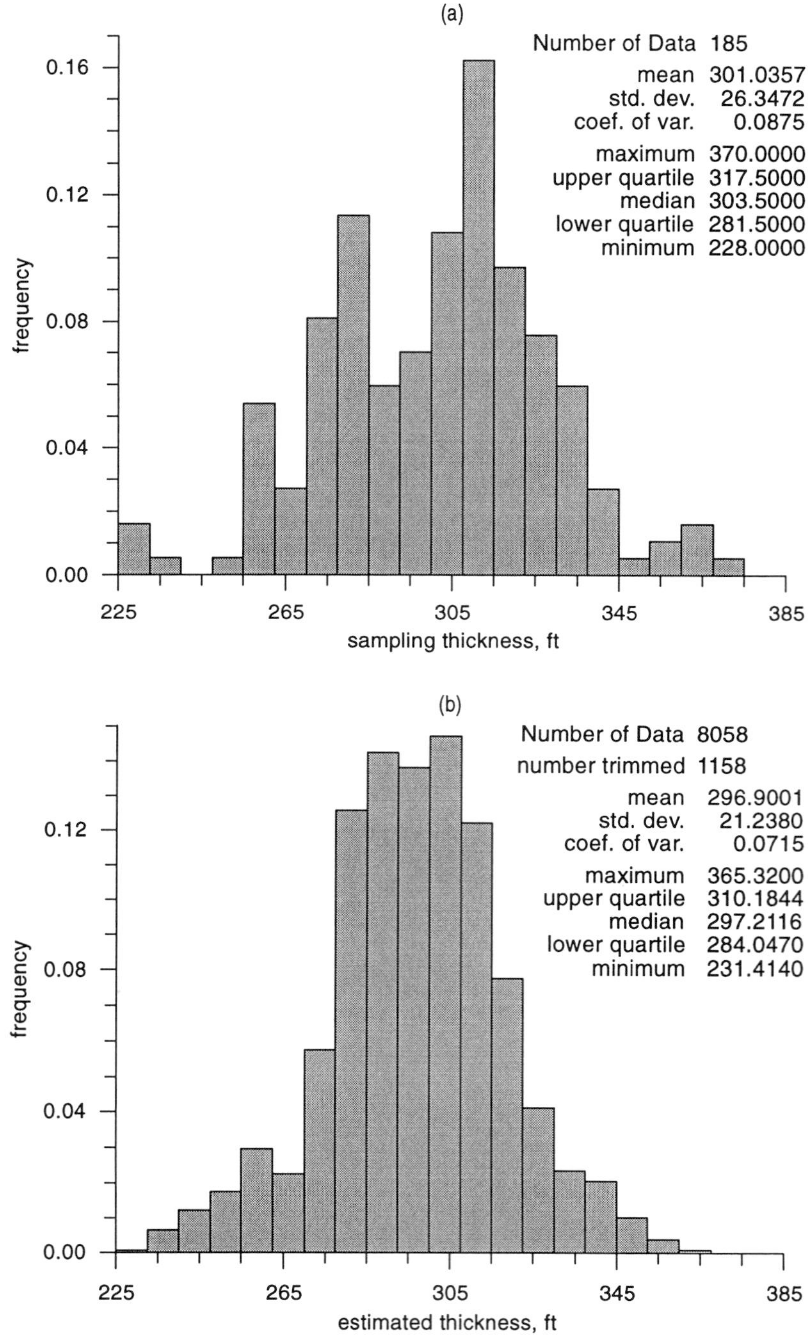

Figure 9.3. Histogram for the sampling used to prepare Figure 9.1 (a) and its ordinary kriging grid (b).

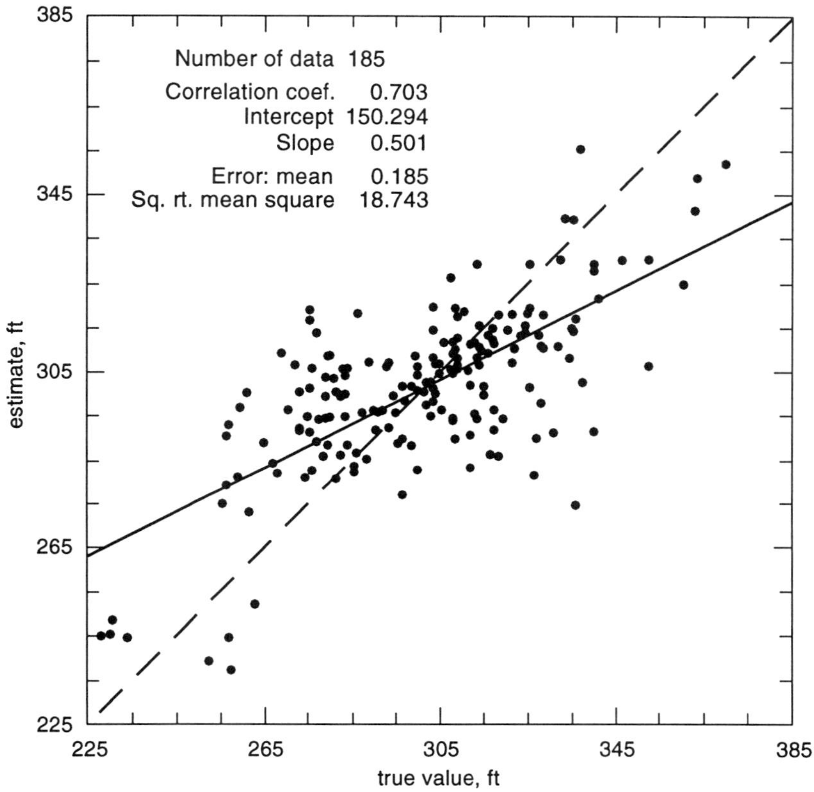

Figure 9.4. Crossvalidation of the formation thickness in Hodgeman County.

Proof

If $z(\boldsymbol{x}_0)$ is the true value of an attribute at location $\boldsymbol{x}_0$, the trivial equation $z(\boldsymbol{x}_0) = z(\boldsymbol{x}_0)$ still holds if one adds and subtracts the value $\hat{z}(\boldsymbol{x}_0)$ from its true value

$$z(\boldsymbol{x}_0) = z(\boldsymbol{x}_0) - \hat{z}(\boldsymbol{x}_0) + \hat{z}(\boldsymbol{x}_0) \ ;$$

but, the difference of the first two terms on the right-hand side of the equation is the estimation error,

$$z(\boldsymbol{x}_0) = e(\boldsymbol{x}_0) + \hat{z}(\boldsymbol{x}_0) \ .$$

If $e(\boldsymbol{x}_0)$ derives from $N\left(0, \sigma^2(\boldsymbol{x}_0)\right)$, then shifting the distribution by $\hat{z}(\boldsymbol{x}_0)$ makes it $N\left(\hat{z}(\boldsymbol{x}_0), \sigma^2(\boldsymbol{x}_0)\right)$. □

Sequential Gaussian simulation is then a form of kriging with feedback. Employing kriging normal scores, the working hypothesis is that each simulated surface is a realization of a multivariate normal process. Although the number of different ways to visit the nodes is limited to $N!$, given a visitation schedule, the number of different realizations still remains infinite because each cumulative frequency distribution function $F(x_i;\ t_i|n+i-1)$ is able to generate an infinite number of different values, independently of the visitation schedule. Practice shows that diversity among realizations is maximized by visiting the nodes at random and preparing a different schedule for every realization.

The conditioning to the data is automatic. The kriging variance at a sampling location is zero, assuring that the only possible drawing is that of the observed value. In practice, however, the data rarely coincide with the nodes. The method is more stable if the data are moved to the closest nodes.

Algorithm 9.1

This is a basic procedure to generate realizations of an attribute that can be regarded as the realization of a multivariate normal random function. Variants abound.

1. In case the sampling is not univariate normal, transform the data to obtain normal scores.

2. Using standard semivariogram modeling techniques, find the best semivariogram model for the transformed data.

3. The standard practice is to arrange the simulation locations regularly at the nodes of a regular grid.

4. Move the data to the closest node, averaging or discarding all but one of those observations that share the same node. Pick at random those without data and stack them in a queue.

5. From the node sequence queue, draw the location $\boldsymbol{x}_i$ of the node due for the generation of a simulated value. Employ kriging to find the estimate $\hat{Z}(\boldsymbol{x}_i)$ and the estimation variance $\sigma^2(\boldsymbol{x}_i)$ using an expanded sampling comprising all data plus all values that may have been simulated already.

6. Employ $\hat{Z}(\boldsymbol{x}_i)$ and $\sigma^2(\boldsymbol{x}_i)$ to define a normal distribution with mean $\hat{Z}(\boldsymbol{x}_i)$ and variance $\sigma^2(\boldsymbol{x}_i)$, $N\left(\hat{Z}(\boldsymbol{x}_i), \sigma^2(\boldsymbol{x}_i)\right)$. The simulated value $z_s(\boldsymbol{x}_i)$ is the number that results from randomly drawing a value from

$$N\left(\hat{Z}(\boldsymbol{x}_i), \sigma^2(\boldsymbol{x}_i)\right).$$

7. Add $z_s(\boldsymbol{x}_i)$ to the expanded sampling.

8. If $\boldsymbol{x}_i$ is not the last node in the queue, go back to Step 5.

9. In case Step 1 was necessary, back transform the values in the multivariate realization to the original space. □

If the total number of observations is n and the number of nodes to simulate is N, then after simulating the last node, the size of the expanded sampling is $n+N-1$. Because typically $n \ll N$, the method needs to discriminate in favor of the original observations to work properly.

EXERCISE 9.1

The semivariogram model of the normal scores for the formation thickness in the sampling in Appendix D is exponential with nugget 0.05, sill 0.75, and range 8,680 ft. Use sequential Gaussian simulation for:

1. Generating three realizations of the formation thickness.
2. Showing how close the semivariograms of the realizations approximate the experimental semivariogram of the sampling.
3. Showing how well the histograms of the realizations compare with the histogram of the original sampling.

Solution

1. Figure 9.5 displays three realizations of formation thickness plus a duplicate of the ordinary kriging estimate in Figure 9.1. Calculations were performed employing program **sgsim** in GSLIB (Deutsch and Journel, 1992) with a 1,000-ft spacing and a maximum search radius of 18,000 ft. As expected, the maximum discrepancies occur at the diagonally opposed corners without control points.

2. Figure 9.6 jointly displays all realization semivariograms plus the model for the attribute. Agreement between the experimental semivariograms of the data and that of a realization theoretically is achieved only in expected value, which in practice is always good.

3. The realization histograms shown in Figure 9.7 are quite similar to that of the sampling. The quartile deviations are now 36.7 ft, 35.6 ft, and 37.5 ft, almost identical to the 36 ft of quartile deviation for the sampling, and much more realistic than the 26.1 ft quartile deviation for ordinary kriging. Perhaps the most noticeable difference is the increase in the proportion of values falling in the class with the smallest values, which can be explained in terms of the difference in sampling density. The sampling has few values at the east end of the area where most of the low values occur, while the realizations have a perfectly regular array of values with an understandably higher proportion of low values denoting thin formation thickness. □

Simulated Annealing

Simulated annealing is one of the newest and most popular simulation methods in a node-by-node basis. The method was initially formulated to solve the statistical mechanics problem of calculating the variation with temperature of properties of substances composed of interacting molecules (Metropolis *et al.*, 1953), such as energy levels in the metallurgical annealing process of prolonged heating and slow cooling of a piece of metal. During this process, as regularity of the crystalline lattice increases, the free energy of the solid decreases. Kirkpatrick *et al.* (1983) were the first to see the potential of the procedure for solving optimization problems such as the design of computer parts and the classical traveling salesman problem. Credit for introducing the method in stochastic simulation goes to Deutsch and Journel (1992).

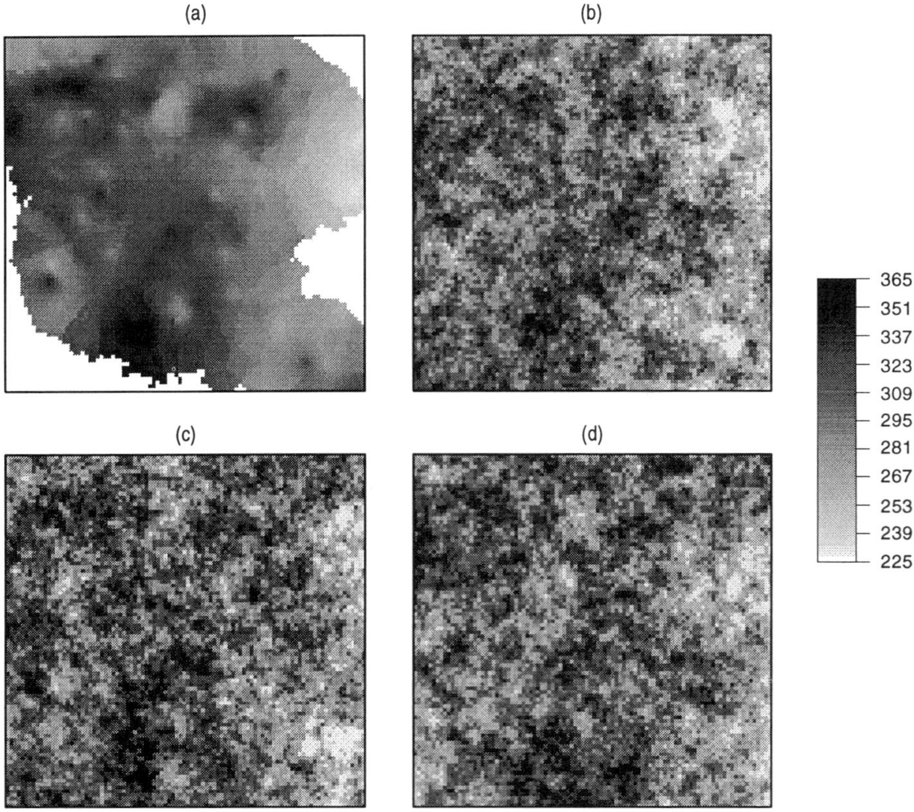

Figure 9.5. Dakota formation thickness (in ft), south-central Hodgeman County, Kansas. For reference, map in upper left corner is a copy of Figure 9.1 and other maps are realizations by sequential Gaussian simulation. Visit Internet site http://www.kgs.ukans.edu/Mathgeo/Books/Geostat/index.html for a color rendition.

In the application of annealing to geostatistics, one deals with values of a spatial attribute instead of molecules. The method presumes that the sampling took place at some of the sites to be considered in the characterization through the partial realization of the random function, typically a grid of regular nodes. If that is not the case, the observations are moved to the closest node. Duplicates are discarded or averaged.

The grid values are assigned in two steps, which commonly are:

i) All nodes coinciding with an observation are given the value of the observation at the node site.

ii) Values are assigned to the remaining nodes by drawing at random from a cumulative frequency distribution function provided by the user— typically the sampling cumulative frequency distribution function.

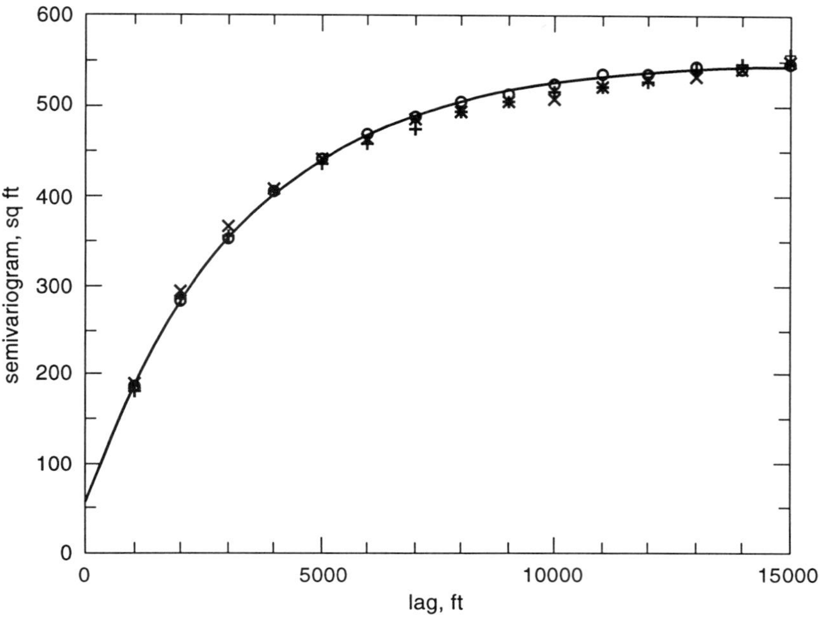

Figure 9.6. Experimental semivariograms for Dakota formation thickness, south-central Hodgeman County, Kansas. Realizations are indicated by circles for 9.5b, crosses for 9.5c, and plus signs for 9.5d. For reference, the solid line is the semivariogram model for the thickness, which is not used in the calculations: exponential with a nugget of 57 sq. ft, range of 9498 ft and (sill – nugget) of 482 sq. ft.

The final solution does not strongly depend on the choice of the initial solution (Aarts and Korst, 1989, p. 75). This peculiar way of defining the node values instantly assures that the realization is a conditional simulation and that the realization has a prespecified histogram. All that is left to simulated annealing is to achieve a match between a function of the original observations and the one for simulated realizations, typically the experimental semivariogram. In such a case, matching is accomplished by reducing the sum of weighted differences,

$$G = \sum_{h} \frac{\left[\gamma_G(\boldsymbol{h}) - \gamma(\boldsymbol{h})\right]^2}{\gamma^2(\boldsymbol{h})},$$

below a small threshold. $\gamma(\boldsymbol{h})$ is the semivariogram model and $\gamma_G(\boldsymbol{h})$ is the semivariogram for the realization. The objective function G plays the same role as the Gibbs free energy function in the physical process of annealing.

The reduction in the objective function is achieved by swapping pairs of values $z_s(\boldsymbol{x}_i)$ and $z_s(\boldsymbol{x}_j)$ chosen at random and observing the effect on the objective function. Recalculation of the objective function can be considerably simplified by correcting the effect of the swapping rather than starting the

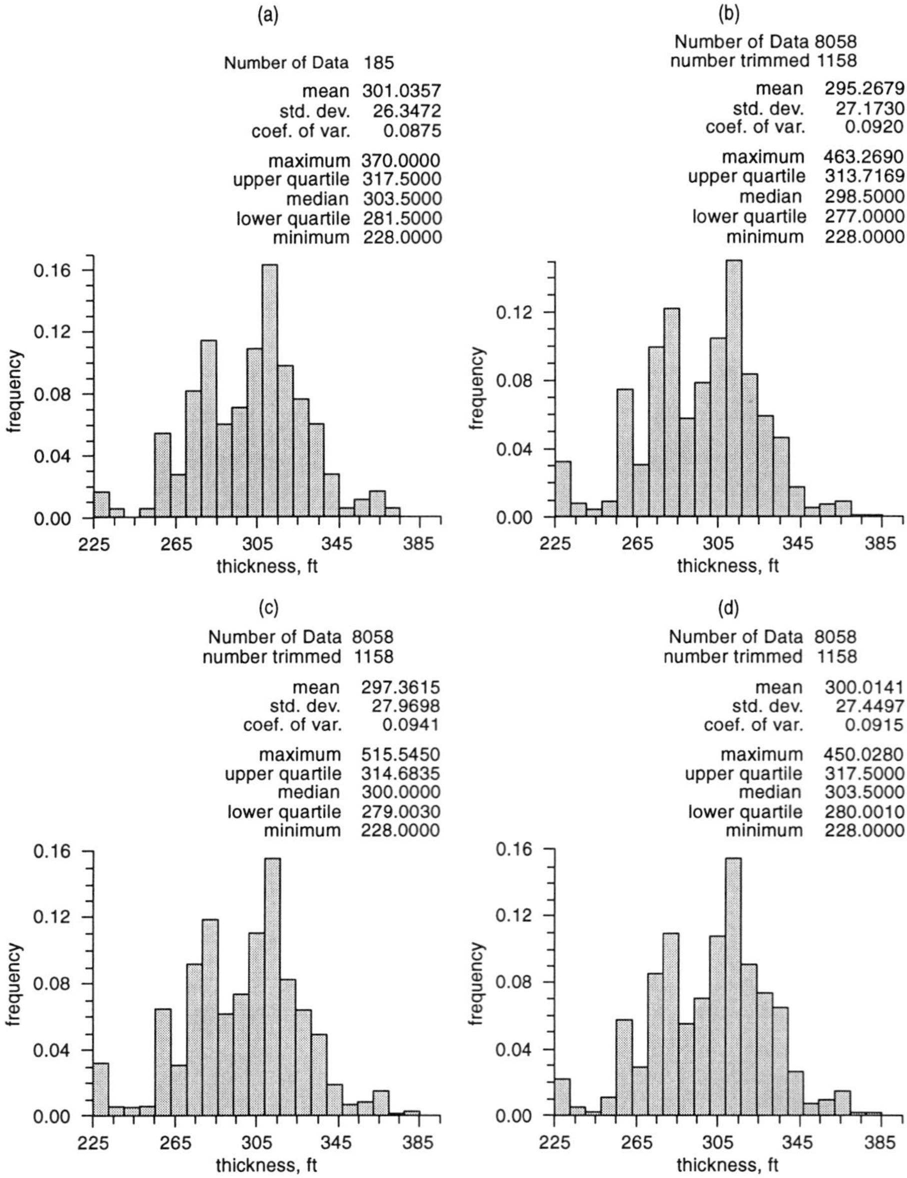

Figure 9.7. Histograms for Dakota formation thickness, south-central Hodgeman County, Kansas. For reference, histogram (a) is the sampling, while the others are those for the realizations in Figure 9.5, in the same sequential order. For a more equitable comparison, the values associated with the nodes in the blanked areas of Figure 9.5a are trimmed off.

recalculation of G from scratch. The equivalent of swapping with true annealing is molecular vibration. A swap is accepted provided that:

i) Neither of the locations involved in the swapping coincides with a sampling site.

ii) There is a decrease in the objective function.

iii) The trickiest and key point in the whole procedure is that not all swaps failing ii) are rejected. Some of them are retained, but the frequency with which those apparently unfavorable swaps are retained decreases with $e^{\frac{G_{old}-G_{new}}{t}}$, where t mimics the temperature parameter in the Boltzmann distribution. Parameter t must be lowered slowly to avoid convergence to local minima.

The specification of how to lower t is called the *annealing schedule.* Its proper selection is critical for the performance of the method (van Laarhoven and Aarts, 1987, p. 98). Successful practice recommends starting at an initial high temperature and lowering it slowly by some multiplicative parameter when enough swaps have been accepted or too many have been tried. In essence the procedure works by trial and error (Deutsch and Cockerham, 1994).

Conceptually the method differs from all geostatistical methods in several respects; for example:

i) Assignment of the initial values of the nodes of a partial realization is accomplished by duplicating the data and drawing numbers from a cumulative distribution rather than by running any form of regression.

ii) The core of the method—optimization by swapping of values—belongs to the realm of operations research. Yet, the method is essentially stochastic. The final product is still viewed as the realization of a random function, the initial stages involve drawing from a cumulative distribution function, and the function to be matched is usually a semivariogram.

Algorithm 9.2

This is a basic procedure for the generation of one partial realization of a random function through simulated annealing.

1. Model the semivariogram $\gamma(\boldsymbol{h})$ using the original sampling.

2. Assign one value to each node. Those nodes coinciding with a sampling site take the value of the corresponding observation. If some or all observations do not coincide with the location of a node, move such observations to the closest node, averaging or discarding duplicates. The remaining values are assigned by drawing values at random from a cumulative distribution function, typically that of the data sampling.

3. Compute the objective function G.

4. Swap pairs of values $z_s(\boldsymbol{x}_i)$ and $z_s(\boldsymbol{x}_j)$, chosen at random, and recalculate the objective function.

5. Keep or reject the swap, depending on the impact on the objective function and the annealing schedule.

6. If G is above tolerance or the number of attempted perturbations is below a stopping limit, go back to Step 4. □

EXERCISE 9.2

Employ the data in Appendix D for the Dakota formation thickness to perform simulated annealing using the following parameters:

Grid spacing: 200 ft
Initial temperature: 1
Temperature reduction factor: 0.12
Maximum attempted perturbations per temperature range:
75 times the number of nodes
Accepted perturbations per temperature range:
8 times the number of nodes
Number of lags for conditioning: 201
Semivariogram: Exponential model with nugget 57 sq. ft;
(sill – nugget) 482 sq. ft; range 9,498 ft.

Analyze the realization and its experimental semivariogram for the following cases:

1. A maximum value 1.0 for the objective function.
2. A maximum value 0.1 for the objective function.
3. A maximum value 0.00001 for the objective function.
4. Compare the histograms of the realization and that of the sampling.

Solution

Figure 9.8 summarizes results obtained using module **sasim** in GSLIB (Deutsch and Journel, 1992). Given the implementation in such a module, a grid spacing of 1,000 ft and 201 conditioning lags permits conditioning of the semivariogram up to a lag of 11,314 ft.

1. Initially the value of the objective function is 1. Because the node values are arranged at random, the experimental semivariogram for the map in Figure 9.8a is pure nugget effect.

2. As the collection of node values becomes more organized, the experimental semivariogram decreases, especially for small arguments. Figures 9.8c and 9.8d illustrate the situation for the requested objective function of 0.1, which requires half a million swaps. Notice the operation of the typical progression of annealing optimization—adjusting to the large features first and delaying accommodation of the small features until toward the end.

3. For a sufficiently small objective function, for all practical purposes, the realization experimental semivariogram exactly matches the experimental semivariogram of the attribute. The optimization reaches the objective function of 0.00001 after 1,100,000 swaps and more than one hour of CPU time on a Sun SPARCstation 5.

4. Figure 9.9 shows the histograms for the realization in Figure 9.8e. Figures 9.8f and 9.9b confirm the capability of simulated annealing to reproduce perfectly the sampling histogram and the sampling semivariogram. □

The grain of the realization in Figure 9.8e is comparable to the grain for those realizations in Figure 9.5 generated by sequential Gaussian simulation.

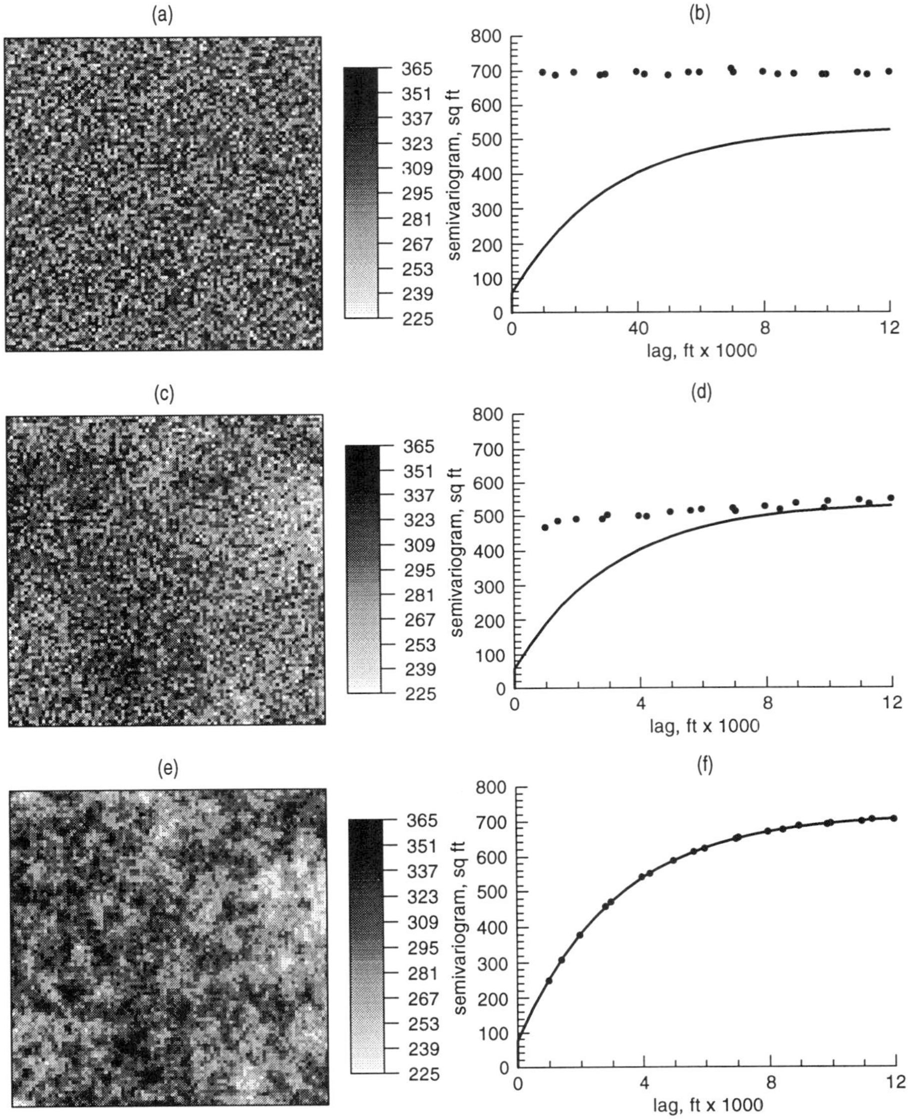

Figure 9.8. Creation of a realization by simulated annealing for Dakota formation thickness (in ft), south-central Hodgeman County, Kansas. Map (a) is the initial random layout whose semivariogram (b) is pure nugget effect. Rendition (c) is an intermediate stage with semivariogram (d), and (e) is the final realization whose semivariogram (f) perfectly fits the semivariogram model for the attribute. For a color rendition, visit Internet site http://www.kgs.ukans.edu/Mathgeo/Books/Geostat/index.html.

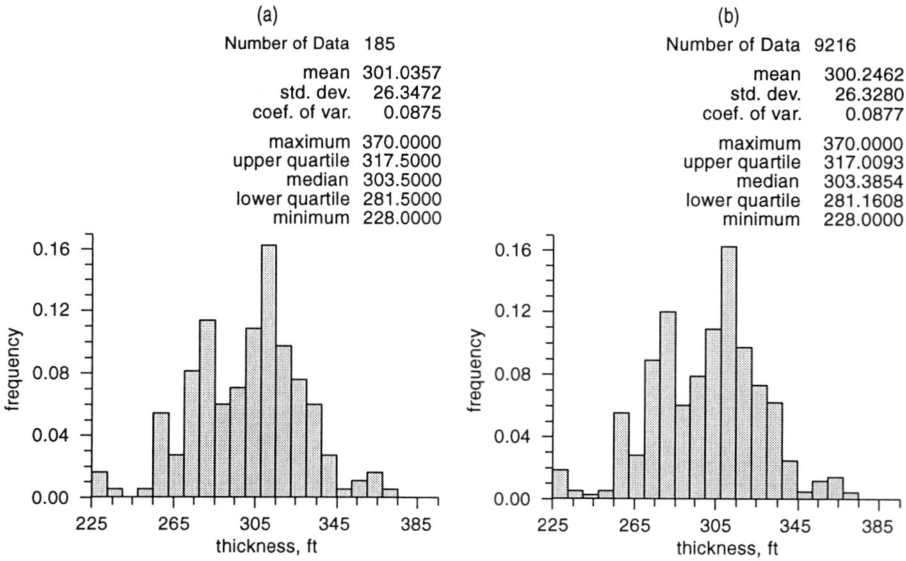

Figure 9.9. Histograms for (a) the sampling used to prepare the simulated annealing realization in Figure 9.8 and (b) for the realization itself.

Deviations from a perfect approximation to the histogram and to the semivariogram by those realizations in Figure 9.5 are too small to translate into mapping properties that can be detected by eye.

Advantages and Disadvantages of Simulated Annealing

On the positive side, simulated annealing is able to generate equiprobable realizations without the need to rely on any distributional assumptions about the parent random function and offers the possibility of handling multiple constraints.

The fact that simulated annealing can match any histogram or semivariogram compatible with the data is both positive and negative. What the user expects from the method is the generation of equiprobable characterizations of reality. Simulated annealing, however, at best matches histograms and semivariograms derived from samplings. To the extent that such samplings are not a good representation of reality, the ability of the method to produce perfect matchups turns into a liability. In the Dakota formation thickness example, for instance, the ability of sequential Gaussian simulation to generate realizations with a higher proportion of low values than the sampling most likely is more realistic than having realizations exactly reproducing the data proportions in the sampling. Although simulated annealing offers the possibility of drawing the values from a cumulative distribution function other than that for the sampling, because preparing alternative cumulative distribution functions cannot be done rigorously, such a practice may lead to dangerously subjective decisions.

The computational side of simulated annealing is clearly its weakest aspect. Depending on the grid spacing, the processing time may be several orders of magnitude larger than that for any other simulation method. If the sampling has not been done at the nodes of a regular grid, a fine grid may be necessary to properly render the surface and to minimize errors in the relocation of sampling sites. As the grid spacing gets finer and finer, however, depending on the implementation of the method, the number of lags required for the proper conditioning of the semivariogram may become larger and larger, in which case the processing time goes up. In addition, although the method is not known to suffer from convergence problems, theoretically there is no assurance of convergence to the global minimum (van Laarhoven and Aarts, 1987, p. 55).

Finally, sequential Gaussian simulation presents no problems for the characterization of attributes with a trend, which current implementations of simulated annealing cannot handle directly, as they do not account for the effect that the deterministic trend has on the semivariogram.

Increasing acceptance of simulated annealing is a clear indication that users are more impressed with its positive aspects than with its drawbacks. In addition, the method is still novel; its repeated use will certainly contribute to its further enhancement.

Lower–Upper (LU) Decomposition

Conditional simulation by LU decomposition is another example of geostatistics borrowing existing techniques independently developed to fulfill other needs. In this case the adaptation was the idea of M.W. Davis (1987).

If **A** is a square matrix, provided that the determinant of **A** is not zero, one can factorize the matrix as the unique product of two triangular matrices

$$\begin{bmatrix} a_{11} & a_{12} & \cdots & a_{1n} \\ a_{21} & a_{22} & \cdots & a_{2n} \\ \cdots & \cdots & \cdots & \cdots \\ a_{n1} & a_{n2} & \cdots & a_{nn} \end{bmatrix} = \begin{bmatrix} l_{11} & 0 & \cdots & 0 \\ l_{21} & l_{22} & \cdots & 0 \\ \cdots & \cdots & \cdots & \cdots \\ l_{n1} & l_{n2} & \cdots & l_{nn} \end{bmatrix} \begin{bmatrix} u_{11} & u_{12} & \cdots & u_{1n} \\ 0 & u_{22} & \cdots & u_{2n} \\ \cdots & \cdots & \cdots & \cdots \\ 0 & 0 & \cdots & u_{nn} \end{bmatrix}$$

called the *LU decomposition* or *LU factorization* (Golub and Van Loan, 1989, p. 96).

All kriging covariance matrices are amenable to LU decomposition. Because a covariance matrix must have a determinant that is not zero and in addition is positive definite and symmetric, one can employ the particular and more efficient Cholesky decomposition that takes advantage of the fact that the upper triangular matrix is equal to the transpose of the lower triangular matrix. The decomposition plus the following pair of theorems are the core of the method.

Theorem 9.3

Let $\mathbf{L}$ be the lower triangular matrix for the sampling of a spatial attribute whose covariance matrix is $\mathbf{K}$, let $\mathbf{g}$ be a vector of independent random numbers drawn from a standard normal distribution, and let the random vector $\mathbf{y}$ be the product $\mathbf{L}\,\mathbf{g}$. Then

$$\mathrm{E}\left[\mathbf{y}\mathbf{y}'\right] = \mathbf{K}\,.$$

Proof

By the definition of the random vector $\mathbf{y}$,

$$\mathrm{E}\left[\mathbf{y}\mathbf{y}'\right] = \mathrm{E}\left[\mathbf{L}\,\mathbf{g}\,(\mathbf{L}\,\mathbf{g})'\right]\,.$$

Expanding the transpose, because the covariance matrix is symmetric,

$$\mathrm{E}\left[\mathbf{y}\mathbf{y}'\right] = \mathrm{E}\left[\mathbf{L}\,\mathbf{g}\mathbf{g}'\,\mathbf{U}\right]$$

and, expanding the expectation,

$$\mathrm{E}\left[\mathbf{y}\mathbf{y}'\right] = \; L\mathrm{E}\left[\mathbf{g}\,\mathbf{g}'\right]\mathbf{U}\,.$$

Because $\mathbf{g}$ is a vector of independent random numbers from standard normal distribution, $\mathrm{E}\left[\mathbf{g}\,\mathbf{g}'\right]$ is equal to the identity matrix. Hence,

$$\mathrm{E}\left[\mathbf{y}\mathbf{y}'\right] = \mathbf{LIU} \qquad \text{or} \qquad \mathrm{E}\left[\mathbf{y}\mathbf{y}'\right] = \mathbf{LU}$$

and the proof follows because the right-hand product is the factorization of the covariance matrix. □

The random vector $\mathbf{y}$ is then a nonconditional simulation of the random function whose partial realization has covariance matrix $\mathbf{K}$. To condition the simulations, it is convenient to distinguish between sampling sites and sites that are unsampled.

Definition 9.1

Consider the set $S = S_n \bigcup S_N$, where S_n is the sampling set of size n and S_N is the set of N unsampled grid nodes completing the definition of a partial realization of the normal scores of a random function. Conveniently ordering the locations in S, one can decompose the covariance matrix of S into four submatrices: $\mathbf{C}_{nn}$, the covariance matrix between the observations; $\mathbf{C}_{nN} = \mathbf{C}'_{Nn}$, the covariance matrices between the observations and the unsampled nodes; and $\mathbf{C}_{NN}$, the covariance matrix among the unsampled points, all in the normal score space. Then $\mathbf{L}_{nn}$, $\mathbf{L}_{Nn}$, and $\mathbf{L}_{NN}$ are matrices in the following factorization of the LU decomposition of the covariance matrix of the normal scores of S:

$$\begin{bmatrix} \mathbf{C}_{nn} & \mathbf{C}_{nN} \\ \mathbf{C}_{Nn} & \mathbf{C}_{NN} \end{bmatrix} = \begin{bmatrix} \mathbf{L}_{nn} & 0 \\ \mathbf{L}_{Nn} & \mathbf{L}_{NN} \end{bmatrix} \begin{bmatrix} \mathbf{U}_{nn} & \mathbf{U}_{nN} \\ 0 & \mathbf{U}_{NN} \end{bmatrix}\,. \quad \square$$

Theorem 9.4

Let $\mathbf{L}_{nn}$, $\mathbf{L}_{Nn}$, and $\mathbf{L}_{NN}$ be the matrices in Definition 9.1; let $\varphi(z_n)$ be the normal scores of the sampling sites; and let $\mathbf{g}_N$ be a vector of N independent random numbers from a standard normal distribution. Then

$$\mathbf{y} = \begin{bmatrix} \varphi(z_n) \\ \mathbf{L}_{Nn}\mathbf{L}_{nn}^{-1}\varphi(z_n) + \mathbf{L}_{NN}\mathbf{g}_N \end{bmatrix}$$

is a partial conditional realization of the random function Z with its same covariance. □

Proof

From Theorem 9.3,

$$\mathbf{y} = \begin{bmatrix} \mathbf{y}_n \\ \mathbf{y}_N \end{bmatrix} = \begin{bmatrix} \mathbf{L}_{nn} & 0 \\ \mathbf{L}_{Nn} & \mathbf{L}_{NN} \end{bmatrix} \begin{bmatrix} \mathbf{g}_n \\ \mathbf{g}_N \end{bmatrix}$$

is an unconditional realization with the same covariance as the parent random function. To condition the realization, $\mathbf{y}$ must honor the data, which implies that $\mathbf{g}_n$ can only be the random vector satisfying

$$\varphi(z_n) = \mathbf{L}_{nn}\mathbf{g}_n \ .$$

Hence, solving for $\mathbf{g}_n$,

$$\mathbf{y} = \begin{bmatrix} L_{nn} & 0 \\ \mathbf{L}_{Nn} & \mathbf{L}_{NN} \end{bmatrix} \begin{bmatrix} \mathbf{L}_{nn}^{-1}\varphi(z_n) \\ \mathbf{g}_N \end{bmatrix} .$$

Multiplication of the matrices proves the theorem. □

The following algorithm is a systematic summary of the findings.

Algorithm 9.3

This is a procedure to generate multiple conditional realizations of a random function using LU decomposition of the covariance matrix:

1. Transform the data to obtain normal scores $\varphi(z_n)$.
2. Using standard semivariogram modeling techniques, find the best covariance model for the transformed data.
3. Calculate matrices $\mathbf{L}_{nn}$, $\mathbf{L}_{Nn}$, and $\mathbf{L}_{NN}$ in Definition 9.1.
4. Generate a vector $\mathbf{g}_N$ of N independent random numbers from a standard normal distribution.
5. Compute realization $\mathbf{y}$:

$$\mathbf{y} = \begin{bmatrix} \varphi(z_n) \\ \mathbf{L}_{Nn}\mathbf{L}_{nn}^{-1}\varphi(z_n) + \mathbf{L}_{NN}\mathbf{g}_N \end{bmatrix} .$$

6. Transform the realization back to the original space.

7. If more realizations are required, return to Step 4. Otherwise, stop. □

In order to generate more than one realization, all that needs to be done is to generate more vectors of random numbers drawn from a standard normal distribution. This is the greatest strength of the method, which makes it extremely fast. Its greatest and fatal weakness, however, is the limitation of the covariance matrix size to no more than 1,000 sites total—data plus nodes (Cressie, 1991, p. 207). Although this is not a theoretical limitation, but an implementation problem, the drawback has severely restricted the application of LU simulation. GSLIB offers the program **lusim** for running an LU simulation (Deutsch and Journel, 1992, p. 163–164).

The Turning Bands Method

Originally formulated by Matheron (1973), the idea behind the turning bands method is a special case of the more general approach using Radon projections and space transformations (Radon, 1917; Cressie, 1991, p. 206-207; Christakos, 1992, p. 215-216). The advantage of the procedure lies in the computational efficiency derived from the reduction in the dimension of the space where the simulation is actually performed. Simulated realizations of a multidimensional random function are obtained by generating a series of independent one-dimensional realizations of a random function that share a common covariance. Although this is the oldest method in use for the simulation of three-dimensional random fields (Deutsch and Journel, 1992, p. 144), generation of conditional realizations is cumbersome and the results are not free from problems, which explains the declining use of the method as more alternatives become available.

Implementation of the algorithm, such as the program **tb3d** in GSLIB (Deutsch and Journel, 1992, p. 160-163), requires that the random function be second order stationary and follow a univariate standard normal distribution.

Algorithm 9.4

The steps involved in the turning bands simulation of a multidimensional random function are:

1. Transform the data to obtain normal scores.

2. Using standard semivariogram modeling techniques, find the best covariance model $\mathrm{Cov}_n\,[\boldsymbol{h}]$ for the anamorphosed data.

3. Derive the covariance $\mathrm{Cov}_1\,[h]$ that linear combinations of independent realizations on the line should have in order to generate realizations of a multidimensional process with covariance $\mathrm{Cov}_n\,[\boldsymbol{h}]$. For the three-dimensional, isotropic case, the unidimensional covariance is

$$\mathrm{Cov}_1\,[h] = \frac{\partial}{\partial h} h\,\mathrm{Cov}_3\,[h] \ .$$

For example, if $\text{Cov}_3[\boldsymbol{h}]$ is the exponential model Ce^{-ah}, then $\text{Cov}_1[h]$ is $C(1-ah)e^{-ah}$, and for the spherical semivariogram,

$$\text{Cov}_3[\boldsymbol{h}] = \begin{cases} \mathbf{C}\left[1 - \frac{3}{2}\frac{\boldsymbol{h}}{\boldsymbol{a}} + \frac{1}{2}\left(\frac{\boldsymbol{h}}{\boldsymbol{a}}\right)^3\right] & \text{for } \boldsymbol{h} < \boldsymbol{a} \\ 0 & \text{for } \boldsymbol{h} \geq \boldsymbol{a} \end{cases}$$

$$\text{Cov}_1[h] = \begin{cases} \frac{C}{2}\left[1 - 3\frac{h}{a} + 2\left(\frac{h}{a}\right)^3\right] & \text{for } h < a \\ 0 & \text{for } h \geq a \end{cases}$$

Paradoxically, the relationship between $\text{Cov}_2[\boldsymbol{h}]$ and $\text{Cov}_1[h]$ is a more complex integral equation whose solution (Brooker, 1985) is:

$$\text{Cov}_1[h] = \text{Cov}_2[0] + \int_0^h h\left(h^2 - t^2\right)\frac{d}{dt}\left[\text{Cov}_2[t]\right]dt\ ,$$

which for the case of the spherical semivariogram turns into

$$\text{Cov}_1[h] = \begin{cases} C\left[1 - \frac{3\pi}{4}\left(\frac{h}{a} - \frac{1}{2}\left(\frac{h}{a}\right)^3\right)\right] \text{for } h < a \\ C\left[1 - \frac{3}{2}\left(\frac{h}{a} - \frac{1}{2}\left(\frac{h}{a}\right)^3\right)\arcsin\left(\frac{a}{h}\right) - \frac{3}{4}\left(\frac{h}{a}\right)^2\sqrt{1 - \left(\frac{a}{h}\right)^2}\right] \text{for } h \geq a \end{cases}$$

4. Generate a series of discrete, independent, unidimensional realizations along lines radiating from the origin of a multidimensional sphere. All realizations must have the same covariance $\text{Cov}_1[h]$. The theory calls for an infinite number of lines, but in practice the recommended number of lines varies with the dimensionality of the space. For a two-dimensional space, Mantoglou and Wilson (1982) recommend 4 to 16 lines, and for three dimensions, Deutsch and Journel (1992, p. 144) use 15 lines, while Tompson *et al.* (1989) observed significant improvements in the quality of the simulation by increasing the number of lines up to 100.

There are several methods for generating realizations on the line that have been around much longer than the turning bands method (Christakos, 1992, p. 316–321), the most common being convolution techniques, autoregressive techniques, and fast Fourier techniques. Program **tb3d** uses convolution, which is based on the property that $\text{Cov}_1[h]$ can be expressed as the convolution of a properly chosen function $p(u)$:

$$\text{Cov}_1(h) = \int_{-\infty}^{\infty} p(u)\,p(u+h)du\ .$$

If, for example, $\text{Cov}_3[\boldsymbol{h}]$ is spherical with variance C and sill a, then

$$p(u) = \begin{cases} \sqrt{\frac{6C}{\pi a^3}}u & \text{for } u \leq \frac{a}{2} \\ 0 & \text{for } u > \frac{a}{2}\ . \end{cases}$$

Values $y_1(x_i)$ are drawn from a uniform distribution and assigned to points x_i at regular intervals Δx on a line. The realization on the line in this case is the set of weighted values $Z_1(x_i)$ given by

$$Z_1(x_i) = \sum_{j=i-n}^{i+n} p \,|\, x_j - x_i \,|\, y(x_j) \ ,$$

where $n = \frac{a - \Delta x}{2\,\Delta x}$.

5. Add the contributions of the line realizations to produce the multidimensional realization $Z_n(\boldsymbol{x}_i)$. The summation is straightforward, but what is added up is not. If multidimensional perpendicular planes are erected in between each simulated value of each of the N realizations along directions $\mathbf{u}_j$, then the space is partitioned in multidimensional bands of width Δx. The partition is easier to visualize in two dimensions, in which case the bands degenerate into strips (Figure 9.10). Given a unidimensional realization, each point in the multidimensional space takes that simulated value $z_{1,i}(x)$ sharing the same band. Of all the values in realization $Z_{1,i}(\boldsymbol{x})$, the only one making a contribution to the sum is the one in the band containing the perpendicular from $\boldsymbol{x}_0$ to the line $\mathbf{u}_i$. In other words, the argument of $z_{1,i}(\cdot)$ is the length of the projection of $\boldsymbol{x}_0$ on $\mathbf{u}_i$. By the central limit theorem, each simulated value is a realization of a multidimensional random function at $\boldsymbol{x}_0$ obtained as the scaled addition of the unidimensional simulated values at the N projected points,

$$z_n(\boldsymbol{x}_0) = \frac{1}{\sqrt{N}} \sum_{i=1}^{N} z_{1,i}(\boldsymbol{x}_0 \cdot \mathbf{u}_i) \ .$$

6. The turning bands method produces values that do not honor the data. If conditioning the data is not required, skip this step.

As shown in Figure 9.11, the conditioning involves the kriging of two different data sets. Let $\hat{z}_d(\boldsymbol{x}_0)$ be a kriging estimate at location $\boldsymbol{x}_0$ based on the original data set of normal scores. Let $z_n(\boldsymbol{x}_0)$ be the nonconditional value produced by Step 5 using the same normal scores. A set of $z_n(\boldsymbol{x}_0)$ values comprising a realization, by construction, has the same covariance as the attribute being simulated. A second data set is generated by employing the same N line realizations in Step 4 to simulate values $z_s(\boldsymbol{x}_i)$ at each of the $\boldsymbol{x}_i$ locations of the original data set. A second kriging is now done, using the $z_s(\boldsymbol{x}_i)$'s as data to obtain $\hat{z}_n(\boldsymbol{x}_0)$. The expression

$$z_c(\boldsymbol{x}_0) = \hat{z}_d(\boldsymbol{x}_0) + z_n(\boldsymbol{x}_0) - \hat{z}_n(\boldsymbol{x}_0)$$

provides the required conditionally simulated value because:

i) by Theorem 2.5, if $\boldsymbol{x}_0$ coincides with a sampling site, $z_n(\boldsymbol{x}_0) = \hat{z}_n(\boldsymbol{x}_0)$ and $\hat{z}_d(\boldsymbol{x}_0) = z(\boldsymbol{x}_0)$, hence $z_c(\boldsymbol{x}_0)$ is an exact interpolator because $z_c(\boldsymbol{x}_0) = z(\boldsymbol{x}_0)$;

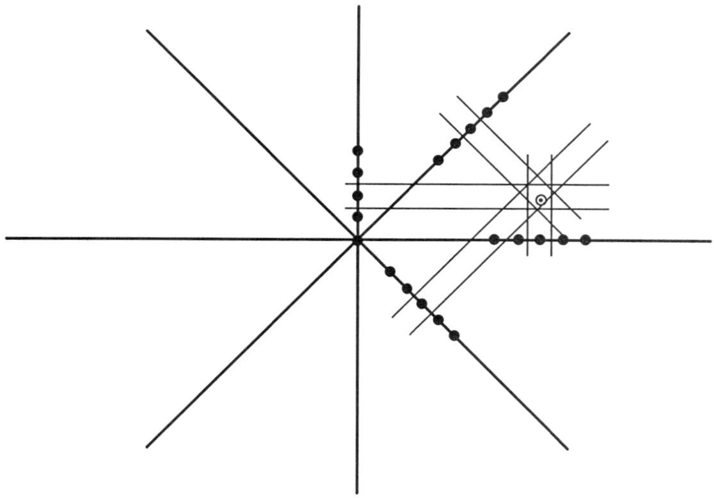

Figure 9.10. Illustration of the turning bands method in two dimensions (after Luster, 1985). Each simulated point on the plane (dot within circle) is the average of all simulated line values whose bands of influence contain the point. For simplicity, only four line simulations are depicted here. Actually 8 or 16 lines would be preferable.

ii) considering that the kriging is an unbiased interpolator (Theorem 2.4), $\mathrm{E}[Z(\boldsymbol{x}_0)] = \mathrm{E}\left[\hat{Z}_d(\boldsymbol{x}_0)\right]$ and $\mathrm{E}[Z_n(\boldsymbol{x}_0)] = \mathrm{E}\left[\hat{Z}_n(\boldsymbol{x}_0)\right]$. Consequently the expected value of $Z_c(\boldsymbol{x})$ is the same as the expected value of the random function of normal scores $Z(\boldsymbol{x})$;

iii) one can decompose a random function into its kriging estimate plus its kriging error $Z(\boldsymbol{x}) = \hat{Z}_d(\boldsymbol{x}) + \left\{Z(\boldsymbol{x}) - \hat{Z}_d(\boldsymbol{x})\right\} = \hat{Z}_d(\boldsymbol{x}) + R(\boldsymbol{x})$, in which case, as the mean of the normal scores is zero,

$$\mathrm{Cov}(\boldsymbol{h}) = \mathrm{E}\left[\left\{\hat{Z}_d(\boldsymbol{x}) + R(\boldsymbol{x})\right\}\left\{\hat{Z}_d(\boldsymbol{x}+\boldsymbol{h}) + R(\boldsymbol{x}+\boldsymbol{h})\right\}\right].$$

Expanding, by the orthogonality of estimates and errors (Theorem 2.6),

$$\mathrm{Cov}(\boldsymbol{h}) = \mathrm{E}\left[\hat{Z}_d(\boldsymbol{x})\,\hat{Z}_d(\boldsymbol{x}+\boldsymbol{h})\right] + E[R(\boldsymbol{x})\,R(\boldsymbol{x}+\boldsymbol{h})].$$

However, as the unknown error $Z(\boldsymbol{x}) - \hat{Z}_d(\boldsymbol{x})$ has the same covariance as the known $Z_n(\boldsymbol{x}) - \hat{Z}_n(\boldsymbol{x}) = R_n(\boldsymbol{x})$, then

$$\mathrm{Cov}(\boldsymbol{h}) = \mathrm{E}\left[\hat{Z}_d(\boldsymbol{x})\,\hat{Z}_d(\boldsymbol{x}+\boldsymbol{h})\right] + E\left[R_n(\boldsymbol{x})\,R_n(\boldsymbol{x}+\boldsymbol{h})\right] = \mathrm{Cov}_{Z_c}(\boldsymbol{h}),$$

because $R_n(\boldsymbol{x})$ is independent of $\hat{Z}_d(\boldsymbol{x})$. Therefore, the covariance of the combination $Z_c(\boldsymbol{x})$ is the same as the covariance of the random function of normal scores $Z(\boldsymbol{x})$.

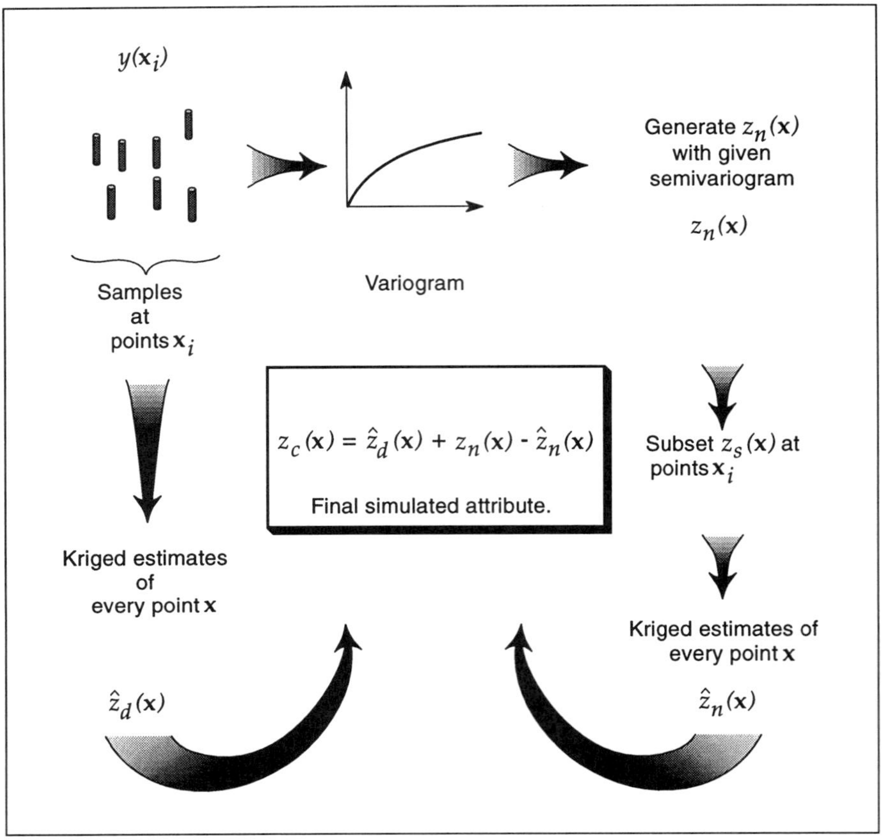

Figure 9.11. Summary of the conditioning process for the turning bands method (after David, 1977, p. 327).

7. In case Step 1 was necessary, back transform the normal score realization to the original space. □

Available software such as program **tb3d** automatically takes care of the details in Steps 3 through 5. Despite its complexity, the nonconditional turning bands method is fast; however, it is unreliable. The realizations may be marred by artifacts due to sensitivity to the number and spacing of the lines (Mantoglou and Wilson, 1982; Tompson *et al.*, 1989). The danger of artifacts plus the cumbersome conditioning step have encouraged users to abandon the method in favor of those mentioned above.

Chapter 10

RELIABILITY

Characteristic of geostatistics and other stochastic methods is the ability to assign confidence intervals to the estimates. The confidence intervals are derived from cumulative distributions of random functions.

Once a cumulative distribution is known, the options in terms of what confidence limits to employ are up to the user and can be any number in the domain extending from 0 to 100%. In geostatistical practice, however, the cases of interest can be grouped into three classes:

i) kth percentile maps, in which for any location on a map the interest is in finding that threshold for which the probability of occurrence of the true value is $100\,k$. For example, on a scale of 0–1, a 90th percentile denotes that with 0.9 probability the true value of the attribute will not exceed the value of that percentile.

ii) Threshold probability maps showing the fluctuations in the probability that the attribute is below a certain value. For example, for a chlorine survey, a threshold map may indicate the probability that the true chlorine concentration remains below 50 ppm.

iii) Interval probability maps displaying the areal fluctuations in the probability that the true value is between two given values, such as the probability that depth to a certain formation is within an oil saturation zone ranging from 5,200 to 5,350 ft below sea level.

There are at least three different ways to generate these reliability maps. The basic difference between the three is in the generation of the random function probability distributions, which may be by:

i) Normality of kriging errors.

ii) Transformation of the original data.

iii) Stochastic simulation.

Kriging Under Normality of Errors

We have seen that if the standard scores conform to a standard normal distribution, then a strong but not uncommon working hypothesis is that the errors are multinormally distributed.

By Theorem 9.2, if the probability distribution for the true value is $N\left(\hat{z}\left(\boldsymbol{x}_0\right), \sigma^2\left(\boldsymbol{x}_0\right)\right)$, then its cumulative distribution is

$$\operatorname{Prob}\left[z \leq t\right]_{\boldsymbol{x}0} = F\left(t\right)_{\boldsymbol{x}0} = \int_{-\infty}^{t} \frac{1}{\sqrt{2\pi}} e^{-\frac{\left(t-\hat{z}\left(\boldsymbol{x}_0\right)\right)^2}{2\sigma^2\left(\boldsymbol{x}_0\right)}} dz \, .$$

The estimate and its estimation variance completely determine the distribution of the variate, from which one can calculate the probability associated to any threshold or interval. For example, if the estimate is 1,000 and the kriging variance is 100, then

$$\operatorname{Prob}\left[z \leq 1020\right]_{\boldsymbol{x}0} = 97.73\%$$

and

$$\begin{aligned} \operatorname{Prob}\left[990 \leq z \leq 1010\right]_{\boldsymbol{x}0} &= \operatorname{Prob}\left[z \leq 1010\right]_{\boldsymbol{x}0} - \operatorname{Prob}\left[z \leq 990\right]_{\boldsymbol{x}0} \\ &= 84.13 - 15.87 \\ &= 68.26\% \, . \end{aligned}$$

The variations on how to denote reliability are many, but the basis of all of them is the normal distribution.

Normality of the errors is by far the most congenial solution to the reliability problem, but it also makes the strongest assumption.

Algorithm 10.1

This is a procedure to generate percentile maps for the kriging estimate of a multivariate random function.

1. Select a percentile.

2. Use the most adequate form of kriging to generate a set of estimates and its corresponding kriging standard deviation.

3. From the standard normal distribution, find the number α of standard deviations above or below the mean corresponding to the percentile in Step 1.

4. Through a grid-to-grid operation, generate the percentile grid as

$$\hat{Z}\left(\boldsymbol{x}\right) + \alpha\sigma\left(\boldsymbol{x}\right) \, . \quad \square$$

If the domain of the attribute is bounded, Algorithm 10.1 may produce unfeasible percentiles, such as negative chemical concentrations.

EXERCISE 10.1

Given the sampling in Exercise 7.2 and the optimal parameters for answer 7.2.3, prepare grids with a 200-ft increment extending over the entire sampling space and then display the results for the following cases:

1. Ordinary kriging and its kriging standard deviation.
2. Elevations assuring that with 90% probability at any location the true elevation will be less then the value on the map.
3. The probability that the true elevation is less than −7,950 ft and the probability that the true elevation is less than −7,850 ft.
4. The probability that the true elevation is between −7,950 and −7,850 ft.

Solution

1. Most geostatistical packages have a routine to generate a grid with values for the estimate and its kriging standard deviation. If the output reports variance instead of standard deviation, a simple utility program can handle the transformation. The tails of small values of kriging variance display better on a map scaled in standard deviations.

Figure 10.1 displays the requested maps. The sampling density is fairly regular and the standard deviation is maximum toward the corners where it reaches 60 ft due to lack of control.

Considering that Exercise 7.2 (5) supports normality of the errors, the map in Figure 10.1a can be regarded as the 50th percentile map, or the elevation for which—at any location—there is equal probability that the true value will be smaller or larger than the value on the map.

Similarly, one can interpret the standard deviation as the length of the estimate-centered confidence interval that, with 68.3% probability, contains the true elevation; or half the length of the estimate-centered confidence interval that, with 95.5% probability, contains the true elevation. Toward the corners, for example, where the standard deviation is as large as 60 ft, the 95.5% probability interval can reach 120 ft—60 ft above and below the kriging estimate.

2. Normal cumulative distributions reach the value of 90% probability for an argument equal to the mean plus 1.28 times the standard deviation. Thus, through a straight grid-to-grid operation, the requested map is that shown in Figure 10.2, which is obtained as the one in Figure 10.1a plus 1.28 times the one in Figure 10.1b.

3. At each node the probabilities result from the direct application of the cumulative distribution function $F(t)_{x_0}$. The corresponding maps are in Figure 10.3.

4. At each node x_0,

$$\text{Prob}\left[-7950 \le z \le -7850\right]_{x_0} = \text{Prob}\left[z \le -7850\right]_{x_0} - \text{Prob}\left[z \le -7950\right]_{x_0} .$$

Therefore the requested map is the difference between the maps in Figure 10.3 shown in Figure 10.4. □

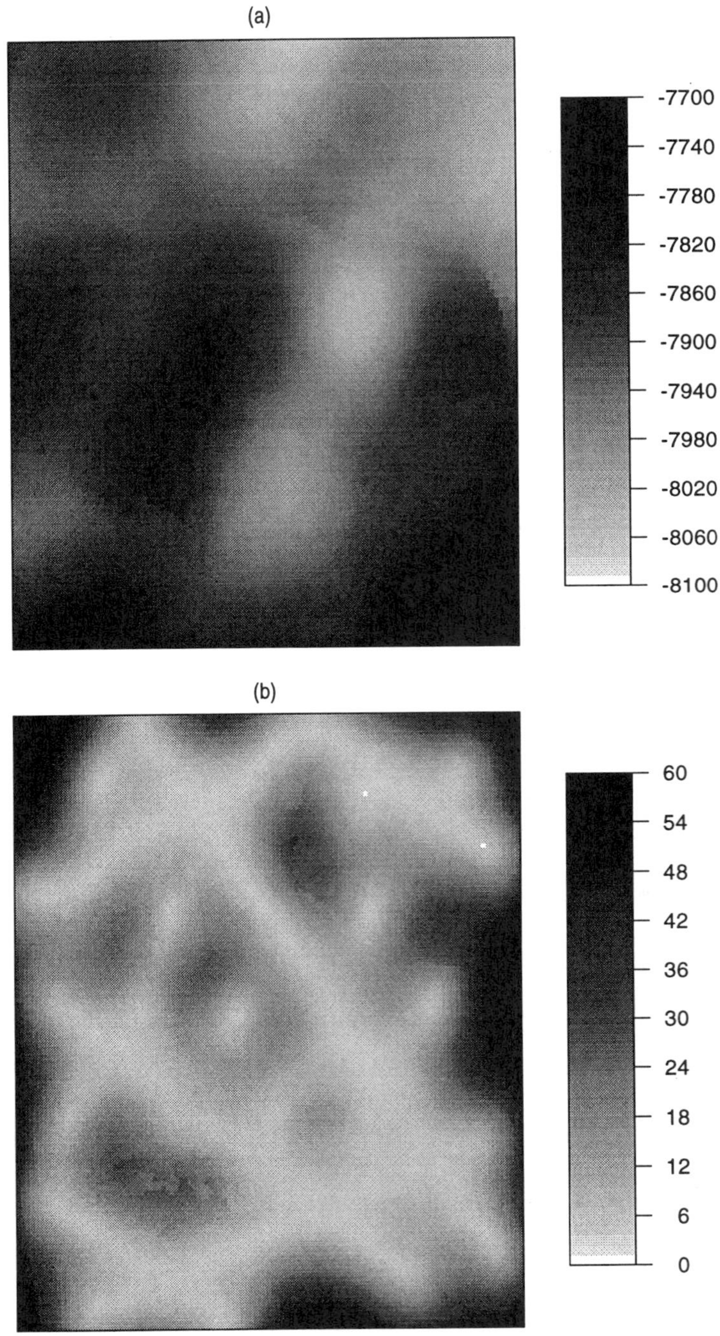
(a)
-7700
-7740
-7780
-7820
-7860
-7900
-7940
-7980
-8020
-8060
-8100
(b)
60
54
48
42
36
30
24
18
12
6
0

[Figure 10.1]

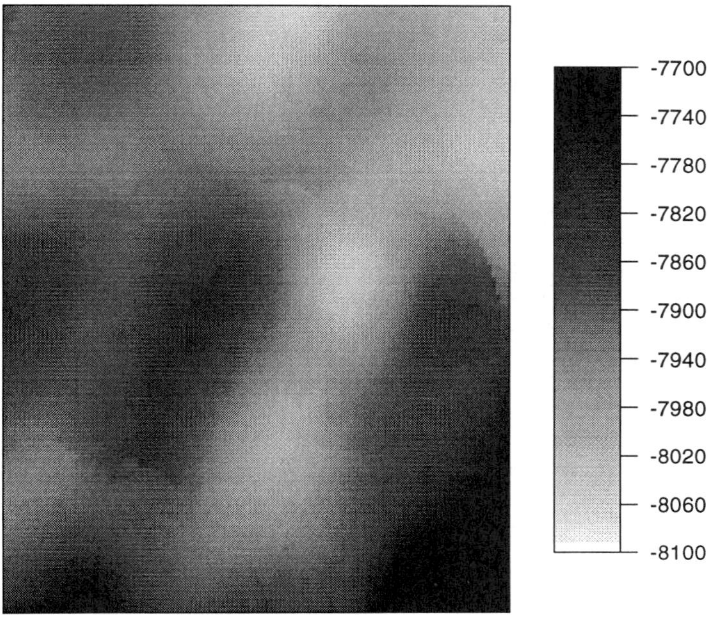

Figure 10.2. The 90th percentile for elevation (ft below sea level) to UNCF based on the normality of ordinary kriging errors. For a color rendition, visit Internet site http://www.kgs.ukans.edu/Mathgeo/Books/Geostat/index.html.

Indicator Kriging

The second solution to the reliability problem is based on a nonlinear transformation of the data to approximate the cumulative distribution numerically.

Definition 10.1

Given a random function $Z(\boldsymbol{x})$, its indicator $I(\boldsymbol{x}, t)$ is the binary transformation

$$I(\boldsymbol{x}, t) = \begin{cases} 0, & \text{if } Z(\boldsymbol{x}) > t \\ 1, & \text{if } Z(\boldsymbol{x}) \leq t\,. \end{cases} \quad \square$$

Such a transformation has the interesting property that its expected value is equal to the cumulative probability of the variate.

[Opposite page, left.]

Figure 10.1. Mapping of ordinary kriging results for elevation to UNCF. (a) The estimate or, in reliability terms, the 50th percentile, in feet below sea level.(b) The kriging standard deviation or length of interval above and below the estimate containing the true value with 68.3% probability, in feet. Visit Internet site http://www.kgs.ukans.edu/Mathgeo/Books/Geostat/index.html for a color rendition.

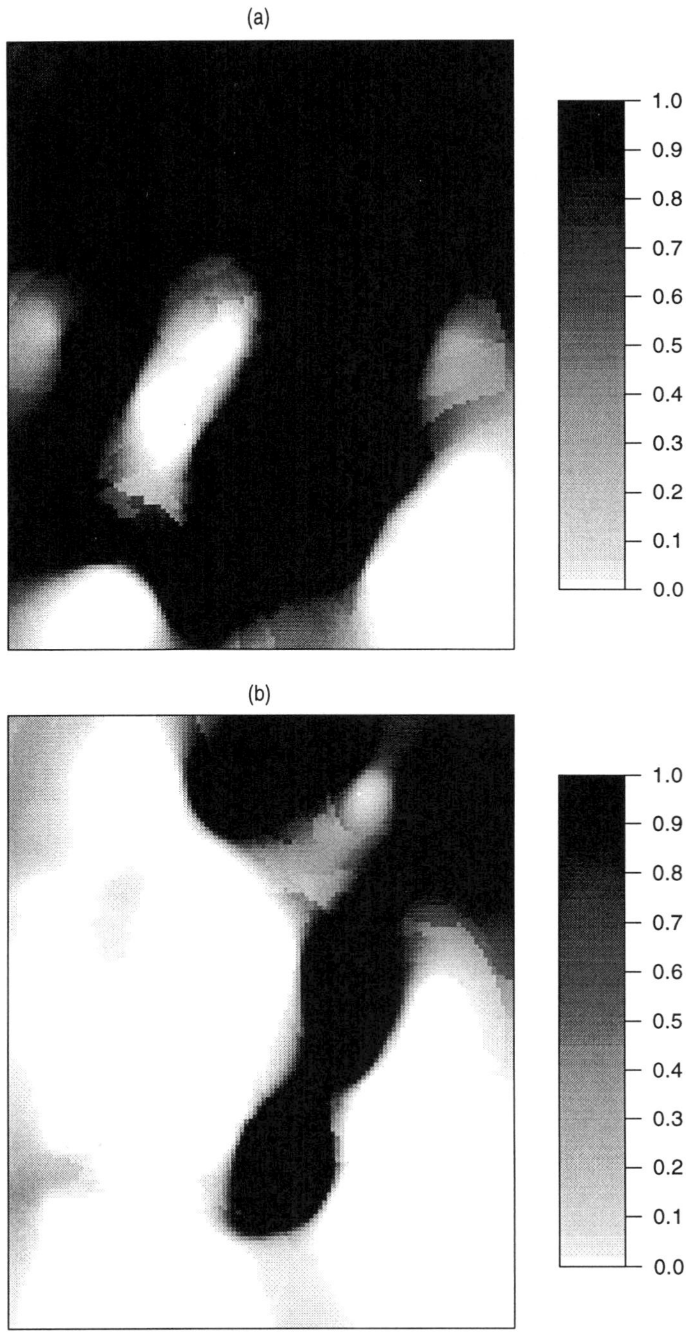

[Figure 10.3]

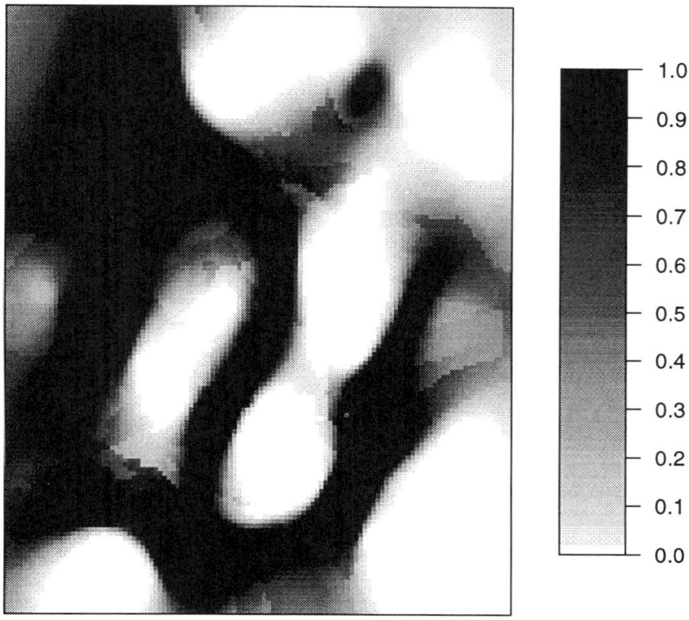

Figure 10.4. Probability that elevation to UNCF is between −7850 and −7950 ft. Results based on the normality of ordinary kriging errors. Visit Internet site http://www.kgs.ukans.edu/Mathgeo/Books/Geostat/index.html for a color rendition.

Theorem 10.1

For any variate of a random function $Z(\boldsymbol{x})$, the cumulative distribution is equal to the expected value of the indicator of the variate

$$F(t)_{\boldsymbol{x}} = \mathrm{E}\left[I(\boldsymbol{x}, t)\right] .$$

Proof

The expected value of an indicator is

$$\mathrm{E}\left[I(\boldsymbol{x}, t)\right] = \int_{-\infty}^{\infty} I(\boldsymbol{x}, t) f(z)\, dz .$$

The indicator of a variate is a step function that is 1 below the threshold and zero from the threshold and up. Hence

[Opposite page, left.]

Figure 10.3. Probability maps for two thresholds for elevation to UNCF based on normality of ordinary kriging errors. (a) Probability that elevation is below −7850 ft. (b) Probability that elevation is below −7950 ft. Visit Internet site http://www.kgs.ukans.edu/Mathgeo/Books/Geostat/index.html for a color rendition.

$$\mathrm{E}\left[I\left(\boldsymbol{x},t\right)\right] = \int_{-\infty}^{t} 1 f\left(z\right) dz + \int_{t}^{\infty} 0 f\left(z\right) dz \, .$$

The second integral vanishes and $\int_{-\infty}^{t} f\left(z\right) dz$ is, by definition, the cumulative distribution function $F\left(t\right)_{\boldsymbol{x}}$:

$$F\left(t\right)_{\boldsymbol{x}} = \mathrm{E}\left[I\left(\boldsymbol{x},t\right)\right] . \quad \square$$

Therefore, mapping of the cumulative probabilities for any threshold reduces to mapping the expected value of the indicators. Conceivably the mapping of several indicators could be used to produce a point-by-point approximation of the cumulative distribution at every location in the area of interest, a task which is customarily accomplished by various forms of kriging of the indicators. The following algorithm is a basic procedure for the approximation of cumulative probabilities.

Algorithm 10.2

This is a procedure to approximate the cumulative distribution of a random function by indicator kriging.

1. Generate several indicator data sets by selecting at least ten thresholds spaced within the extreme values of the attribute.

2. Model the semivariograms. Because of the dependence among indicator semivariograms, the slope of an indicator semivariogram model must be at least as steep as that of a linear model, which according to Figure 5.6 precludes the Gaussian, the cubic, and the sine hole effect models as permissible functions.

3. Perform ordinary kriging of each indicator data set separately, which generates as many grids as thresholds.

4. Numerically approximate the cumulative distribution function. The set of values for the same node in the different grids contains the necessary information. □

Again, after obtaining the cumulative probability of a variate, one can produce the same results as those obtained from the cumulative function presented in the previous section considering normality of the errors.

As ingenious as indicator kriging may be, it suffers from some theoretical and practical problems. On the theoretical side:

(a) Because kriging is a non-convex combination, its domain is the entire set of real numbers, while a cumulative distribution is restricted to the interval (0, 1). Potentially unbounded kriging estimates can be less than zero or larger than 1, which values are unfeasible for cumulative probabilities.

(b) Kriging cannot account for a second property of cumulative distributions, which must be monotonic. Cumulative distributions derived from indicator kriging sometimes fail this order-relationship requirement and decrease in value as the variate increases.

On the practical side:

(a) To produce the same type of results, the effort demanded by indicator kriging is higher than that required by the approach based on the normality of errors, when the latter is applicable. In the simple approach in Algorithm 10.2, the semivariogram must be modeled and kriging must be run as many times as there are thresholds. Some variations to kriging indicators simplify the work, others make it worse, but none eliminate the theoretical drawbacks above.

(b) The semivariogram for very high or very low thresholds is customarily difficult to model, which adds concern about the optimality of the estimates.

Despite all these problems, indicator kriging offers one advantage that makes it a popular alternative to reliability mapping, even if the data can be regarded as a partial realization of a multivariate random function. If the errors are, indeed, normally distributed, indicator kriging still works, of course. The favorable side of indicator kriging is that it allows the extraction of some information from imprecise data. For example, suppose that at a given site the value of the attribute is not known exactly, but there are grounds to believe that it is less than 10 units. This type of information, which ordinary kriging cannot handle at all, sometimes results in indicator values that can be used with indicator kriging to improve knowledge about the attribute. In the case of the example, the imprecise information will be of no use for thresholds below 10, but for any threshold larger than 10, the indicator will be 1.

EXERCISE 10.2

Given the sampling in Exercise 7.2, prepare grids with a 200-ft increment extending over the entire sampling space. Then display the results for the following operations:

1. Generate indicators for −7,850 and −7,950 ft and model their omnidirectional semivariograms.

2. Generate maps of the probability that the elevation is less than −7,950 ft and of the probability that it is less than −7,850 ft. Compare results with Exercise 10.1 (3).

3. Generate a map of the probability that the elevation is between −7,950 and −7,850. Compare results with Exercise 10.1 (4).

Solution

1. Figure 10.5 contains the experimental semivariograms and the models.

2. Figure 10.6 shows the maps for $\text{Prob}[z < -7850]$ and $\text{Prob}[z < -7950]$. Compared to the same maps produced taking advantage of the normality of errors, it may be observed that the maps are generally comparable and that there is a systematic tendency by indicator kriging to reduce the areas associated to both extreme probability values of zero and 1.

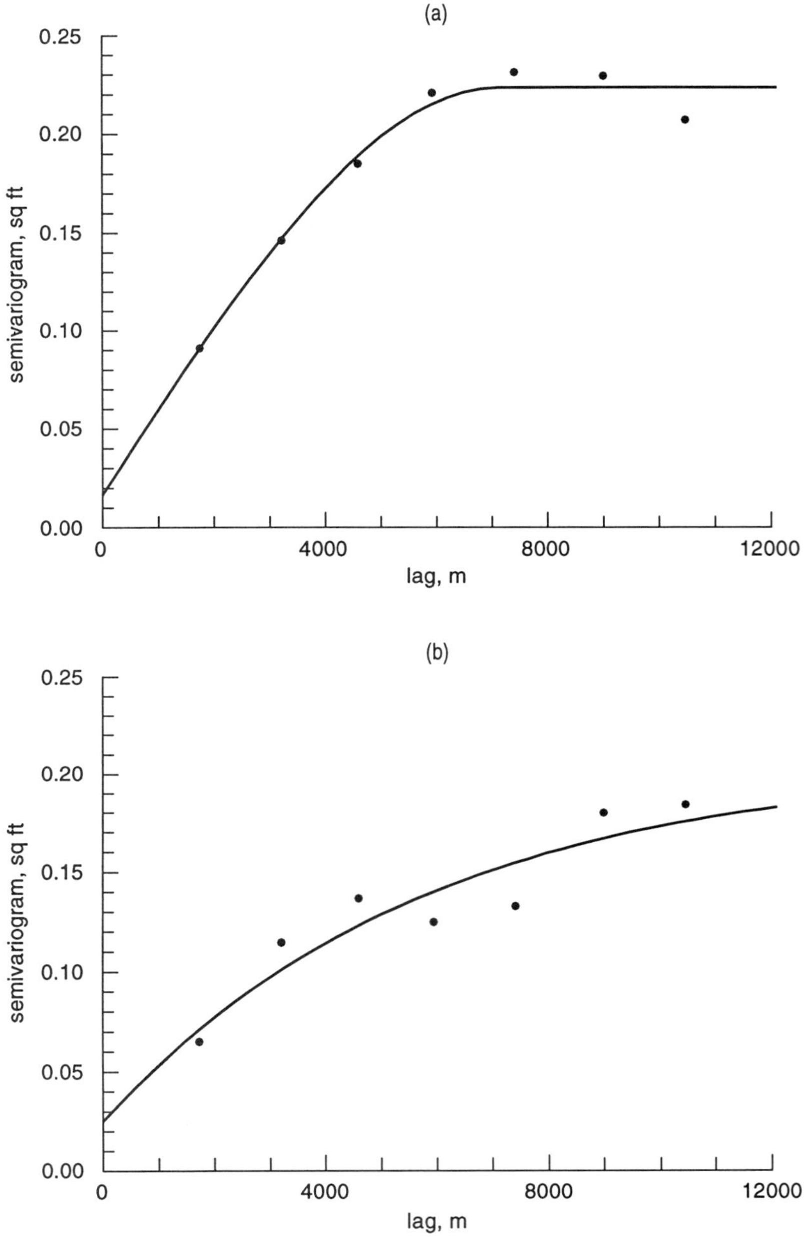

Figure 10.5. Omnidirectional indicator semivariogram for elevation to UNCF. Dots denote experimental values and lines denote semivariogram models: (a) experimental semivariogram for threshold −7950 ft and spherical model with nugget of 0.016 sq. ft, (sill – nugget) of 0.207 sq. ft, and range of 7119 m; (b) experimental semivariogram for threshold −7850 ft and exponential model with nugget of 0.025 sq. ft, (sill – nugget) of 0.181 sq. ft, and range of 17659 m.

The blank areas highlight negative probabilities. There are also probabilities above 1, but the graphics package **cscale** (Deutsch and Journel, 1992) used in the grid display does not show them. Values below the minimum of zero are blank, but those above the maximum of 1.0 are black instead of also being blank. In Figure 10.6a, 0.26% or 43 node values are above 1.0 and in Figure 10.6b, 197 values are above 1.0 and 385 below zero.

3. The map for Prob$[-7950 < z < -7850]$ is shown in Figure 10.7. It emphasizes the differences between maps in Figures 10.3 and 10.6. Besides the areas with negative probabilities, the mean difference is the discrepancy on the areal extension of the areas with probability above 0.9, which in Figure 10.4 are contiguous and in Figure 10.7 are split into three separate areas. In addition, the narrow strip in the center of the map is completely missing, allowing the merging of two low probability areas that in Figure 10.4 are separate.

All the blank areas in Figure 10.7 are the result of problems with order relationships. □

Stochastic Simulation

Like indicator kriging, most stochastic simulation methods also model the cumulative distribution numerically.

In Chapter 9, we saw that the map of any stochastic simulation is a display of an equiprobable realization of the random function. Each realization contributes with one point to the cumulative distribution of any individual variate. Above a bare minimum of 20 realizations, the set of realizations provides all the information necessary to approximate the distribution of any variate at a node in the grids. A safer rule of thumb is to generate 100 realizations.

Algorithm 10.3

This is a procedure for the numerical approximation of the cumulative distribution of a random function through conditional stochastic simulation.

1. Select a conditional simulation method and its parameters that provide the best characterization of the attribute of interest

2. Generate at least 20 realizations.

3. From the different realizations, group the values by node.

4. For each node, rank the values. At any node the cumulative probability of $F(t)$ is equal to k/n, where k is the number of simulated values at such a node that is smaller or equal to t and n is the number of realizations. □

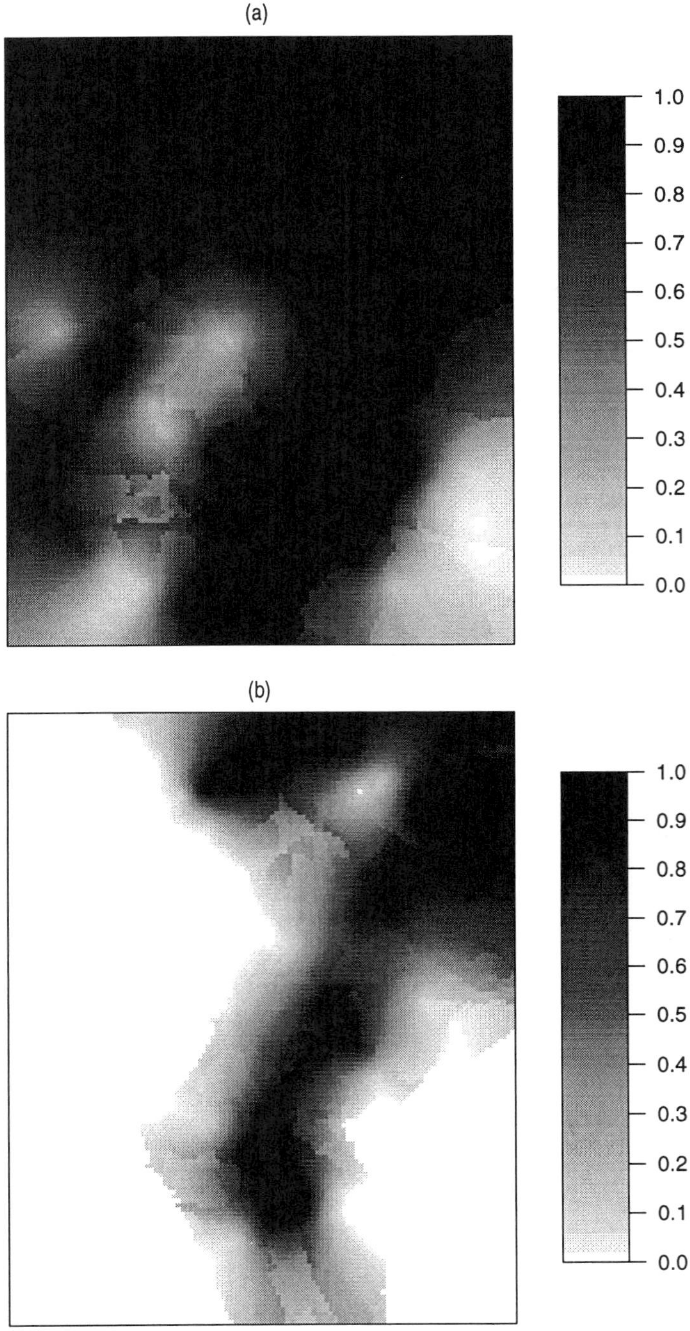

Figure 10.6. Probability maps for two thresholds for elevation to UNCF based on indicator kriging. (a) Probability elevation is below −7850 ft. (b) Probability elevation is below −7950 ft. Visit Internet site for color rendition.

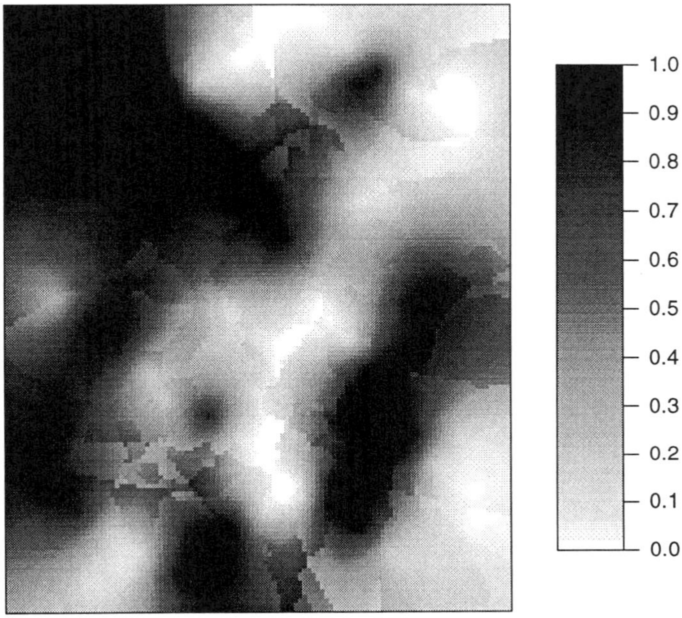

Figure 10.7. Probability that elevation to UNCF is between −7850 and −7950 ft. Results based on indicator kriging. For a color rendition, visit Internet site http://www.kgs.ukans.edu/Mathgeo/Books/Geostat/index.html.

EXERCISE 10.3

Given the sampling in Exercise 7.2, using simulated annealing and the following parameters, generate 20 realizations extending over the entire sampling space.

Grid spacing: 200 ft
Initial temperature: 1
Temperature reduction factor: 0.12
Maximum attempted perturbations per temperature range:
75 times the number of nodes
Accepted perturbations per temperature range:
8 times the number of nodes
Maximum value for the objective function: 0.0001.

Display the results for the following cases:

1. The probability that the elevation is less than −7,950 ft.
2. The probability that the elevation is less than −7,850 ft.
3. The probability that the elevation is between −7,950 and −7,850 ft.

Solution

1. Figure 10.8 shows the maps for Prob[$z < -7850$] and Prob[$z < -7950$].
2. The map for Prob[$-7950 < z < -7850$] is given in Figure 10.9. □

(a)

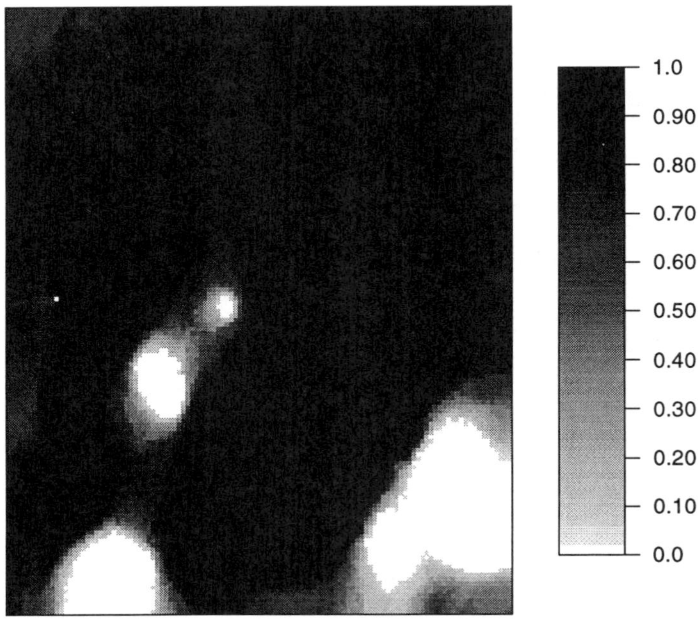

(b)

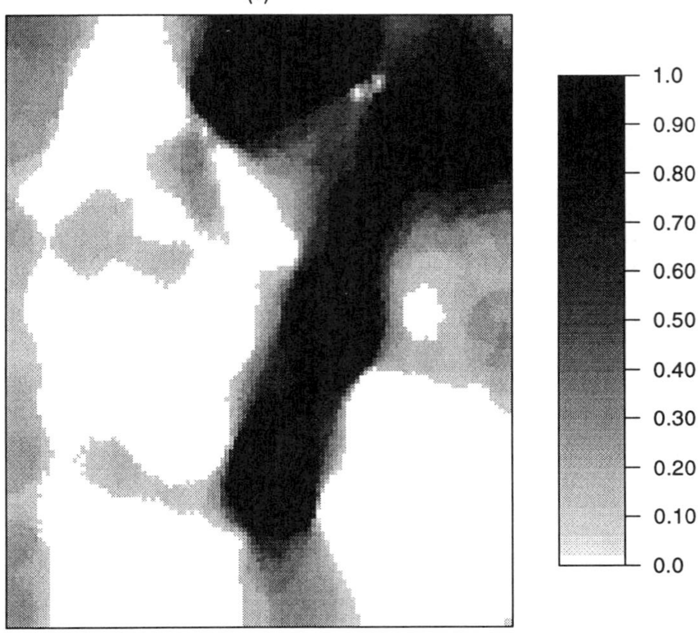

Figure 10.8. Probability maps based on simulated annealing for two thresholds for elevation to UNCF. (a) Probability elevation is below −7850 ft. (b) Probability elevation is below −7950 ft. Visit Internet site for color rendition.

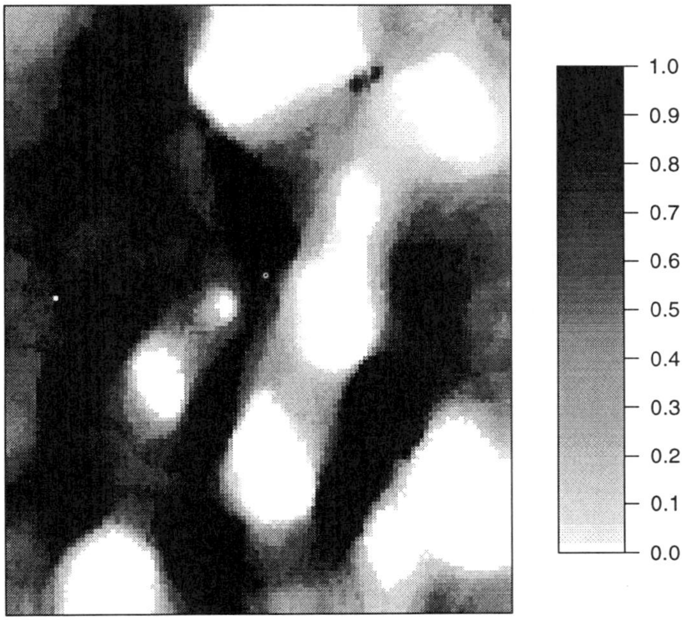

Figure 10.9. Probability that elevation to UNCF is between −7850 and −7950 ft. Results based on simulated annealing. For a color rendition, visit Internet site http://www.kgs.ukans.edu/Mathgeo/Books/Geostat/index.html.

Comparisons

Although there is no analytical way to compare the confidence intervals obtained by multinormal kriging, indicator kriging, or simulations, experience and the exercises in this chapter tend to support the following pragmatic generalizations:

1. Simulated annealing requires the minimal assumptions to generate the results. In addition:

a) Each realization is equally probable.
b) The sampling histogram is a good approximation of the true histogram.
c) The sampling provides a good approximation to the true function or functions used in the objective function—commonly the semivariogram, which is the case in Exercise 10.3.

2. On the other hand, by assuming multinormality, kriging makes the strongest assumption. Multinormality is not only a strong assumption, but one that cannot be verified; at most, it cannot be rejected. Figures 10.4 and 10.9, however, show that when multinormality is likely, kriging is able to produce results comparable to those from simulated annealing.

3. Kriging is the most efficient approach, in the sense that it requires the minimum number of steps and processing time. Simulated annealing is always orders of magnitude more demanding in terms of CPU time.

4. Cumulative functions produced by kriging or simulated annealing are monotonic and vary within the correct 0–1 range.

5. Indicator kriging can make use of imprecise data.

Chapter 11

CUMULATIVE DISTRIBUTION ESTIMATORS

In the previous chapter we have seen different ways to generate cumulative distributions, including a method based on the kriging estimate, when the assumption of multinormality is accepted.

The converse, generating an estimate from a cumulative distribution, is also possible. Once a cumulative distribution has been produced, the expected value, the median, or some other statistic of the cumulative distribution can be used to generate an estimate. The cases of widest practical interest are the conditional expectation estimator, also called the E-type estimator, and those derived from minimization of the impact that errors have in decision-making.

Let us begin with E-type estimates, of which only the stochastic simulation and indicator probability maps are of interest. For multivariate normal kriging, the kriging estimate is directly the E-type estimate.

Simulation E-type Estimator

Each realization of a conditional stochastic simulation honors the data and is equiprobable. If several realizations are all partial realizations at the same nodes of a grid, each realization contributes exactly one outcome to the discrete cumulative distribution at each node.

Theorem 11.1

Given n_s stochastic realizations of a random function $Z(\boldsymbol{x})$, the expected value for a variate at site $\boldsymbol{x}$ is given by

$$\hat{Z}_E(\boldsymbol{x}) = \frac{1}{n_s} \sum_{i=1}^{n_s} z_i(\boldsymbol{x}) ,$$

where $z_i(\boldsymbol{x})$ are outcomes at site $\boldsymbol{x}$.

Proof

The expected value for a discrete variate is

$$\hat{Z}_E(\boldsymbol{x}) = \sum_{i=1}^{n_s} z_i(\boldsymbol{x}) f_i(\boldsymbol{x}) .$$

But because the outcomes are equiprobable,

$$\hat{Z}_E(\boldsymbol{x}) = \frac{1}{n_s} \sum_{i=1}^{n_s} z_i(\boldsymbol{x}) . \quad \square$$

Algorithm 11.1

This is a procedure to produce a conditional expectation estimator from equiprobable realizations of a random function. Rather than producing a single estimate, the algorithm generates an array of estimates at the nodes of a grid.

1. Select a conditional simulation method and the parameters providing the best characterization of the attribute of interest.
2. Generate at least 20 realizations.
3. Group the different outcomes by node.
4. At each node compute the mean and the standard deviation. The mean provides E-type estimates and the corresponding standard deviations are their standard errors. □

EXERCISE 11.1

Given the sampling in Exercise 7.2, employ the parameters in Exercise 10.3 to produce:

1. A map for the E-type estimation.
2. A map of the standard error.
3. Compare the results to those for Exercise 10.1, 1.

Solution

1. Figure 11.1a is the E-type estimate for the elevation to UNCF resulting from the processing of the same realizations employed to generate Figure 10.8.

2. Figure 11.1b is the standard error map.

3. There is a remarkable similarity between equivalent maps in Figures 10.1 and 11.1. Except for the eastern edge, the elevation maps are almost identical.

The kriging standard deviation map shows the same tendencies as the standard error map, but as shown on Figure 11.2, on average the kriging standard deviation is only 62.7% of the E-type standard deviation at the same node. Figure 11.3 confirms the same tendency on the basis of the individual cumulative distribution function of four nodes. □

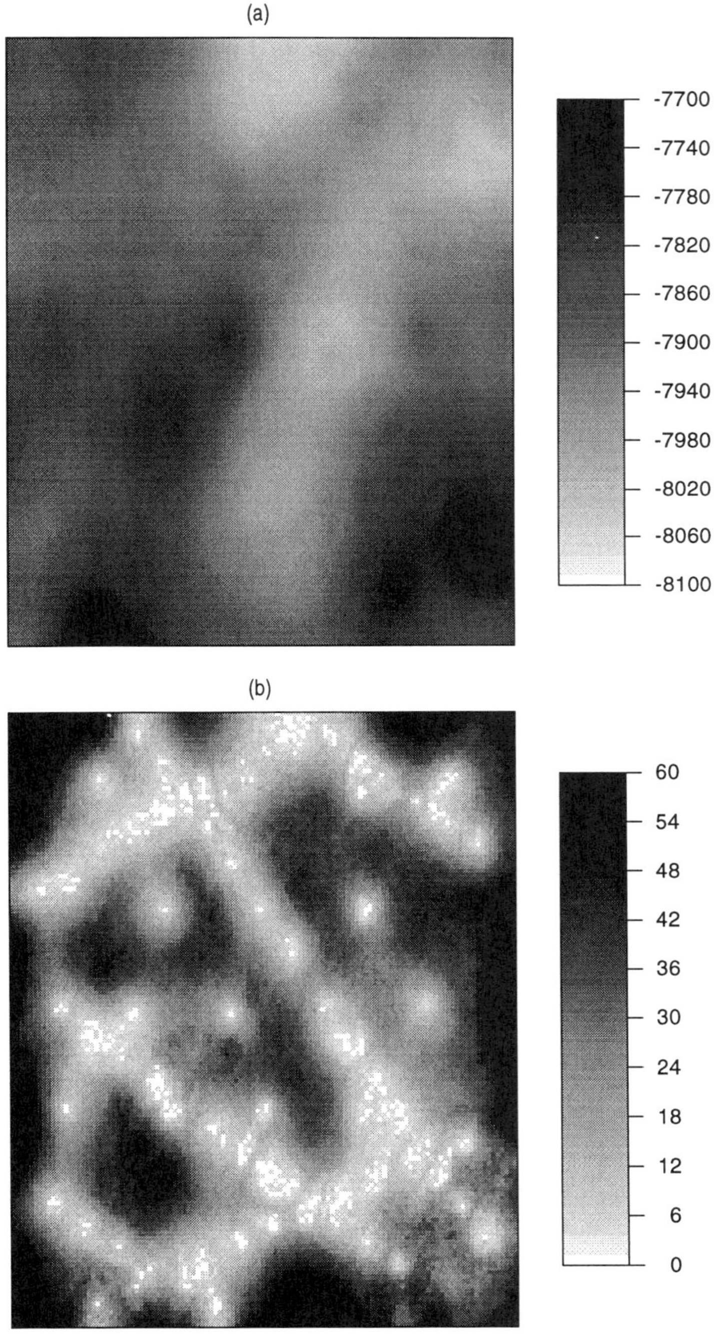

Figure 11.1. Conditional expectation estimation by simulated annealing for UNCF unconformity. (a) Estimated surface, feet below sea level. (b) Standard deviation for estimate, feet. A color rendition of the maps is available on the Internet: //www.kgs.ukans.edu/Mathgeo/Books/Geostat/index.html.

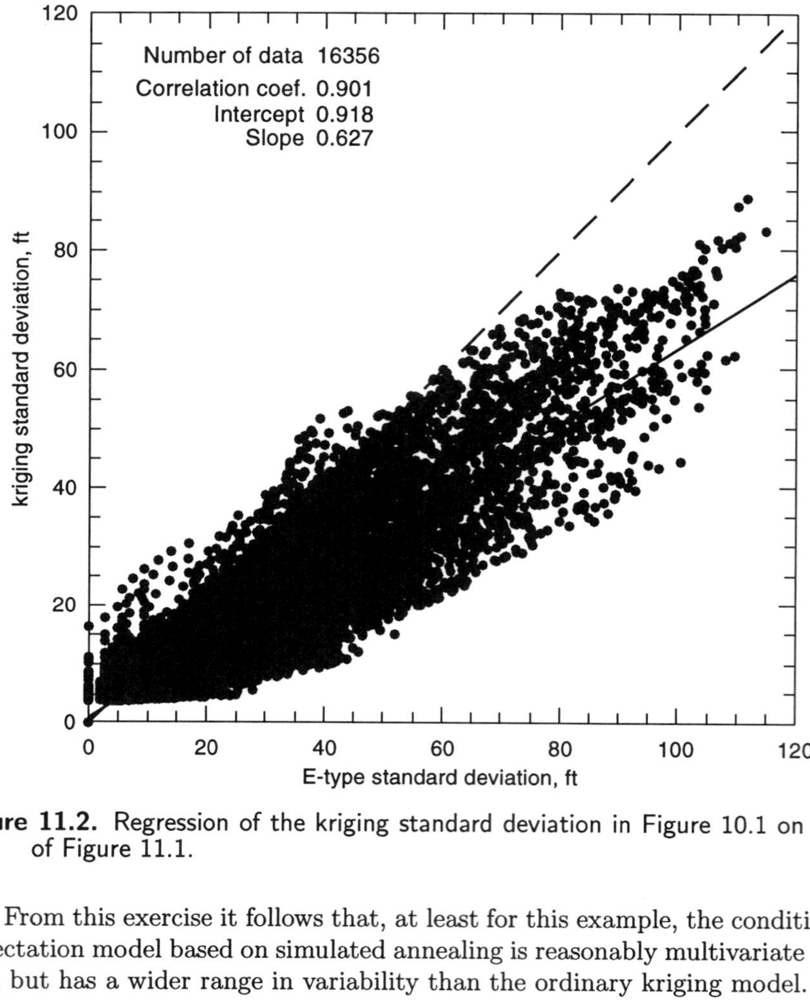

Figure 11.2. Regression of the kriging standard deviation in Figure 10.1 on that of Figure 11.1.

From this exercise it follows that, at least for this example, the conditional expectation model based on simulated annealing is reasonably multivariate normal, but has a wider range in variability than the ordinary kriging model.

Indicator E-type Estimator

Conceptually, both a set of stochastic realizations and a set of probability maps for different cutoffs allow the generation of cumulative distributions. Practical considerations, however, make use of the indicator cumulative distributions less suitable for the calculation of E-type estimates. Cumulative distributions that can be generated from stochastic realizations are equiprobable. Thus, for the cumulative distribution, data are spaced apart at regular probability increments at every node. The increments of cumulative probabilities from the indicator approach are irregularly spaced, however, denying indicator E-type estimates the convenience and accuracy of Theorem 11.1, for reasons that become apparent in the following theorem.

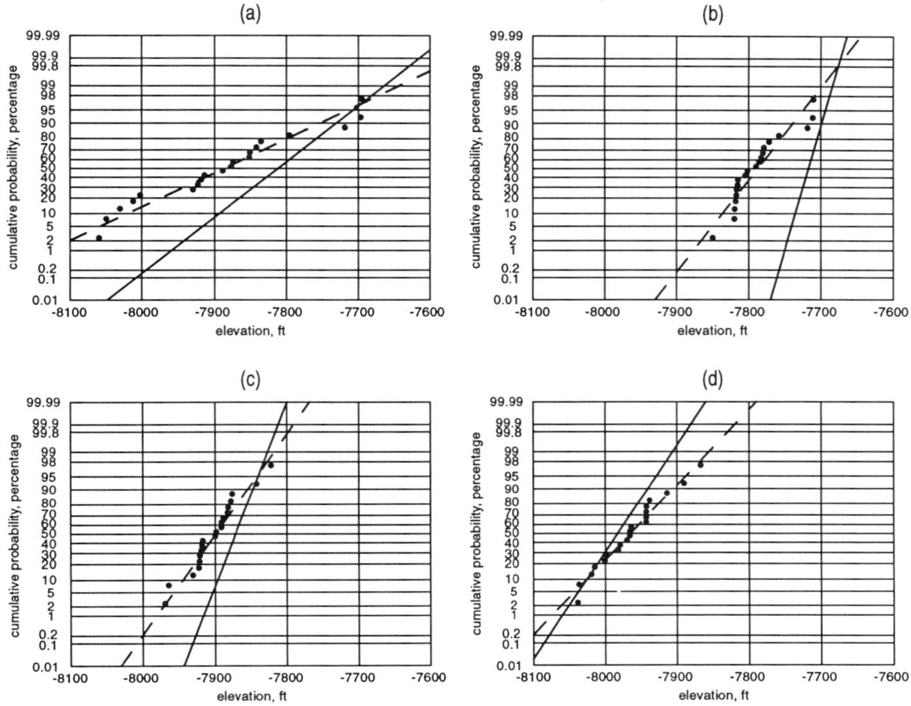

Figure 11.3. Comparison of some nodal cumulative distribution functions for the 116 by 141 node grids in Figures 10.1 a and 11.1 a. The solid line is the normal distribution with mean equal to the kriging estimate and standard deviation equal to the square root of the kriging variance; the dots are the values of the realizations used to produce Figure 11.1. The segmented line is the normal distribution with the same mean and variance as the dot distribution. (a) Node (1,1) in the southwest corner, with a maximum discrepancy, D, between the cumulative distribution function of the dots and the segmented line of 0.11. (b) Node (93,10) in the southeast with D = 0.16. (c) Node (38,86) in the northwest, $D = 0.13$. (d) Node (97,98) in the northeast, $D = 0.12$.

Theorem 11.2

Let $F_{\boldsymbol{x}}(z_i)$ be the probability at $\boldsymbol{x}$ obtained from the kriging of indicator values derived from cutoff z_i, $i = 1,\ 2,\ \ldots,\ n$. The conditional expectation estimator $\hat{Z}_E(\boldsymbol{x})$ that can be derived from such a set of n indicator cutoffs is

$$\hat{Z}_E(\boldsymbol{x}) \approx \sum_{i=1}^{n+1} \bar{z}_i \left[F_{\boldsymbol{x}}(z_i) - F_{\boldsymbol{x}}(z_{i-1})\right],$$

where $\bar{z}_i$ is the mean value of the class $(z_{i-1},\ z_i)$, with z_0 representing the lowest sampling value and z_{n+1} the largest.

Proof

The conditional expectation estimator is given by the Stieltjes integral,

$$\hat{Z}_E(\boldsymbol{x}) = \int_{-\infty}^{\infty} z dF_{\boldsymbol{x}}(z) \, ,$$

which can be decomposed into

$$\hat{Z}_E(\boldsymbol{x}) = \int_{-\infty}^{z_1} z dF_{\boldsymbol{x}}(z) + \sum_{i=1}^{n-1} \int_{z_i}^{z_{i+1}} z dF_{\boldsymbol{x}}(z) + \int_{z_n}^{\infty} z dF_{\boldsymbol{x}}(z) \, .$$

The proof follows by approximating each integral by the product of the probability class width times its mean. □

Algorithm 11.2

This is a procedure to produce a grid of conditional expectation estimators from a set of indicator kriging grids.

1. Select a kriging method and generate probability maps for at least 10 different cutoffs.
2. Group the probabilities by node.
3. At each node, calculate an E-type estimate by the approximation,

$$\hat{Z}_E(\boldsymbol{x}) \approx \sum_{i=1}^{n+1} \bar{z}_i \left[F_{\boldsymbol{x}}(z_i) - F_{\boldsymbol{x}}(z_{i-1}) \right] \, ,$$

justified in Theorem 11.2. □

Loss Functions

It is commonly necessary to use the result of a model, such as the map in Figure 11.1a, in decision-making processes that are sensitive to estimation errors, $e(\boldsymbol{x})$. If we assume that there is a loss function $L\big(e(\boldsymbol{x})\big)$ assessing the negative impact of each discrepancy, its expected value is given by

$$\mathrm{E}\big[L\big(e(\boldsymbol{x})\big)\big] = \int_{-\infty}^{\infty} L\big(e(\boldsymbol{x})\big) \, dF_{\boldsymbol{x}}(Z) \, .$$

Different scenarios require different loss functions, leading to different estimates. The linear and quadratic loss functions are the ones of greatest interest.

Theorem 11.3

Let $L\big(e(\boldsymbol{x})\big)$ be the linear, asymmetric loss function

$$L\big(e(\boldsymbol{x})\big) = \begin{cases} \alpha_1 \left| e(\boldsymbol{x}) \right| , & \text{for } e(\boldsymbol{x}) < 0 \\ \alpha_2 e(\boldsymbol{x}) , & \text{for } e(\boldsymbol{x}) \geq 0 \end{cases}$$

for random function $Z(\boldsymbol{x})$, where both coefficients α_1 and α_2 are positive. Then the estimator that minimizes the expected loss function is the $\alpha_1/(\alpha_1+\alpha_2)$ quantile.

Proof

From the definition of expected loss function and for the special case of the linear asymmetric case above,

$$\mathrm{E}\big[L(e(\boldsymbol{x}))\big] = \int_{-\infty}^{\hat{Z}(\boldsymbol{x})} \alpha_2 \left(\hat{Z}(\boldsymbol{x}) - Z(\boldsymbol{x})\right) dF_{\boldsymbol{x}}(Z) - \int_{\hat{Z}(\boldsymbol{x})}^{\infty} \alpha_1 \left(\hat{Z}(\boldsymbol{x}) - Z(\boldsymbol{x})\right) dF_{\boldsymbol{x}}(Z)\,.$$

Differentiating with respect to the unknown $\hat{Z}(\boldsymbol{x})$,

$$\mathrm{E}\big[L(e(\boldsymbol{x}))\big] = \alpha_2 \int_{-\infty}^{\hat{Z}(\boldsymbol{x})} dF_{\boldsymbol{x}}(Z) - \alpha_1 \int_{\hat{Z}(\boldsymbol{x})}^{-\infty} dF_{\boldsymbol{x}}(Z)\,.$$

The optimal estimator is the one making the derivative equal to zero. Considering that the cumulative distribution is bounded between zero and 1,

$$\int_{\hat{Z}(\boldsymbol{x})}^{\infty} dF_{\boldsymbol{x}}(Z) = 1 - \int_{-\infty}^{\hat{Z}(\boldsymbol{x})} dF_{\boldsymbol{x}}(Z)\,;$$

then,

$$\alpha_2 \int_{-\infty}^{\hat{Z}(\boldsymbol{x})} dF_{\boldsymbol{x}}(Z) - \alpha_1 \left(1 - \int_{-\infty}^{\hat{Z}(\boldsymbol{x})} dF_{\boldsymbol{x}}(Z)\right) = 0\,,$$

which is equivalent to the following implicit equation on $\hat{Z}(\boldsymbol{x})$:

$$\int_{-\infty}^{\hat{Z}(\boldsymbol{x})} dF_{\boldsymbol{x}}(Z) = \frac{\alpha_1}{\alpha_1 + \alpha_2}\,.$$

The proof follows because this is the mathematical definition of the $\alpha_1/(\alpha_1+\alpha_2)$ quantile. □

Corollary 11.1

Let the loss function be the linear, symmetric expression $\alpha\,|e(\boldsymbol{x})|$. Then the estimator that minimizes the expected loss function is the median.

Proof

The proof follows directly from Theorem 11.3 by making $\alpha_1 = \alpha_2$. □

If underestimation has more severe consequences that overestimation, then $\alpha_1 > \alpha_2$ and the estimate that minimized the loss function is above the median, biased toward the upper tail of the cumulative density function. This is typically the case in environmental remediation in which medical problems that may develop because of underrating a contaminated site are more serious than would be the wasting of resources by overcleaning. Conversely, higher importance of overestimation implies that $\alpha_1 < \alpha_2$, which is typical of the case in mining ventures in which optimistic preliminary estimates may lead to ruinous investments. The median is the best estimate in the less frequent case of indifference.

Theorem 11.4

Let $L(e(\boldsymbol{x}))$ be the quadratic loss function $\alpha e^2(\boldsymbol{x})$ for random function $Z(\boldsymbol{x})$. Then the estimator that minimizes the expected loss function is the conditional expectation estimator.

Proof

If the loss function is equal to $\alpha e^2(\boldsymbol{x})$,

$$\mathrm{E}\left[L(e(\boldsymbol{x}))\right] = \int_{-\infty}^{\infty} \alpha \left(\hat{Z}(\boldsymbol{x}) - Z(\boldsymbol{x})\right)^2 dF_{\boldsymbol{x}}(Z) \; .$$

Differentiating with respect to the unknown $\hat{Z}(\boldsymbol{x})$,

$$\frac{\partial \mathrm{E}\left[L(e(\boldsymbol{x}))\right]}{\partial \hat{Z}(\boldsymbol{x})} = \int_{-\infty}^{\infty} 2\alpha \left(\hat{Z}(\boldsymbol{x}) - Z(\boldsymbol{x})\right) dF_{\boldsymbol{x}}(Z) \; .$$

The optimal estimator is the one making the derivative equal to zero. Dividing both sides of the equation by α and expanding

$$\hat{Z}(\boldsymbol{x}) \int_{-\infty}^{\infty} dF_{\boldsymbol{x}}(Z) + \int_{-\infty}^{\infty} Z(\boldsymbol{x})\, dF_{\boldsymbol{x}}(Z) = 0 \, ,$$

but from the properties of the cumulative distribution function, the first integral is equal to 1. Adding $\hat{Z}(\boldsymbol{x})$ to both sides of the equation,

$$\hat{Z}_{\mathrm{E}}(x) = \int_{-\infty}^{\infty} Z(x)\, dF_{\boldsymbol{x}}(Z),$$

which is the definition of the conditional expectation estimator. □

A quadratic loss function is more sensitive to large errors and is symmetric, but its relative simplicity accounts for its use, even in approximate situations for which the implications of underestimation may not be exactly the same as those of overestimation.

Chapter 12

BLOCK KRIGING

In the Introduction I mentioned that kriging was formulated from the outset to generalize linear regression, partly in order to consider different supports for the estimate and the sampling. In Chapter 5 we defined support as the shape, size, and orientation of the volume associated with any observation. So far we have not used that potential of kriging.

For practical and pedagogic reasons we have been dealing with point estimates or estimates whose supports are small enough at the scale of the sampling domain that one can regard such small supports—cores, plugs, jugs—as points. The practical reasons are obvious. We have been able to cover ten chapters of applications devoted to point formulations. The pedagogic reasons will be clear at the end of this chapter, but in a nutshell, the generalization to estimates with non-point supports does not come without a price. The models are more complex than those for point kriging. Hopefully, at this stage of the exposition, the reader has understood the fundamentals of point kriging and the generalization will be easier to grasp than if we had followed the historical sequence.

Block kriging is the generic name given to any form of kriging in which the interest lies in estimation of a linear average for an attribute inside supports that are intermediate in size between the support of the sampling and the sampling domain.

Definition 12.1

Let $Z(\boldsymbol{x})$ be a point support random function in D and let $S(\boldsymbol{x}_0) \subset D$ be a support of finite size S centered around $\boldsymbol{x}_0$. The true block average $Z_S(\cdot)$ at location $\boldsymbol{x}_0$ is

$$Z_S(\boldsymbol{x}_0) = \frac{1}{S} \int_{S(\boldsymbol{x}_0)} Z(\boldsymbol{x})\, d\boldsymbol{x}\,. \quad \square$$

When Matheron formulated kriging, he presented it in the form of block kriging (Matheron, 1963) partly for mathematical generality but mostly because his immediate concern was estimation of *in situ* resources, that is, estimates of the average ore content of mining blocks, a problem conceptually similar to more recent concerns in environmental remediation or to the estimation of block values for use in fluid flow simulations in hydrology and petroleum engineering.

Historically and conceptually block kriging is the mother of all forms of kriging for the characterization of a single attribute, in the sense that any form of point kriging is a limiting form of some kind of block kriging, which has as many variations as point kriging. I will restrict attention, however, to ordinary block kriging under second order quasi-stationarity, which is the most widely applied form of block kriging.

The Estimator

The purpose of block kriging is to produce direct block averages from point or quasi-point measurements, not from known block averages.

Definition 12.2

Let $Z(\boldsymbol{x}_i)$ be k variates of a random function inside a neighborhood $d \subset D$ of the sampling domain and let $S \subset d$ be the support for the estimator $\hat{Z}_S(\boldsymbol{x}_0)$ at site $\boldsymbol{x}_0$. Then the block kriging estimator $\hat{Z}_S(\boldsymbol{x}_0)$ is

$$\hat{Z}_S(\boldsymbol{x}_0) = \sum_{i=1}^{k} \lambda_i Z(\boldsymbol{x}_i) ,$$

subject to $\sum_{i=1}^{k} \lambda_i = 1$. □

Block kriging requires judicious application. Although the model is able to process any numbers, indiscriminate use will result in some values that may be useless and others that will be invalid. While it is appropriate to run block kriging to estimate the amount of manganese nodules per square kilometer on the ocean floor, the average water table depth per square kilometer is of no interest; and employing block kriging to investigate the permeability of blocks of sedimentary rocks is improper, because the physics of fluid flow dictates that the linear average permeability of a block is not necessarily the same as the equivalent permeability of the block.

Assumptions

Like any form of kriging, derivation of the normal equations and estimation of moments require some definitions and assumptions. For an organized presentation, it is better to state them up front rather than to address them as needed. Let us start with two definitions.

Definition 12.3

Let $\text{Cov}(\boldsymbol{x}_0, \boldsymbol{x})$ be the covariance between the variate $Z(\boldsymbol{x}_0)$ and all the variates $Z(\boldsymbol{x})$ in the block S. Then the covariance $\text{Cv}(\boldsymbol{x}_0, S)$ between a site and the block is

$$\text{Cv}(\boldsymbol{x}_0, S) = \frac{1}{S}\int_S \text{Cov}(\boldsymbol{x}_0, \boldsymbol{x}) d\boldsymbol{x} .$$

Definition 12.4

Let $\text{Cov}(\boldsymbol{x}, \mathbf{y})$ be the covariance between a variate $Z(\boldsymbol{x})$ in block S_i and variate $Z(\mathbf{y})$ in the block S_j. Then the block covariance $\text{CV}(S_i, S_j)$ between two blocks is

$$\text{CV}(S_i, S_j) = \frac{1}{S_i S_j}\int_{S_i}\int_{S_j} \text{Cov}(\boldsymbol{x}, \mathbf{y}) d\boldsymbol{x}\, d\mathbf{y} .$$

The minimum set of assumptions to formulate and apply block kriging follows.

Assumption 12.1

A sampling is a partial realization of a point random function. □

Assumption 12.2

The point random function is second order stationary inside the neighborhood d in Definition 12.2

$$\left.\begin{array}{l} \text{E}[Z(\boldsymbol{x})] = m \\ \text{Cov}[Z(\boldsymbol{x}), Z(\boldsymbol{x}+\boldsymbol{h})] = \text{Cov}(\boldsymbol{h}) \end{array}\right\} \forall\, \boldsymbol{x},\ \boldsymbol{x}+\boldsymbol{h} \text{ in } d ,$$

where m is a constant, $\boldsymbol{h}$ is an Euclidean distance, and $\text{Cov}(\cdot)$ is the point covariance. □

It can be proved that second order quasi-stationarity of the point random function implies quasi-stationarity of the point-to-block covariances and second order quasi-stationarity of the block averages.

Lemma 12.1

If a point random function is second order quasi-stationary, its point-to-block covariance $\text{Cv}(\boldsymbol{x}_0, S)$ is quasi-stationary.

Proof

By Definition 12.3,

$$\text{Cv}(\boldsymbol{x}_0, S) = \frac{1}{S}\int_S \text{Cov}(\boldsymbol{x}_0, \boldsymbol{x}) d\boldsymbol{x} .$$

Because the point covariance is quasi-stationary,

$$\mathrm{Cv}\,(\boldsymbol{x}_0, S) = \frac{1}{S}\int_S \mathrm{Cov}\,(\boldsymbol{x}_0 - \boldsymbol{x})d\boldsymbol{x}\,,$$

which is a function of the Euclidean distance and not of position. □

Lemma 12.2

If a point random function is second order quasi-stationary, its block average is second order quasi-stationary.

Proof (a)

By Definition 12.1,

$$\mathrm{E}\,[Z_S\,(\boldsymbol{x}_0)] = \mathrm{E}\left[\frac{1}{S}\int_S Z\,(\boldsymbol{x})\,d\boldsymbol{x}\right].$$

By the commutative property of summation and expectation,

$$\mathrm{E}\,[Z_S\,(\boldsymbol{x}_0)] = \frac{1}{S}\int_S \mathrm{E}\,[Z\,(\boldsymbol{x})]\,d\boldsymbol{x}\,;$$

but, because the point random function is second order quasi-stationary, its mean is constant:

$$\begin{aligned}\mathrm{E}\,[Z_S\,(\boldsymbol{x}_0)] &= \frac{1}{S}\int_S m\,d\boldsymbol{x}\\ &= m\,.\end{aligned}$$

Hence, the point and the block random function have the same constant mean.

Proof (b)

From Definition 12.4,

$$\mathrm{CV}\,(S_i, S_j) = \frac{1}{S_i S_j}\int_{S_i}\int_{S_j} \mathrm{Cov}\,(\boldsymbol{x}, \mathbf{y})d\boldsymbol{x}\,d\mathbf{y}\,.$$

Considering that the point random function is second order quasi-stationary, its covariance is a function of distance and not of location:

$$\mathrm{CV}\,(S_i, S_j) = \frac{1}{S_i S_j}\int_{S_i}\int_{S_j} \mathrm{Cov}\,(\boldsymbol{x} - \mathbf{y})d\boldsymbol{x}\,d\mathbf{y}\,.$$

Proofs (a) and (b) prove the lemma. □

Estimation Error

In the line of generalized linear regression, block kriging employs minimization of the mean square estimation error as the optimization criterion. Although detrimental to the continuity of the exposition, first proving two lemmas is a convenient way to derive the objective function.

Lemma 12.3

Let $Z_S(\cdot)$ be the true block average at location $\boldsymbol{x}_0$ for a point random function $Z(\boldsymbol{x})$ that is second order quasi-stationary . Then

$$\mathrm{E}\left[Z(\boldsymbol{x}) - Z_S(\boldsymbol{x}_0)\right] = 0 .$$

Proof

The proof is a direct consequence of Assumption 12.2 and Lemma 12.2. □

Lemma 12.4

Let Cov $(\cdot)$ be the covariance in Assumption 12.2, let Cv $(\cdot)$ be the point-to-block covariance in Definition 12.3, and let CV $(\cdot)$ be the block-to-block covariance in Definition 12.4. Then,

$$\begin{aligned}&\mathrm{Cov}\big(Z(\boldsymbol{x}_i) - Z_S(\boldsymbol{x}_0), Z(\boldsymbol{x}_j) - Z_S(\boldsymbol{x}_0)\big) = \\ &\qquad \mathrm{CV}(S,S) - \mathrm{Cv}(\boldsymbol{x}_i, S) - \mathrm{Cv}(\boldsymbol{x}_j, S) + \mathrm{Cov}(\boldsymbol{x}_i, \boldsymbol{x}_j) .\end{aligned}$$

Proof

By the definition of covariance and Lemma 12.3,

$$\begin{aligned}&\mathrm{Cov}\big(Z(\boldsymbol{x}_i) - Z_S(\boldsymbol{x}_0), Z(\boldsymbol{x}_j) - Z_S(\boldsymbol{x}_0)\big) = \\ &\qquad \mathrm{E}\left[\{Z(\boldsymbol{x}_i) - Z_S(\boldsymbol{x}_0)\}\{Z(\boldsymbol{x}_j) - Z_S(\boldsymbol{x}_0)\}\right] ,\end{aligned}$$

which by expansion, after adding and subtracting twice the square of the mean m, gives

$$\begin{aligned}\mathrm{Cov}\big(Z(\boldsymbol{x}_i) - Z_S(\boldsymbol{x}_0), Z(\boldsymbol{x}_j) - Z_S(\boldsymbol{x}_0)\big) &= \left\{\mathrm{E}\left[Z_S(\boldsymbol{x}_0), Z_S(\boldsymbol{x}_0)\right] - m^2\right\} \\ &\quad - \left\{\mathrm{E}\left[Z(\boldsymbol{x}_i), Z_S(\boldsymbol{x}_0)\right] - m^2\right\} \\ &\quad - \left\{\mathrm{E}\left[Z(\boldsymbol{x}_j), Z_S(\boldsymbol{x}_0)\right] - m^2\right\} \\ &\quad + \left\{\mathrm{E}\left[Z(\boldsymbol{x}_i), Z(\boldsymbol{x}_j)\right] - m^2\right\} .\end{aligned}$$

Terms in the first bracket are equivalent to CV (S, S) by Definition 12.4 and Lemma 12.2, those in the second and third are equal and equivalent to Cv $(\boldsymbol{x}, S)$ according to Definition 12.3 and Lemma 12.2, and the last bracket is the definition of Cov $(\boldsymbol{x}_i, \boldsymbol{x}_j)$. All these equivalences prove the lemma. □

Theorem 12.1

Let λ_i be the weights in Definition 12.2, let Cov $(\cdot)$ be the covariance in Assumption 12.2, let Cv $(\cdot)$ be the point-to-block covariance in Definition 12.3, and let CV $(\cdot)$ be the block-to-block covariance in Definition 12.4. Then the estimation variance $\sigma^2(\boldsymbol{x}_0)$ is given by

$$\sigma^2(\boldsymbol{x}_0) = \mathrm{CV}(S, S) - 2\sum_{i=1}^{k} \lambda_i \mathrm{Cv}(\boldsymbol{x}_i, S) + \sum_{i=1}^{k}\sum_{j=1}^{k} \lambda_i \lambda_j \mathrm{Cov}(\boldsymbol{x}_i, \boldsymbol{x}_j).$$

Proof

The variance of the estimation error is

$$\sigma^2(\boldsymbol{x}_0) = \mathrm{Var}\left[\hat{Z}_S(\boldsymbol{x}_0) - Z_S(\boldsymbol{x}_0)\right],$$

which by Definition 12.2 and the constraint in the weights in the same definition is equivalent to

$$\sigma^2(\boldsymbol{x}_0) = \mathrm{Var}\left[\sum_{i=1}^{k} \lambda_i \left\{Z(\boldsymbol{x}_i) - Z_S(\boldsymbol{x}_0)\right\}\right].$$

Considering that the difference of two variates is another variate, by Lemma 2.4,

$$\sigma^2(\boldsymbol{x}_0) = \sum_{i=1}^{k}\sum_{j=1}^{k} \lambda_i \lambda_j \mathrm{Cov}\big(Z(\boldsymbol{x}_i) - Z_S(\boldsymbol{x}_0), Z(\boldsymbol{x}_j) - Z_S(\boldsymbol{x}_0)\big)$$

and by Lemma 12.4,

$$\sigma^2(\boldsymbol{x}_0) = \sum_{i=1}^{k}\sum_{j=1}^{k} \lambda_i \lambda_j \left\{\mathrm{Cov}(S, S) - \mathrm{Cov}(\boldsymbol{x}_i, S) - \mathrm{Cov}(\boldsymbol{x}_j, S) + \mathrm{Cov}(\boldsymbol{x}_i, \boldsymbol{x}_j)\right\}.$$

Expanding, because of the constraint $\sum_{i=1}^{k} \lambda_i = 1$,

$$\sigma^2(\boldsymbol{x}_0) =$$

$$\mathrm{Cov}(S, S) - \sum_{i=1}^{k} \lambda_i \left\{\mathrm{Cov}(\boldsymbol{x}_j, S)\right\} - \sum_{j=1}^{k} \lambda_j \left\{\mathrm{Cov}(\boldsymbol{x}_j, S)\right\} + \sum_{i=1}^{k}\sum_{j=1}^{k} \lambda_i \lambda_j \mathrm{Cov}(\boldsymbol{x}_i, \boldsymbol{x}_j).$$

Equality of the second and third terms in the right-hand side of the equation proves the theorem. □

Normal Equations

Now that we have an expression for the mean square error we can attempt to minimize it. From Theorem 12.2, the estimation variance is a function of the covariance and the estimation weights. The covariance is an intrinsic property of the random function that one cannot modify in order to reduce the estimation error. The suitable terms for an optimal selection are the weights λ_i in the expression for $\sigma^2(\boldsymbol{x}_0)$. Yet, by Definition 12.2, these weights cannot vary freely. They have to honor the constraint of summing to 1, which makes the search for the optimal weights a constrained optimization problem that can be solved as a pseudo-unconstrained problem by employing the Lagrange method of multipliers (Hillier and Lieberman, 1995, p. 961).

Definition 12.5

Let λ_i be the weights in Definition 12.2, let $\text{Cov}(\cdot)$ be the covariance in Assumption 12.2, let $\text{Cv}(\cdot)$ be the point-to-block covariance in Definition 12.3, let $\text{CV}(\cdot)$ be the block-to-block covariance in Definition 12.4, and let μ be a Lagrange multiplier. From the expression for the estimation error variance in Theorem 12.1, the Lagrangian function for block kriging is

$$L(\lambda_i, \mu) =$$

$$\text{CV}(S, S) - 2\sum_{i=1}^{k} \lambda_i \text{Cv}(\boldsymbol{x}_i, S) + 2\mu\left(\sum_{i=1}^{k} \lambda_i - 1\right) + \sum_{i=1}^{k}\sum_{j=1}^{k} \lambda_i \lambda_j \text{Cov}(\boldsymbol{x}_i, \boldsymbol{x}_j). \quad \square$$

This is the objective function of the pseudo-unconstrained optimization problem. The arbitrary factor 2 in front of μ was selected for further convenience. The Lagrangian function is a quadratic expression in the unknown weights involving a linear constraint. Consequently, if the covariance is positive definite—that is if the quadratic term is always positive—then:

(a) the optimum is a global optimum,

(b) the solution is unique,

(c) the objective function for the optimum weights is a minimum, and

(d) the value of the objective function—the kriging variance—is zero or positive.

Theorem 12.2

Let λ_i be the weights in Definition 12.2, let μ be the Lagrange multiplier in Definition 12.5, let $\text{Cov}(\cdot)$ be the point covariance in Assumption 12.2, and let $\text{Cv}(\cdot)$ be the point-to-block covariance in Definition 12.3. Then the weights that produce the minimum estimation variance are the solution to

$$\left.\begin{array}{l}\sum_{i=1}^{k}\lambda_i\mathrm{Cov}\left(\boldsymbol{x}_i,\boldsymbol{x}_1\right)+\mu=\mathrm{Cv}\left(\boldsymbol{x}_1,S\right)\\ \sum_{i=1}^{k}\lambda_i\mathrm{Cov}\left(\boldsymbol{x}_i,\boldsymbol{x}_2\right)+\mu=\mathrm{Cv}\left(\boldsymbol{x}_2,S\right)\\ \cdots\cdots\cdots\cdots\cdots\cdots\cdots\cdots\cdots\cdots\\ \sum_{i=1}^{k}\lambda_i\mathrm{Cov}\left(\boldsymbol{x}_i,\boldsymbol{x}_k\right)+\mu=\mathrm{Cv}\left(\boldsymbol{x}_k,S\right)\\ \sum_{i=1}^{k}\lambda_i=1\end{array}\right\}.$$

Proof

Differentiating the Lagrangian function with respect to the unknowns,

$$\frac{\partial L\left(\lambda_i,\mu\right)}{\partial\lambda_i}=-2\mathrm{Cv}\left(\boldsymbol{x}_i,S\right)+2\mu+2\sum_{j=1}^{k}\lambda_j\mathrm{Cov}\left(\boldsymbol{x}_i,\boldsymbol{x}_j\right)$$

$$\frac{\partial L\left(\lambda_i,\mu\right)}{\partial\mu}=\sum_{i=1}^{k}\lambda_i-1\,.$$

The optimal weights are those that make these derivatives equal to zero. The proof follows after moving terms around and dividing both sides of the first k equations by 2. □

The system of equations in Theorem 12.2 contains the normal equations for block kriging. Cancellation of all the coefficients whose value is 2 in the normal equations demonstrates the convenience of introducing a factor of 2 in Definition 12.5. Notice the remarkable similarity to the normal equation for point ordinary kriging in Theorem 4.4. The only difference is in the nature of the covariance on the right-hand side of the equation.

It is also of interest to have the value of the mean square error for the set of optimal weights. Let us first prove a lemma.

Lemma 12.5

Let λ_i be the weights in Definition 12.2, let μ be the Lagrange multiplier in Definition 12.5, let $\mathrm{Cov}\left(\cdot\right)$ be the point covariance in Assumption 12.2, and let $\mathrm{Cv}\left(\cdot\right)$ be the point-to-block covariance in Definition 12.3. Then,

$$\sum_{i=1}^{k}\sum_{j=1}^{k}\lambda_i\lambda_j\mathrm{Cov}\left(\boldsymbol{x}_i,\boldsymbol{x}_j\right)=\sum_{i=1}^{k}\lambda_i\mathrm{Cv}\left(\boldsymbol{x}_i,S\right)-\mu\,.$$

Proof

From Theorem 12.2,

$$\sum_{j=1}^{k} \lambda_j \mathrm{Cov}\left(\boldsymbol{x}_i, \boldsymbol{x}_j\right) = \mathrm{Cv}\left(\boldsymbol{x}_i, S\right) - \mu\,, \text{ for } i = 1,\ 2,\ \ldots,\ k\,.$$

Each of the k expressions is still true if one multiplies both sides of the ith equation by λ_i. Adding them up,

$$\sum_{i=1}^{k} \lambda_i \sum_{j=1}^{k} \lambda_j \mathrm{Cov}\left(\boldsymbol{x}_i, \boldsymbol{x}_j\right) = \sum_{i=1}^{k} \lambda_i \mathrm{Cv}\left(\boldsymbol{x}_i, S\right) - \mu \sum_{i=1}^{k} \lambda_i\,.$$

The proof follows by remembering that, by Definition 12.2, the sum of the weights is 1, and by introducing λ_i under the second summation. □

Theorem 12.3

Let Cov $(\cdot)$ be the point covariance in Assumption 12.2, let Cv $(\cdot)$ be the point-to-block covariance in Definition 12.3, let CV $(\cdot)$ be the block covariance in Definition 12.4, let λ_i be the weights in Definition 12.2, and let μ be the Lagrange multiplier in Definition 12.5. Then the minimum mean square error for block kriging, $\sigma^2_{BK}\left(\boldsymbol{x}_0\right)$, is

$$\sigma^2_{BK}\left(\boldsymbol{x}_0\right) = \mathrm{CV}\left(S, S\right) - \sum_{i=i}^{k} \lambda_i \mathrm{Cv}\left(\boldsymbol{x}_i, S\right) - \mu\,.$$

Proof

From Theorem 12.1,

$$\sigma^2\left(\boldsymbol{x}_0\right) = \mathrm{CV}\left(S, S\right) - 2\sum_{i=1}^{k} \lambda_i \mathrm{Cv}\left(\boldsymbol{x}_i, S\right) + \sum_{i=1}^{k}\sum_{j=1}^{k} \lambda_i \lambda_j \mathrm{Cov}\left(\boldsymbol{x}_i, \boldsymbol{x}_j\right),$$

and after replacing the last term by its equivalent expression in Lemma 12.5,

$$\sigma^2_{BK}\left(\boldsymbol{x}_0\right) = \mathrm{CV}\left(S, S\right) - 2\sum_{i=1}^{k} \lambda_i \mathrm{Cv}\left(\boldsymbol{x}_i, S\right) + \sum_{i=1}^{k} \lambda_i \mathrm{Cv}\left(\boldsymbol{x}_i, S\right) - \mu\,.$$

Cancellation of the $\sum_{i=1}^{k} \lambda_i \mathrm{Cv}\left(\boldsymbol{x}_i, S\right)$ terms proves the theorem. □

Let us summarize the finding by defining some matrices and formulating an algorithm.

Definition 12.6

Let $\boldsymbol{x}_i$ be the locations of k variates inside the neighborhood $d \subset D$ in Definition 12.2 and let $\text{Cov}(\boldsymbol{x}_i, \boldsymbol{x}_j)$ be the covariance in Assumption 12.2. Then $\mathbf{B}$ is the matrix

$$\mathbf{B} = \begin{bmatrix} \text{Cov}(\boldsymbol{x}_1, \boldsymbol{x}_1) & \text{Cov}(\boldsymbol{x}_2, \boldsymbol{x}_1) & \cdots & \text{Cov}(\boldsymbol{x}_k, \boldsymbol{x}_1) & 1 \\ \text{Cov}(\boldsymbol{x}_1, \boldsymbol{x}_2) & \text{Cov}(\boldsymbol{x}_2, \boldsymbol{x}_2) & \cdots & \text{Cov}(\boldsymbol{x}_k, \boldsymbol{x}_2) & 1 \\ \cdots & \cdots & \cdots & \cdots & \cdots \\ \text{Cov}(\boldsymbol{x}_1, \boldsymbol{x}_k) & \text{Cov}(\boldsymbol{x}_2, \boldsymbol{x}_k) & \cdots & \text{Cov}(\boldsymbol{x}_k, \boldsymbol{x}_k) & 1 \\ 1 & 1 & \cdots & 1 & 0 \end{bmatrix} . \ \square$$

Definition 12.7

Let λ_i be the optimal weights for the estimator in Definition 12.2, let μ be the Lagrange multiplier in Definition 12.5, and let the prime stand for the transpose of the matrix. Then $\mathbf{L}$ is the matrix

$$\mathbf{L} = [\, \lambda_1 \quad \lambda_2 \quad \cdots \quad \lambda_k \quad \mu]' . \ \square$$

Definition 12.8

Let $\boldsymbol{x}_i$ be the locations of k variates inside the neighborhood $d \subset D$ in Definition 12.2, let $S \subset d$ be the support in the same definition, let $\boldsymbol{x}_0$ be the estimation location, and let $\text{Cv}(\cdot)$ be the point-to-block covariance in Definition 12.3. Then $\mathbf{b}$ is the matrix

$$\mathbf{b} = [\, \text{Cv}(\boldsymbol{x}_1, S) \quad \text{Cv}(\boldsymbol{x}_2, S) \quad \cdots \quad \text{Cv}(\boldsymbol{x}_k, S) \quad 1]' . \ \square$$

Definition 12.9

Let $\boldsymbol{x}_i$ be the locations of k variates inside the neighborhood $d \subset D$ in Definition 12.2. Then $\mathbf{Z}$ is the matrix

$$\mathbf{Z} = [\, Z(\boldsymbol{x}_1) \quad Z(\boldsymbol{x}_2) \quad \cdots \quad Z(\boldsymbol{x}_k) \quad 0]' . \ \square$$

Algorithm 12.1

This is a procedure for ordinary block kriging estimation of attributes honoring Assumptions 12.1 and 12.2.

1. Calculate each term in matrix $\mathbf{B}$ in Definition 12.6.
2. Calculate each term in matrix $\mathbf{b}$ in Definition 12.8.
3. Solve the system of equations

$$\mathbf{BL} = \mathbf{b} ,$$

where **L** is the matrix in Definition 12.7.

4. Compute the estimate

$$\hat{Z}_{BK}(\boldsymbol{x}_0) = \mathbf{Z}'\mathbf{L} = \mathbf{Z}'\mathbf{B}^{-1}\mathbf{b},$$

where **Z** is the matrix in Definition 12.9.

5. Calculate the ordinary kriging estimation variance

$$\sigma^2_{BK}(\boldsymbol{x}_0) = \mathrm{CV}(S, S) - \mathbf{b}'\mathbf{L} = \mathrm{CV}(S, S) - \mathbf{b}'\mathbf{B}^{-1}\mathbf{b}. \quad \square$$

Note that applying the algorithm produces one estimate and its kriging variance for a single site. In general kriging, several sites, such as the nodes in a grid, require applying the algorithm as many times as there are sites. The exception is highly regular sampling patterns. But increasingly fast computers are making it more attractive to ignore duplicate sampling-estimate configurations and blindly run the algorithm for every required estimation location, regardless of potential duplication in the solution of the normal equations.

Covariance Modeling

As in other forms of kriging, performance of block kriging presumes knowledge of some covariance terms. But the situation is now more critical, because three different types of covariance are being dealt with here instead of one, as in any other form of point kriging for one attribute.

The solution to this conundrum is no different than that which we have seen before, *i.e.*, model the covariances and ignore the effects that the use of models rather than the true covariances have in the normal equations.

Modeling the point-to-point covariance proceeds in exactly the same way as it does for point kriging: First, model a semivariogram $\hat{\gamma}(\boldsymbol{h})$ and then use Lemma 4.3 to convert the semivariogram to a covariance,

$$\mathrm{Cov}(\boldsymbol{h}) = \hat{C} - \hat{\gamma}(\boldsymbol{h}),$$

where $\hat{C}$ is the sill of the semivariogram model. The semivariogram must be transitive and permissible to allow the conversion and to assure that the covariance models are positive definite.

Once the point-to-point covariance is known, Definitions 12.3 and 12.4 are used to calculate the point-to-block covariance and the block variance. In the 1960s, in the early days of kriging, solution of these integrals was critical for the applicability of the method. There was great fascination in solving the integrals by exact analytical methods for any imaginable shape of block, and summarizing the results graphically in the form of charts. Today the prevailing practice is to introduce further approximations in the estimation process and calculate the integrals by numerical approximations.

If one subdivides the support into smaller pieces that can be regarded as points, the point-to-block covariance is the average of all the point covariances between the observation and the center of the subdivisions. Isaaks and Srivastava (1989, p. 330) recommend subdividing each side of a rectangular support into four smaller sides, resulting in a partition of 4^D subdivisions, where $D = 2, 3$. Table 12.1 shows the maximum subdivisions, depending upon the dimensionality of the support, as recommended by Journel and Huijbregts (1978, p. 97).

Table 12.1. Maximum number of subdivisions in calculation of block covariances.

Dimension	Subdivisions
1	10
2	6×6
3	$4 \times 4 \times 4$

EXERCISE 12.1

Let the point covariance be the isotropic model $e^{-\frac{h}{0.8}}$ and let the supports be squares 0.25 miles on each side. Then

1. Show a step-by-step calculation of $\mathrm{Cv}\,(0.3)$.
2. Prepare a graph showing both $\mathrm{Cov}\,(h)$ and $\mathrm{Cv}\,(h)$.
3. Calculate the block variance.

Solution

1. According to rules of thumb, a tessellation of the squares into 25 smaller squares 0.05 miles on a side is a convenient subdivision intermediate between the criteria given above. As shown in Figure 12.1, averaging the 25 values for the point covariance between a point 0.3 miles from the center of the 0.25-mile square and the centers of the 0.05-mile squares gives an estimate of 0.682 for $\mathrm{Cv}\,(0.3)$.

2. Repetition of the calculations for other lags provides the following table of values (Table 12.2) that were used to prepare Figure 12.2. In this instance, beyond 0.5 miles the difference from point-to-point and block-to-point covariance is negligible. In the averaging process, the negative discrepancies between the point covariance for the central subdivision and those for distant subdivisions tend to exactly compensate the positive discrepancies between the point covariance for the central subdivision and those for close subdivisions.

3. Calculation of the block variance $\mathrm{CV}\,(S, S)$ involves calculating the point-to-block covariance between the point at the center of each small square and the large square, which in this case involves averaging 25 point-to-block covariances, or equivalently, 625 point covariances. Figure 12.3 shows the point covariance values for all possible configurations. Symmetry of the subdivision dictates

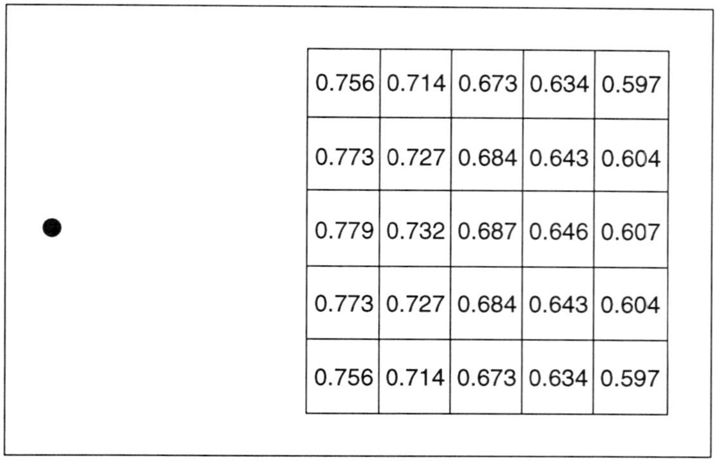

Figure 12.1. Numerical approximation of point-to-block covariance Cv$(\cdot)$ for the random function whose partial realization is given in Appendix A. The point covariance is $e^{-\frac{h}{0.8}}$, the side of the square is 0.25 miles, and the distance from the point on the left to the center of the square is 0.3 miles. The values show the point covariance between the point on the left and the center of the 0.05-mile squares of the subdivision. Averaging the 25 numbers gives a value of 0.682 for Cv(0.3).

that any of the other 19 configurations can be obtained from the basic ones in Figure 12.3. Averaging the point-to-block configurations,

$$\begin{aligned}\mathrm{CV}(S,S) &= \frac{0.890 + 4\cdot(0.823+0.854+0.872+0.881)+8\cdot 0.846}{25}\\ &= 0.855\,. \quad \square\end{aligned}$$

When using computer software the user does not have to worry about calculating point-to-block covariances or block variances. Block kriging routines perform all these calculations automatically.

Table 12.2. Covariance values for the sampling in Appendix A.

Lag (mi.)	Cov (·)	Cv (·)
0.00	1.000	0.890
0.05	0.934	0.881
0.10	0.882	0.854
0.20	0.779	0.769
0.30	0.687	0.682
0.40	0.607	0.604
0.50	0.535	0.534
0.60	0.472	0.472
0.80	0.368	0.368
1.00	0.287	0.287

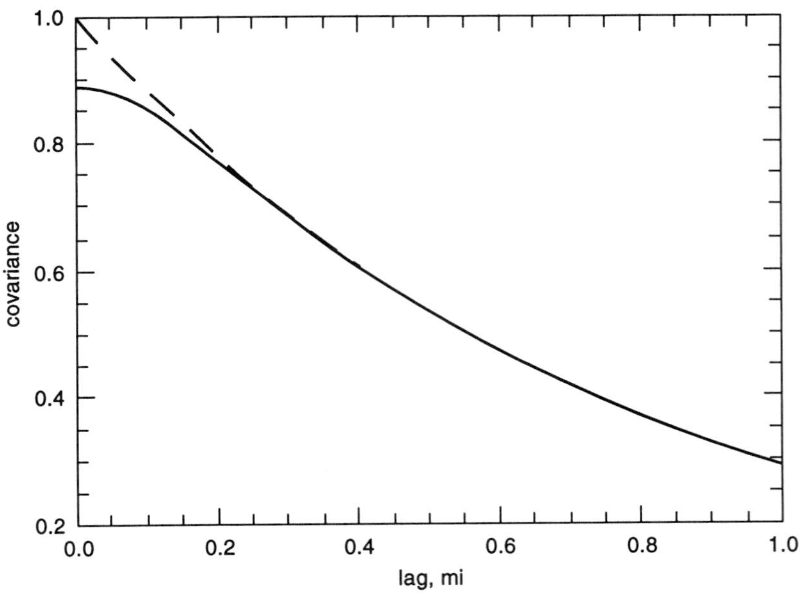

Figure 12.2. Point-to-point (dashed line) and point-to-block covariance (solid line) for the random function whose partial realization is given in Appendix A, when the support is a square 0.25 miles on a side.

0.779	0.773	0.756	0.732	0.702
0.829	0.821	0.798	0.767	0.732
0.882	0.870	0.838	0.798	0.756
0.939	0.915	0.870	0.821	0.773
●	0.939	0.882	0.829	0.779

0.823

0.773	0.779	0.773	0.756	0.732
0.821	0.829	0.821	0.798	0.767
0.870	0.882	0.870	0.838	0.798
0.915	0.939	0.915	0.870	0.821
0.939	●	0.939	0.882	0.829

0.846

0.756	0.773	0.779	0.773	0.756
0.798	0.821	0.829	0.821	0.798
0.838	0.870	0.882	0.870	0.838
0.870	0.915	0.939	0.915	0.870
0.882	0.939	●	0.939	0.882

0.854

0.821	0.829	0.821	0.798	0.767
0.870	0.882	0.870	0.838	0.798
0.915	0.939	0.915	0.870	0.821
0.939	●	0.939	0.882	0.829
0.915	0.939	0.915	0.870	0.821

0.872

0.798	0.821	0.829	0.821	0.798
0.828	0.870	0.882	0.870	0.838
0.870	0.915	0.939	0.915	0.870
0.882	0.939	●	0.939	0.882
0.970	0.915	0.939	0.915	0.870

0.881

0.838	0.870	0.882	0.870	0.838
0.870	0.915	0.939	0.915	0.870
0.882	0.939	●	0.939	0.882
0.870	0.915	0.939	0.915	0.870
0.838	0.870	0.882	0.870	0.838

0.890

Figure 12.3. Point-to-block covariances for all configurations required in the numerical approximation of the block covariance for the random function whose partial realization is given in Appendix A. The support is a square 0.25 miles on a side. The dot denotes the location of the point, where the point covariance, Cov(0), in this case is always 1.0. The number under the square is the point-to-block covariance.

EXERCISE 12.2

As seen in Exercise 3.1, the sampling in Appendix A does not approximate a normal distribution. Hence, a normal score transformation should produce better estimates than a direct estimation in the original data space. The normal scores are regarded as isotropic and their point covariance is $e^{-\frac{h}{0.8}}$. For this sampling, considering squares a quarter-mile on each side,

1. Show a detailed block kriging calculation for the estimator and the kriging variance at (1.85, 2.15) employing the six closest wells.
2. Prepare maps showing the point and block fluctuations of the attribute and compare them.
3. Do the same for the kriging standard deviation.

Solution

1. The six closest wells are:

ID	Easting mi.	Northing mi.	Porosity %	Normal score
60	1.864	2.127	13.9	1.150
52	1.862	2.379	11.3	−0.250
59	1.616	2.127	10.5	−0.924
61	2.092	2.1241	10.6	−0.782
68	1.866	1.880	11.7	0.250
67	1.676	1.937	11.2	−0.354

From Exercise 12.1 and Algorithm 12.1,

$$\mathbf{L} = \begin{bmatrix} 1.0 & & & & & & \\ 0.730 & 1.0 & & & & & \\ 0.733 & 0.644 & 1.0 & & & & \\ 0.752 & 0.651 & 0.552 & 1.0 & & & \\ 0.734 & 0.536 & 0.644 & 0.660 & 1.0 & & \\ 0.716 & 0.549 & 0.780 & 0.565 & 0.780 & 1.0 & \\ 1.0 & 1.0 & 1.0 & 1.0 & 1.0 & 1.0 & 0.0 \end{bmatrix}^{-1} \begin{bmatrix} 0.887 \\ 0.743 \\ 0.738 \\ 0.731 \\ 0.707 \\ 0.704 \\ 1.0 \end{bmatrix}$$

$$= [0.552 \quad 0.151 \quad 0.118 \quad 0.093 \quad 0.048 \quad 0.038 \quad 0.006]'$$

Hence the estimate of block porosity is

$$\hat{z}_{BK}(1.85,\ 2.15) = [1.150 \ -0.250 \ -0.924 \ -0.782 \ -0.250 \ -.0354 \quad 0] \begin{bmatrix} 0.552 \\ 0.151 \\ 0.118 \\ 0.093 \\ 0.048 \\ 0.038 \\ 0.006 \end{bmatrix}$$

$$= 0.414,$$

which after back transformation represents 12.04% porosity. The kriging variance in the normal score space is

$$\sigma_{BK}^2(\boldsymbol{x}_0) = 0.855 - [0.887 \quad 0.743 \quad 0.738 \quad 0.731 \quad 0.707 \quad 0.704 \quad 1] \begin{bmatrix} 0.552 \\ 0.151 \\ 0.118 \\ 0.093 \\ 0.048 \\ 0.038 \\ 0.006 \end{bmatrix}$$

$$= 0.032.$$

2. Figure 12.4 shows the point and block porosity maps. The block kriging maps were trimmed to facilitate comparison.

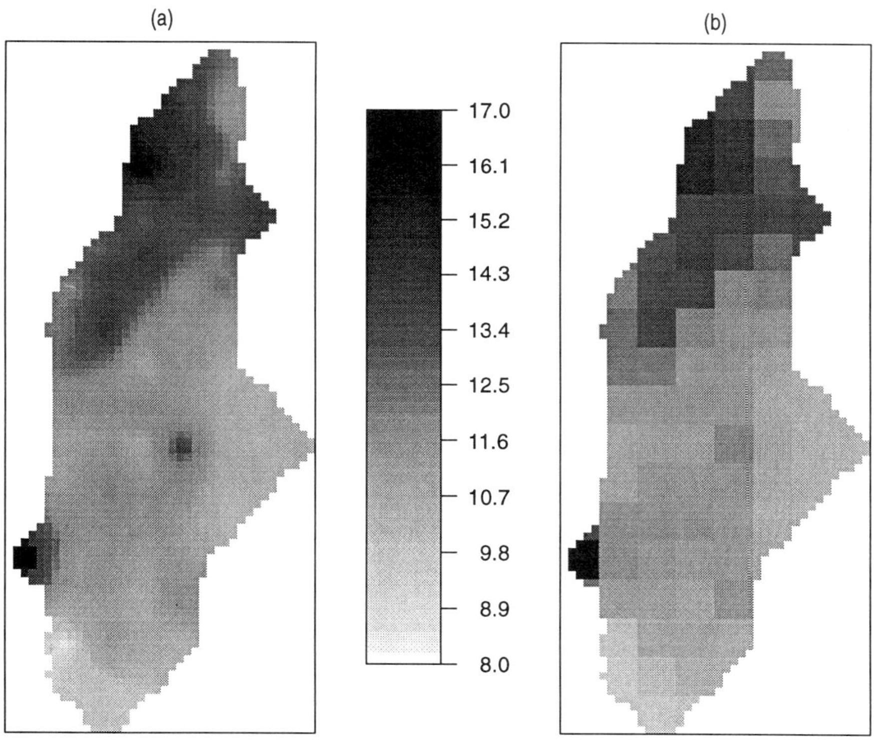

Figure 12.4. Well porosity (in percent) in West Lyons field, Kansas: (a) ordinary kriging rendition; (b) block kriging for 0.25 × 0.25-mile squares. Visit Internet site http://www.kgs.ukans.edu/Mathgeo/Books/Geostat/index.html for a color rendition.

The grid in Figure 12.4a was prepared using ordinary kriging, employing a grid spacing of 0.05 miles, an origin of (0.75, 0.30), search radius of 0.45 miles, and between three and six observations per estimation without octant search.

Block kriging is an even smoother version of reality than the already smoothed point kriging version, which is clearly revealed in the block kriging map by the absence of the extreme values that appear in the ordinary kriging map.

3. As shown in Figure 12.5, when the sampling control is good, the block kriging variance is smaller and smoother in its fluctuations than the point kriging variance. This result is in agreement with the higher continuity of the block averages.

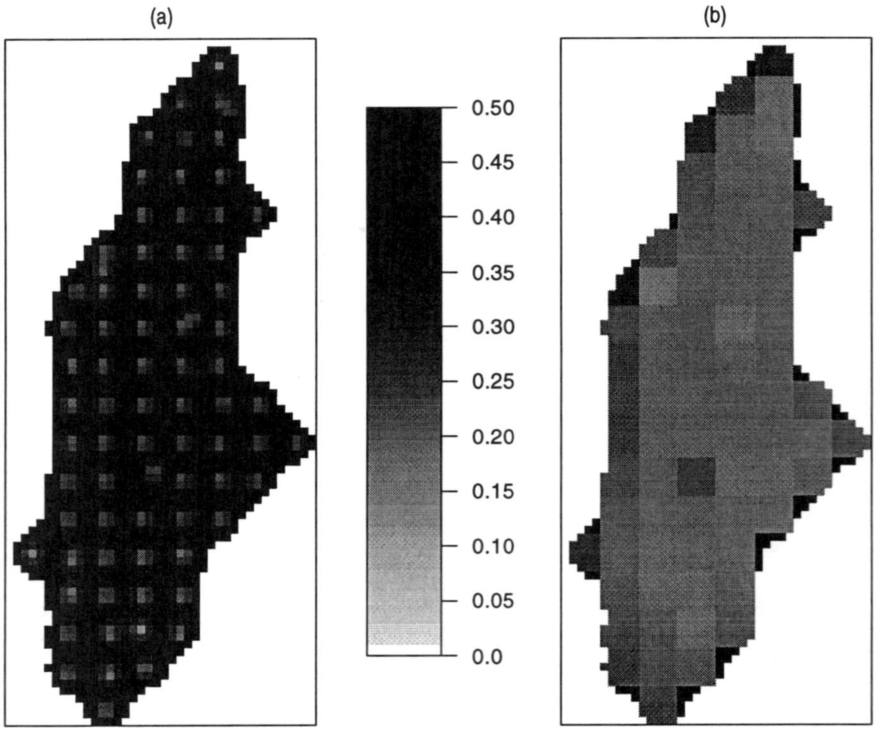

Figure 12.5. Kriging standard deviation in normal score space for well porosity (in percent) in West Lyons field, Kansas: (a) ordinary kriging rendition; (b) block kriging for 0.25 × 0.25-mile squares. For a color rendition, visit Internet site http://www.kgs.ukans.edu/Mathgeo/Books/Geostat/index.html.

Remarks

By construction, block kriging estimates are linear in their weights, unbiased, and have the minimum mean square error.

Block kriging shares with point kriging its declustering power, screen effect, the refusal to accept more than one measurement per site, insensitivity to translation of the Cartesian system, independence of the kriging variance from individual measurements, and dependence on the sampling pattern. An additional similarity to point kriging is demonstrated by the fact that a change of scale in the covariance does not change the block kriging estimate and multiplication of the covariance by a factor changes the estimation variance by the same factor.

If the data support is different than that for the estimates, one cannot talk about exact interpolation. Figure 12.4 and Exercise 12.3 (3) make it evident that, in general, the block average for a support centered at a point observation is different than the value of such an observation.

The following theorem demonstrates that point kriging can produce the same estimates as block kriging, although at a great sacrifice in computational efficiency.

Theorem 12.4

Let λ_i be the weights in Definition 12.2 for the block kriging estimate $\hat{Z}_S(\boldsymbol{x}_0)$ based on k variates $Z(\boldsymbol{x}_i)$ of a point random function and let $\lambda_i(\boldsymbol{x})$ be the weights for the same variates when making point kriging estimates inside the support S. Then,

$$\lambda_i = \frac{1}{S}\int_S \sum_{i=1}^{k} \lambda_i(\boldsymbol{x}) d\boldsymbol{x}, \quad i = 1,\ 2,\ \ldots,\ k\,.$$

Proof

From Definition 12.2,

$$\hat{Z}_S(\boldsymbol{x}_0) = \sum_{i=1}^{k} \lambda_i Z(\boldsymbol{x}_i)\ .$$

From Definition 12.1, the same results are achieved by integrating all possible point estimates $\hat{Z}(\boldsymbol{x})$ inside the support S:

$$\hat{Z}_S(\boldsymbol{x}_0) = \frac{1}{S}\int_S \hat{Z}(\boldsymbol{x})\, d\boldsymbol{x}\,.$$

Hence,

$$\sum_{i=1}^{k} \lambda_i Z(\boldsymbol{x}_i) = \frac{1}{S}\int_S \hat{Z}(\boldsymbol{x})\, d\boldsymbol{x}\,.$$

Considering ordinary or universal kriging estimates,

$$\sum_{i=1}^{k} \lambda_i Z(\boldsymbol{x}_i) = \frac{1}{S}\int_S \sum_{i=1}^{k} \lambda_i(\boldsymbol{x}) Z(\boldsymbol{x}_i)\, d\boldsymbol{x}\,,$$

which is true only if

$$\lambda_i = \frac{1}{S}\int_S \sum_{i=1}^{k} \lambda_i(\boldsymbol{x}) d\boldsymbol{x}, \quad i = 1,\ 2,\ \ldots,\ k\,. \quad \square$$

Consequently, one can reach the same estimates using either block kriging or point kriging. There are at least two disadvantages to employing point kriging:

i) In point kriging, the estimation is more computationally intensive. According to the rule of thumb, block kriging of three-dimensional supports requires $4^3 = 64$ point estimates per block. When block kriging was introduced, charts and calculators prevailed over computers. The savings of 63 estimations per block offered by block kriging meant that the point evaluation of a mine could be reduced from a year to a week by switching to block kriging. At a time of faster and faster computers, the advantages of block kriging are fading. Although the savings ratio is still the same, the difference now is the more subtle alternative of having a processing time of a few seconds instead of a few minutes.

ii) When point estimates are averaged to produce the block estimate, the possibility of having an estimation variance associated with each block is lost. It is not true that the block estimation variance is equal to the mean of the point estimation variances. Thus, use of the point kriging alternative precludes the use of the kriging variance for assessing any kind of reliability of the estimates.

Because of the kriging smoothing property addressed at the beginning of Chapter 9, point kriging is not the best way to obtain information about what is going on inside the supports. Simulation is the most appropriate approach to learning about characteristics such as the variance or the cumulative distributions inside individual blocks. In particular, if the user is willing to make assumptions of multivariate normality, the LU decomposition method in Chapter 9 is an exceedingly fast way to accomplish such tasks, provided that the simulation is done one block at a time.

EXERCISE 12.3

Given the sampling in Appendix A and a square support of $S = 0.25$ miles on a side,

1. Calculate the block estimate for $S(1.85, 2.15)$ as the average of ordinary kriging estimates at the center of 25 smaller squares tessellating S. Employ the same six wells as in Exercise 12.2 and compare the results with the maps in Figure 12.4.

2. Using the wells in Exercise 12.2 to estimate $S(1.85, 2.15)$, calculate the average of the weights per well and compare them with the block kriging weights in the same exercise.

3. Estimate, by block kriging, $S(1.864, 2.127)$, and compare the estimate with the value of Well 60 at the exact same location as the center of the support.

Solution

1. Table 12.3 is a tabulation of ordinary kriging estimates using Algorithm 4.2 and the results from Exercise 12.1, starting from the southwest corner and

progressing eastward. The estimates in the third column were obtained using the same wells as in Exercise 12.2; those in the fourth column are the grid values of Figure 12.4a inside $S\,(1.85, 2.15)$. Discrepancies are the result of differences in the selection of the six closest wells inside the search neighborhood. While the estimates in the third column always employed the same wells, kriging of the fourth column searched for the closest wells individually for every estimate. To make the discrepancies more evident, the absence of a value in the fourth column of a row indicates that the value is the same as in the third column of the same row.

The average of the estimates in the third column of Table 12.3 is 0.414, which is identical to the block kriging estimate in Exercise 12.2 and Figure 12.4b. The average for the grid values is 0.419. Although the values are close, the difference in the searching keeps them from being identical. Hence, depending on how one implements point kriging, the averaging may or may not duplicate exactly the results that can be obtained employing block kriging.

2. From Table 12.4, the average weights are 0.151, 0.118, 0.552, 0.093, 0.038, and 0.048. As predicted by Theorem 12.4, such weights are identical to the block kriging weights in Exercise 12.2.

3. Employing Algorithm 12.1, one learns that $\hat{z}_{BK}\,(1.864,\ 2.127) = 0.450$, which is less than 40% of the value of $z\,(1.864,\ 2.127) = 1.15$. The weight of Well 60 is only 0.561. □

Table 12.3. Point estimates for the 0.25 × 0.25-mile square $S\,(1.85, 2.15)$.

Column	Row	Fixed sampling estimate	Fig. 12.4a estimate
1	1	0.138	
2	1	0.458	
3	1	0.654	0.636
4	1	0.608	0.558
5	1	0.362	0.253
1	2	0.180	0.147
2	2	0.587	
3	2	0.925	
4	2	0.806	
5	2	0.425	0.361
1	3	0.158	0.165
2	3	0.562	
3	3	0.911	
4	3	0.779	
5	3	0.390	0.327

(cont.)

Table 12.3. concluded

Column	Row	Fixed sampling estimate	Fig. 12.4a estimate
1	4	0.007	0.148
2	4	0.369	0.450
3	4	0.553	0.594
4	4	0.497	0.478
5	4	0.249	0.214
1	5	−0.005	0.154
2	5	0.151	0.274
3	5	0.259	0.309
4	5	0.230	0.218
5	5	0.008	0.002

Table 12.4. Weights for the point estimates of the 0.25 × 0.25-mile square $S(1.85, 2.15)$.

Column	Row	52	59	60	61	67	68
1	1	−0.008	0.249	0.371	−0.024	0.302	0.109
2	1	−0.018	0.132	0.519	−0.007	0.201	0.174
3	1	−0.031	0.044	0.612	0.047	0.105	0.223
4	1	−0.030	−0.005	0.584	0.161	0.037	0.253
5	1	−0.022	−0.021	0.458	0.325	0.002	0.258
1	2	0.032	0.337	0.453	−0.032	0.164	0.046
2	2	0.018	0.174	0.657	−0.025	0.111	0.066
3	2	−0.012	0.042	0.841	0.011	0.046	0.072
4	2	−0.003	−0.015	0.767	0.144	0.008	0.099
5	2	0.018	−0.025	0.555	0.339	−0.008	0.121
1	3	0.115	0.384	0.467	−0.035	0.066	0.003
2	3	0.108	0.205	0.671	−0.027	0.042	0.001
3	3	0.079	0.058	0.864	0.007	0.004	−0.012
4	3	0.090	−0.006	0.782	0.142	−0.014	0.008
5	3	0.107	−0.022	0.563	0.339	−0.017	0.030
1	4	0.239	0.388	0.410	−0.030	0.011	−0.018
2	4	0.252	0.224	0.554	−0.012	0.004	−0.023
3	4	0.252	0.096	0.646	0.044	−0.010	−0.028
4	4	0.250	0.020	0.611	0.160	−0.019	−0.022
5	4	0.243	−0.012	0.477	0.325	−0.019	−0.013
1	5	0.387	0.360	0.314	−0.026	−0.012	−0.023
2	5	0.425	0.218	0.399	−0.003	−0.012	−0.028
3	5	0.444	0.104	0.444	0.055	−0.016	−0.030
4	5	0.436	0.030	0.425	0.156	−0.018	−0.029
5	5	0.405	−0.007	0.351	0.294	−0.017	−0.027

Chapter 13

ORDINARY COKRIGING

Simple, ordinary, and universal kriging are not multivariate models in the usual statistical sense of the term. Despite the fact that they employ a random function model comprising an infinite number of random variables, they are all used for the modeling of a single attribute.

Cokriging may be described as a true multivariate estimation procedure because the model deals with two or more attributes within the same field. In geostatistics, two or more regionalized variables defined over a random field are called a coregionalization. Data amenable for cokriging are measurements of two or more attributes at the same sampling location, such as those from a geochemical survey comprising several sites and consisting of several measurements of concentrations of different elements per site. Missing values in general do not lead to problems. On the contrary, cokriging typically works at its best when some of the attributes are undersampled.

Cokriging is basically a generalization of kriging using vectorial random functions, one per attribute. All kriging types discussed in the previous chapters can be regarded as special cases of cokriging for one attribute.

The Estimator

In order to characterize a coregionalization, it is possible to generalize any form of kriging for the modeling of one attribute into its equivalent cokriging form. All the foregoing chapters could be rewritten to generalize results into cokriging. However, we will restrict our attention to the most prevalent form of cokriging, the one that results from the generalization of ordinary kriging under second order stationarity. Other extensions which will not be covered here include universal cokriging (Myers, 1982; Helterbrand and Cressie, 1994), a generalization of universal kriging.

With the help of matrix notation, the chore of generalizing ordinary kriging into ordinary cokriging is largely reduced to replacing variables by matrices.

Definition 13.1

Let $Z_1(\boldsymbol{x}), Z_2(\boldsymbol{x}), \ldots, Z_k(\boldsymbol{x}), \ldots, Z_p(\boldsymbol{x})$ be a set of p random functions defined over the same domain. The vectorial random function $\mathbf{Z}(\boldsymbol{x})$ is the vectorial matrix

$$\mathbf{Z}(\boldsymbol{x}) = [\, Z_1(\boldsymbol{x}) \quad Z_2(\boldsymbol{x}) \quad \cdots \quad Z_p(\boldsymbol{x})\,]' . \quad \square$$

Definition 13.2

$\boldsymbol{\Lambda}_i$ is the square matrix

$$\boldsymbol{\Lambda}_i = \begin{bmatrix} \lambda_{11}^i & \lambda_{12}^i & \cdots & \lambda_{1p}^i \\ \lambda_{21}^i & \lambda_{22}^i & \cdots & \lambda_{2p}^i \\ \cdots & \cdots & \cdots & \cdots \\ \lambda_{p1}^i & \lambda_{p2}^i & \cdots & \lambda_{pp}^i \end{bmatrix},$$

where λ_{jk}^i are real numbers and p is the number of attributes to be modeled. $\square$

Definition 13.3

Let $\mathbf{Z}(\boldsymbol{x})$ be the vectorial random function in Definition 13.1. The cokriging estimator of $\mathbf{Z}(\boldsymbol{x})$, $\hat{\mathbf{Z}}(\boldsymbol{x}_0) = [\, \hat{Z}_1(\boldsymbol{x}_0) \quad \hat{Z}_2(\boldsymbol{x}_0) \quad \cdots \quad \hat{Z}_p(\boldsymbol{x}_0)\,]'$, is the vector

$$\begin{aligned} \hat{\mathbf{Z}}(\boldsymbol{x}_0) &= [\boldsymbol{\Lambda}_1 \quad \boldsymbol{\Lambda}_2 \quad \ldots \quad \boldsymbol{\Lambda}_n] \quad [\mathbf{Z}(\boldsymbol{x}_1) \quad \mathbf{Z}(\boldsymbol{x}_2) \quad \ldots \quad \mathbf{Z}(\boldsymbol{x}_n)]' \\ &= \sum_{i=1}^{n} \boldsymbol{\Lambda}_i \mathbf{Z}(\boldsymbol{x}_i), \end{aligned}$$

where $\boldsymbol{\Lambda}_i$ is the matrix of weights λ_{jk}^i in Definition 13.2 for the estimation of the jth random function using the kth random function at site $\boldsymbol{x}_i$. $\square$

Assumptions

First let us define additional matrices.

Definition 13.4

Let $\mathbf{Z}(\boldsymbol{x})$ be the vectorial random function in Definition 13.1. Then

$$\boldsymbol{m} = [\, m_1 \quad m_2 \quad \cdots \quad m_p\,]' ,$$

where the m_k are scalar constants. $\square$

Definition 13.5

Let $\mathbf{Z}(\boldsymbol{x})$ be the vectorial random function in Definition 13.1 and let $\boldsymbol{h}$ be a vectorial distance in the sampling space. Then its covariance is

$\mathbf{Cov}(\boldsymbol{h}) =$

$$\begin{bmatrix} \mathrm{Cov}\big(Z_1(\boldsymbol{x}), Z_1(\boldsymbol{x}+\boldsymbol{h})\big) & \mathrm{Cov}\big(Z_1(\boldsymbol{x}), Z_2(\boldsymbol{x}+\boldsymbol{h})\big) & \cdots & \mathrm{Cov}\big(Z_1(\boldsymbol{x}), Z_p(\boldsymbol{x}+\boldsymbol{h})\big) \\ \mathrm{Cov}\big(Z_2(\boldsymbol{x}), Z_1(\boldsymbol{x}+\boldsymbol{h})\big) & \mathrm{Cov}\big(Z_2(\boldsymbol{x}), Z_2(\boldsymbol{x}+\boldsymbol{h})\big) & \cdots & \mathrm{Cov}\big(Z_2(\boldsymbol{x}), Z_p(\boldsymbol{x}+\boldsymbol{h})\big) \\ \cdots & \cdots & \cdots & \cdots \\ \mathrm{Cov}\big(Z_p(\boldsymbol{x}), Z_1(\boldsymbol{x}+\boldsymbol{h})\big) & \mathrm{Cov}\big(Z_p(\boldsymbol{x}), Z_2(\boldsymbol{x}+\boldsymbol{h})\big) & \cdots & \mathrm{Cov}\big(Z_p(\boldsymbol{x}), Z_p(\boldsymbol{x}+\boldsymbol{h})\big) \end{bmatrix}$$

where

$$\mathrm{Cov}\big(Z_i(\boldsymbol{x}), Z_j(\boldsymbol{x}+\boldsymbol{h})\big) = \mathrm{E}\big[\big(Z_i(\boldsymbol{x}) - m_i\big)\big(Z_j(\boldsymbol{x}+\boldsymbol{h}) - m_j\big)\big],$$

with $\mathrm{E}[\cdot]$ denoting expected value. □

The assumptions necessary to formulate and run ordinary cokriging follow.

Assumption 13.1

A sampling is a partial realization of a vectorial random function. □

Assumption 13.2

The vectorial random function is second order stationary over the sampling domain, which implies that

$$\begin{aligned} &\mathrm{E}\big[\mathbf{Z}(\boldsymbol{x})\big] = \boldsymbol{m} \\ &\mathrm{Cov}\big(\mathbf{Z}(\boldsymbol{x}), \mathbf{Z}(\boldsymbol{x}+\boldsymbol{h})\big) = \mathbf{Cov}(\boldsymbol{h}), \end{aligned}$$

where $\boldsymbol{m}$ is the vector in Definition 13.4 and $\mathbf{Cov}(\boldsymbol{h})$ is the matrix in Definition 13.5. □

It is always true that $\mathbf{Cov}(\boldsymbol{h}) = \mathbf{Cov}'(-\boldsymbol{h})$; but, in general, $\mathbf{Cov}(\boldsymbol{h}) \neq \mathbf{Cov}'(\boldsymbol{h})$.

Assumption 13.3

None of the variables is a linear combination of the others.

$$Z_k(\boldsymbol{x}) \neq a_0 - \sum_{j \neq k} a_j Z_j(\boldsymbol{x}), \text{ for } k = 1, 2, \ldots, p. \quad \square$$

The exclusion of compositional data is the most significant result of this last assumption. The set of assumptions is the minimum necessary to calculate the weights in such a way that the estimator is unbiased with minimum mean square error. The following is an important corollary.

Corollary 13.1

Let $\mathbf{Cov}(\boldsymbol{h})$ be the covariance matrix in Definition 13.5. Then

$$\mathbf{Cov}(\boldsymbol{h}) = \mathbf{Cov}'(-\boldsymbol{h}) .$$

Proof

Because $\mathbf{Cov}(\boldsymbol{h})$ is invariant under translation, shifting all locations by $-\boldsymbol{h}$ in Definition 13.5,

$$\mathbf{Cov}(\boldsymbol{h}) =$$

$$\begin{bmatrix} \mathrm{Cov}(Z_1(\boldsymbol{x}-\boldsymbol{h}), Z_1(\boldsymbol{x})) & \mathrm{Cov}(Z_1(\boldsymbol{x}-\boldsymbol{h}), Z_2(\boldsymbol{x})) \cdots & \mathrm{Cov}(Z_1(\boldsymbol{x}-\boldsymbol{h}), Z_p(\boldsymbol{x})) \\ \mathrm{Cov}(Z_2(\boldsymbol{x}-\boldsymbol{h}), Z_1(\boldsymbol{x})) & \mathrm{Cov}(Z_2(\boldsymbol{x}-\boldsymbol{h}), Z_2(\boldsymbol{x})) \cdots & \mathrm{Cov}(Z_2(\boldsymbol{x}-\boldsymbol{h}), Z_p(\boldsymbol{x})) \\ \cdots & \cdots \quad \cdots & \cdots \\ \mathrm{Cov}(Z_p(\boldsymbol{x}-\boldsymbol{h}), Z_1(\boldsymbol{x})) & \mathrm{Cov}(Z_p(\boldsymbol{x}-\boldsymbol{h}), Z_2(\boldsymbol{x})) \cdots & \mathrm{Cov}(Z_p(\boldsymbol{x}-\boldsymbol{h}), Z_p(\boldsymbol{x})) \end{bmatrix}.$$

The proof follows because $\mathrm{Cov}\big(Z_i(\boldsymbol{x}), Z_j(\boldsymbol{x}+\boldsymbol{h})\big) = \mathrm{Cov}\big(Z_j(\boldsymbol{x}+\boldsymbol{h}), Z_i(\boldsymbol{x})\big)$ for any variate and location. □

Unbiasedness

Before deriving the expression for the mean square error, let us find the unbiasedness condition.

Theorem 13.1

Let $\boldsymbol{\Lambda}_i$ be the weight matrix in Definition 13.2 and let $\mathbf{I}$ be an identity matrix of order p—a matrix of 1's along the main diagonal and zeros for all the other elements. Then the cokriging estimator is unbiased if and only if

$$\sum_{i=1}^{n} \boldsymbol{\Lambda}_i = \mathbf{I} .$$

Proof (a)

The relationship $\sum\limits_{i=1}^{n} \boldsymbol{\Lambda}_i = \mathbf{I}$ does not change by multiplying both sides by the vector in Definition 13.4,

$$\sum_{i=1}^{n} \boldsymbol{\Lambda}_i \, \boldsymbol{m} = \boldsymbol{m} .$$

From Assumption 13.2,

$$\sum_{i=1}^{n} \boldsymbol{\Lambda}_i \, \mathrm{E}\big[\mathbf{Z}(\boldsymbol{x})\big] = \mathrm{E}\big[\mathbf{Z}(\boldsymbol{x})\big] .$$

By the linearity of the expectation, one can interchange summation and expectation:

$$\mathrm{E}\left[\sum_{i=1}^{n} \boldsymbol{\Lambda}_i \mathbf{Z}(\boldsymbol{x})\right] = \mathrm{E}\big[\mathbf{Z}(\boldsymbol{x})\big].$$

But by Definition 13.3,

$$\mathrm{E}\big[\hat{\mathbf{Z}}(\boldsymbol{x})\big] = \mathrm{E}\big[\mathbf{Z}(\boldsymbol{x})\big],$$

which is the mathematical way of saying that an estimator is unbiased.

Proof (b)

If $\mathrm{E}\big[\hat{\mathbf{Z}}(\boldsymbol{x})\big] = \mathrm{E}\big[\mathbf{Z}(\boldsymbol{x})\big]$, from Definition 13.3,

$$\mathrm{E}\left[\sum_{i=1}^{n} \boldsymbol{\Lambda}_i \mathbf{Z}(\boldsymbol{x})\right] = \mathrm{E}\big[\mathbf{Z}(\boldsymbol{x})\big].$$

Reversing expectation and summation,

$$\mathrm{E}\big[\mathbf{Z}(\boldsymbol{x})\big] \sum_{i=1}^{n} \boldsymbol{\Lambda}_i = \mathrm{E}\big[\mathbf{Z}(\boldsymbol{x})\big]$$

and Proof (b) follows by multiplying both sides of the equation by $\mathrm{E}\big[\mathbf{Z}(\boldsymbol{x})\big]^{-1}$. Proofs (a) and (b) prove the theorem. □

Estimation Error

The prevailing practice is to consider that the cokriging mean square estimation error is the sum of the error variances per attribute.

Definition 13.6

Let $Z_k(\boldsymbol{x})$ be the random function in Definition 13.1 and let $\hat{Z}_k(\boldsymbol{x})$ be its estimator in Definition 13.3. Then the cokriging estimation variance is

$$\sigma^2(\boldsymbol{x}_0) = \sum_{k=1}^{p} \mathrm{Var}\left[\hat{Z}_k(\boldsymbol{x}_0) - Z_k(\boldsymbol{x}_0)\right]. \quad \square$$

To express the estimation variance in terms of covariances, let us first add one more definition and prove five lemmas.

Definition 13.7

The trace of a square matrix is the sum of the elements in the main diagonal

$$\mathrm{Tr}\left\{\begin{bmatrix} a_{11} & a_{12} & \cdots & a_{1n} \\ a_{21} & a_{22} & \cdots & a_{2n} \\ \cdots & \cdots & \cdots & \cdots \\ a_{n1} & a_{n2} & \cdots & a_{nn} \end{bmatrix}\right\} = \sum_{i=1}^{n} a_{ii}. \quad \square$$

Lemma 13.1

Let $\sigma^2(\boldsymbol{x}_0)$ be the estimation variance in Definition 13.6. Then

$$\sigma^2(\boldsymbol{x}_0) = \operatorname{Tr}\left\{\operatorname{Cov}\left(\hat{\mathbf{Z}}(\boldsymbol{x}_0) - \mathbf{Z}(\boldsymbol{x}_0), \hat{\mathbf{Z}}(\boldsymbol{x}_0) - \mathbf{Z}(\boldsymbol{x}_0)\right)\right\}.$$

Proof

By Definition 13.6,

$$\sigma^2(\boldsymbol{x}_0) = \sum_{k=1}^{p} \operatorname{Var}\left[\hat{Z}_k(\boldsymbol{x}_0) - Z_k(\boldsymbol{x}_0)\right]$$

or

$$\sigma^2(\boldsymbol{x}_0) = \sum_{k=1}^{p} \operatorname{Cov}\left(\hat{Z}_k(\boldsymbol{x}_0) - Z_k(\boldsymbol{x}_0), \hat{Z}_k(\boldsymbol{x}_0) - Z_k(\boldsymbol{x}_0)\right),$$

which by Definition 13.7 is equivalent to the trace of

$$\operatorname{Cov}\left(\hat{\mathbf{Z}}(\boldsymbol{x}_0) - \mathbf{Z}(\boldsymbol{x}_0), \hat{\mathbf{Z}}(\boldsymbol{x}_0) - \mathbf{Z}(\boldsymbol{x}_0)\right). \quad \square$$

Lemma 13.2

Let $\mathbf{Z}(\boldsymbol{x})$ be the vectorial random function in Definition 13.1 and let $\boldsymbol{\Lambda}_i$ be the weights in Definition 13.2. Then,

$$\operatorname{Cov}\left(\sum_{i=1}^{n}\boldsymbol{\Lambda}_i\mathbf{Z}(\boldsymbol{x}_i), \sum_{j=1}^{n}\boldsymbol{\Lambda}_j\mathbf{Z}(\boldsymbol{x}_j)\right) = \sum_{i=1}^{n}\sum_{j=1}^{n}\boldsymbol{\Lambda}_i\operatorname{Cov}(\mathbf{Z}(\boldsymbol{x}_i), \mathbf{Z}(\boldsymbol{x}_j))\boldsymbol{\Lambda}'_j.$$

Proof

By the definition of covariance,

$$\begin{aligned}\operatorname{Cov}\left(\sum_{i=1}^{n}\boldsymbol{\Lambda}_i\mathbf{Z}(\boldsymbol{x}_i), \sum_{j=1}^{n}\boldsymbol{\Lambda}_j\mathbf{Z}(\boldsymbol{x}_j)\right) &= \operatorname{E}\left[\sum_{i=1}^{n}\boldsymbol{\Lambda}_i\mathbf{Z}(\boldsymbol{x}_i)\left(\sum_{j=1}^{n}\boldsymbol{\Lambda}_j\mathbf{Z}(\boldsymbol{x}_j)\right)'\right] \\ &\quad - \operatorname{E}\left[\sum_{i=1}^{n}\boldsymbol{\Lambda}_i\mathbf{Z}(\boldsymbol{x}_i)\right]\operatorname{E}\left[\sum_{j=1}^{n}(\boldsymbol{\Lambda}_j\mathbf{Z}(\boldsymbol{x}_j))'\right].\end{aligned}$$

Expanding, by the distributive property of the expectation demonstrated in Lemmas 2.3 and 2.4,

$$\begin{aligned}\operatorname{Cov}\left(\sum_{i=1}^{n}\boldsymbol{\Lambda}_i\mathbf{Z}(\boldsymbol{x}_i), \sum_{j=1}^{n}\boldsymbol{\Lambda}_j\mathbf{Z}(\boldsymbol{x}_j)\right) &= \sum_{i=1}^{n}\sum_{j=1}^{n}\boldsymbol{\Lambda}_i\operatorname{E}\left[\mathbf{Z}(\boldsymbol{x}_i)\mathbf{Z}'(\boldsymbol{x}_j)\right]\boldsymbol{\Lambda}'_j \\ &\quad - \sum_{i=1}^{n}\sum_{j=1}^{n}\boldsymbol{\Lambda}_i\operatorname{E}[\mathbf{Z}(\boldsymbol{x}_i)]\operatorname{E}\left[\mathbf{Z}'(\boldsymbol{x}_i)\right]\boldsymbol{\Lambda}'_j\end{aligned}$$

and factoring,

$$\mathrm{Cov}\left(\sum_{i=1}^{n}\mathbf{\Lambda}_i\mathbf{Z}(\boldsymbol{x}_i),\sum_{j=1}^{n}\mathbf{\Lambda}_j\mathbf{Z}(\boldsymbol{x}_j)\right)=$$
$$\sum_{i=1}^{n}\sum_{j=1}^{n}\mathbf{\Lambda}_i\left\{\mathrm{E}\left[\mathbf{Z}(\boldsymbol{x}_i)\mathbf{Z}'(\boldsymbol{x}_j)\right]-\mathrm{E}\big[\mathbf{Z}(\boldsymbol{x}_i)\big]\mathrm{E}\left[\mathbf{Z}'(\boldsymbol{x}_i)\right]\right\}\mathbf{\Lambda}_j'.$$

The difference on the right-hand side of the equation is equal to

$$\mathrm{Cov}\big(\mathbf{Z}(\boldsymbol{x}_i),\mathbf{Z}(\boldsymbol{x}_j)\big),$$

which proves the lemma. □

Lemma 13.3

Let $\hat{\mathbf{Z}}(\boldsymbol{x}_0)$ be the estimator in Definition 13.3. If the estimator is unbiased, then

$$\mathrm{E}\big[\mathbf{Z}(\boldsymbol{x}_0)-\hat{\mathbf{Z}}(\boldsymbol{x}_0)\big]=0.$$

Proof

From Theorem 13.1,

$$\mathrm{E}\big[\mathbf{Z}(\boldsymbol{x}_0)\big]=\mathrm{E}\big[\hat{\mathbf{Z}}(\boldsymbol{x}_0)\big].$$

Subtracting $\mathrm{E}\big[\hat{\mathbf{Z}}(\boldsymbol{x}_0)\big]$ from both sides of the equation and taking both terms under the expectation prove the lemma. □

Lemma 13.4

Let $\mathbf{Z}(\boldsymbol{x})$ be the vector in Definition 13.1 and let $\hat{\mathbf{Z}}(\boldsymbol{x}_0)$ be the estimate in Definition 13.3. Then,

$$\mathrm{Cov}\big(\mathbf{Z}(\boldsymbol{x}_i)-\mathbf{Z}(\boldsymbol{x}_0),\mathbf{Z}(\boldsymbol{x}_j)-\mathbf{Z}(\boldsymbol{x}_0)\big)=$$
$$\mathrm{Cov}\big(\mathbf{Z}(\boldsymbol{x}_0),\mathbf{Z}(\boldsymbol{x}_0)\big)-\mathrm{Cov}\big(\mathbf{Z}(\boldsymbol{x}_i),\mathbf{Z}(\boldsymbol{x}_0)\big)-$$
$$\mathrm{Cov}'\big(\mathbf{Z}(\boldsymbol{x}_j),\mathbf{Z}(\boldsymbol{x}_0)\big)+\mathrm{Cov}\big(\mathbf{Z}(\boldsymbol{x}_i),\mathbf{Z}(\boldsymbol{x}_j)\big).$$

Proof

By the definition of covariance,

$$\mathrm{Cov}\big(\mathbf{Z}(\boldsymbol{x}_i)-\mathbf{Z}(\boldsymbol{x}_0),\mathbf{Z}(\boldsymbol{x}_j)-\mathbf{Z}(\boldsymbol{x}_0)\big)=$$
$$\mathrm{E}\big[[\mathbf{Z}(\boldsymbol{x}_i)-\mathbf{Z}(\boldsymbol{x}_0)]\,[\mathbf{Z}(\boldsymbol{x}_j)-\mathbf{Z}(\boldsymbol{x}_0)]'\big]-$$
$$\mathrm{E}\big[\mathbf{Z}(\boldsymbol{x}_i)-\mathbf{Z}(\boldsymbol{x}_0)\big]\mathrm{E}\left[[\mathbf{Z}(\boldsymbol{x}_j)-\mathbf{Z}(\boldsymbol{x}_0)]'\right].$$

By Lemma 13.3 the product of the expectations is zero. Expanding, and adding and subtracting twice the square of the vectorial mean in Definition 13.4,

$$\begin{aligned}\mathrm{Cov}\left(\mathbf{Z}\left(\boldsymbol{x}_i\right)-\mathbf{Z}\left(\boldsymbol{x}_0\right),\mathbf{Z}\left(\boldsymbol{x}_j\right)-\mathbf{Z}\left(\boldsymbol{x}_0\right)\right) = &\left\{\mathrm{E}\left[\mathbf{Z}\left(\boldsymbol{x}_0\right),\mathbf{Z}'\left(\boldsymbol{x}_0\right)\right]-\boldsymbol{m}'\boldsymbol{m}\right\}\\ &-\left\{\mathrm{E}\left[\mathbf{Z}\left(\boldsymbol{x}_i\right)\mathbf{Z}'\left(\boldsymbol{x}_0\right)\right]-\boldsymbol{m}'\boldsymbol{m}\right\}\\ &-\left\{\mathrm{E}\left[\mathbf{Z}\left(\boldsymbol{x}_0\right)\mathbf{Z}'\left(\boldsymbol{x}_j\right)\right]-\boldsymbol{m}'\boldsymbol{m}\right\}\\ &+\left\{\mathrm{E}\left[\mathbf{Z}\left(\boldsymbol{x}_i\right)\mathbf{Z}'\left(\boldsymbol{x}_j\right)\right]-\boldsymbol{m}'\boldsymbol{m}\right\}\end{aligned}$$

and the proof follows directly from the definition of covariance. □

Theorem 13.2

Let $\boldsymbol{\Lambda}_i$ be the weights in Definition 13.2 and let $\mathbf{Cov}(\cdot)$ be the matrix in Assumption 13.2. Then the estimation variance is

$$\begin{aligned}\sigma^2\left(\boldsymbol{x}_0\right)=\mathrm{Tr}\Bigg\{&\mathbf{Cov}\left(0\right)-\sum_{i=1}^{n}\boldsymbol{\Lambda}_i\mathbf{Cov}\left(\boldsymbol{x}_0-\boldsymbol{x}_i\right)\\ &-\sum_{i=1}^{n}\mathbf{Cov}\left(\boldsymbol{x}_0-\boldsymbol{x}_i\right)\boldsymbol{\Lambda}_i'+\sum_{i=1}^{n}\sum_{j=1}^{n}\boldsymbol{\Lambda}_i\mathbf{Cov}\left(\boldsymbol{x}_j-\boldsymbol{x}_i\right)\boldsymbol{\Lambda}_j'\Bigg\}.\end{aligned}$$

Proof

From Lemma 13.1 and Definition 13.3,

$$\sigma^2\left(\boldsymbol{x}_0\right)=\mathrm{Tr}\left\{\mathrm{Cov}\left(\sum_{i=1}^{n}\boldsymbol{\Lambda}_i\mathbf{Z}\left(\boldsymbol{x}_i\right)-\mathbf{Z}\left(\boldsymbol{x}_0\right),\sum_{j=1}^{n}\boldsymbol{\Lambda}_j\mathbf{Z}\left(\boldsymbol{x}_j\right)-\mathbf{Z}\left(\boldsymbol{x}_0\right)\right)\right\}$$

and by the restriction $\sum\limits_{i=1}^{n}\boldsymbol{\Lambda}_i=\mathbf{I}$ in the weights in Theorem 13.1,

$$\sigma^2\left(\boldsymbol{x}_0\right)=\mathrm{Tr}\left\{\mathrm{Cov}\left(\sum_{i=1}^{n}\boldsymbol{\Lambda}_i\big(\mathbf{Z}\left(\boldsymbol{x}_i\right)-\mathbf{Z}\left(\boldsymbol{x}_0\right)\big),\sum_{j=1}^{n}\boldsymbol{\Lambda}_j\big(\mathbf{Z}\left(\boldsymbol{x}_j\right)-\mathbf{Z}\left(\boldsymbol{x}_0\right)\big)\right)\right\},$$

which by Lemma 13.2 is equivalent to

$$\sigma^2\left(\boldsymbol{x}_0\right)=\mathrm{Tr}\left\{\sum_{i=1}^{n}\sum_{j=1}^{n}\boldsymbol{\Lambda}_i\mathrm{Cov}\big(\mathbf{Z}\left(\boldsymbol{x}_i\right)-\mathbf{Z}\left(\boldsymbol{x}_0\right),\mathbf{Z}\left(\boldsymbol{x}_j\right)-\mathbf{Z}\left(\boldsymbol{x}_0\right)\big)\boldsymbol{\Lambda}_j'\right\}.$$

By Lemma 13.4,

$$\sigma^2\left(\boldsymbol{x}_0\right)=\mathrm{Tr}\Bigg\{\sum_{i=1}^{n}\sum_{j=1}^{n}\boldsymbol{\Lambda}_i\Big[\mathrm{Cov}\big(\mathbf{Z}\left(\boldsymbol{x}_0\right),\mathbf{Z}\left(\boldsymbol{x}_0\right)\big)-\mathrm{Cov}\big(\mathbf{Z}\left(\boldsymbol{x}_i\right),\mathbf{Z}\left(\boldsymbol{x}_0\right)\big)$$

$$- \operatorname{Cov}'\big(\mathbf{Z}(\boldsymbol{x}_j), \mathbf{Z}(\boldsymbol{x}_0)\big) + \operatorname{Cov}\big((\boldsymbol{x}_i), \mathbf{Z}(\boldsymbol{x}_j)\big)\Big]\boldsymbol{\Lambda}_j'\Bigg\}.$$

Expanding, because of the stationarity Assumption 13.2,

$$\sigma^2(\boldsymbol{x}_0) = \operatorname{Tr}\left\{\sum_{i=1}^{n}\boldsymbol{\Lambda}_i\sum_{j=1}^{n}\boldsymbol{\Lambda}_j'\mathbf{Cov}(0) - \sum_{i=1}^{n}\boldsymbol{\Lambda}_i\mathbf{Cov}(\boldsymbol{x}_0 - \boldsymbol{x}_i)\sum_{j=1}^{n}\boldsymbol{\Lambda}_j' \right.$$
$$\left. - \sum_{i=1}^{n}\boldsymbol{\Lambda}_i\sum_{j=1}^{n}\mathbf{Cov}(\boldsymbol{x}_0 - \boldsymbol{x}_j)\boldsymbol{\Lambda}_j' + \sum_{i=1}^{n}\sum_{j=1}^{n}\boldsymbol{\Lambda}_i\mathbf{Cov}(\boldsymbol{x}_j - \boldsymbol{x}_i)\boldsymbol{\Lambda}_j'\right\}. \quad \square$$

The unbiasedness constraint in the weights $\sum_{i=1}^{n}\boldsymbol{\Lambda}_i = \mathbf{I}$ in Theorem 13.1 proves the theorem. □

Optimization

The Lagrange method of multipliers (Hillier and Lieberman, 1995, p. 961) is the most convenient way to solve the constrained optimization problem.

Definition 13.8

The ordinary cokriging Lagrangian function is:

$$L(\boldsymbol{\Lambda}_1, \boldsymbol{\Lambda}_2, \ldots, \boldsymbol{\Lambda}_n; \boldsymbol{\mu}) = \sigma^2(\boldsymbol{x}_0) + 2\boldsymbol{\mu}\left(\sum_{k=1}^{n}\boldsymbol{\Lambda}_k - \mathbf{I}\right),$$

where $\sigma^2(\boldsymbol{x}_0)$ is the error variance in Theorem 13.2 and $\boldsymbol{\mu}$ is a square matrix of Lagrange multipliers of the same order p as the number of attributes in the coregionalization. □

From Theorem 13.2, $\sigma^2(\boldsymbol{x}_0)$ is quadratic in the weights to be optimized. Hence, in order to have

(a) a global optimum,

(b) a unique solution,

(c) a minimum for the objective function, and

(d) a nonnegative error variance,

a necessary condition is that $\mathbf{Cov}(\boldsymbol{h})$ be positive definite.

Theorem 13.3

Let $\mathbf{\Lambda}_k$ be the weights in Definition 13.2 and let $\mathbf{Cov}(\cdot)$ be the matrix in Assumption 13.2. The optimal weights are the solution to

$$\left.\begin{array}{l}
\sum\limits_{j=1}^{n} \mathbf{\Lambda}_1' \mathbf{Cov}(\boldsymbol{x}_1 - \boldsymbol{x}_j) + \boldsymbol{\mu} = \mathbf{Cov}(\boldsymbol{x}_1 - \boldsymbol{x}_0) \\
\sum\limits_{j=1}^{n} \mathbf{\Lambda}_2' \mathbf{Cov}(\boldsymbol{x}_2 - \boldsymbol{x}_j) + \boldsymbol{\mu} = \mathbf{Cov}(\boldsymbol{x}_2 - \boldsymbol{x}_0) \\
\cdots\cdots\cdots\cdots\cdots\cdots\cdots\cdots\cdots \\
\sum\limits_{j=1}^{n} \mathbf{\Lambda}_n' \mathbf{Cov}(\boldsymbol{x}_n - \boldsymbol{x}_j) + \boldsymbol{\mu} = \mathbf{Cov}(\boldsymbol{x}_n - \boldsymbol{x}_0) \\
\sum\limits_{i=1}^{n} \mathbf{\Lambda}_i = \mathbf{I}
\end{array}\right\}.$$

Proof

The optimal weights $\mathbf{\Lambda}$ are those that make the first derivatives of the Lagrangian function zero with respect to the unknown $\mathbf{\Lambda}_i$ and $\boldsymbol{\mu}$. From Definition 13.8 and the rules on partial derivatives that apply to matrices (Rogers, 1980, chapter 7):

$$\frac{\partial L(\mathbf{\Lambda}_1, \mathbf{\Lambda}_2, \ldots, \mathbf{\Lambda}_n; \boldsymbol{\mu})}{\partial \mathbf{\Lambda}_i} = \sum_{j=1}^{n} \mathbf{\Lambda}_j \mathbf{Cov}(\boldsymbol{x}_i - \boldsymbol{x}_j) + \sum_{j=1}^{n} \{\mathbf{Cov}(\boldsymbol{x}_j - \boldsymbol{x}_i) \mathbf{\Lambda}_j'\}'$$
$$- \mathbf{Cov}(\boldsymbol{x}_0 - \boldsymbol{x}_i) - \mathbf{Cov}(\boldsymbol{x}_0 - \boldsymbol{x}_i) + 2\boldsymbol{\mu}, \text{ for } i = 1, 2, \ldots, n$$

$$\frac{\partial L(\mathbf{\Lambda}_1, \mathbf{\Lambda}_2, \ldots, \mathbf{\Lambda}_n; \boldsymbol{\mu})}{\partial \boldsymbol{\mu}} = \sum_{i=1}^{n} \mathbf{\Lambda}_i - \mathbf{I},$$

but by Corollary 13.1,

$$\frac{\partial L(\mathbf{\Lambda}_1, \mathbf{\Lambda}_2, \ldots, \mathbf{\Lambda}_n; \boldsymbol{\mu})}{\partial \mathbf{\Lambda}_i} = 2\sum_{j=1}^{n} \mathbf{\Lambda}_j \mathbf{Cov}(\boldsymbol{x}_i - \boldsymbol{x}_j)$$
$$- 2\mathbf{Cov}(\boldsymbol{x}_i - \boldsymbol{x}_0) + 2\boldsymbol{\mu}, \text{ for } i = 1, 2, \ldots, n$$

$$\frac{\partial L(\mathbf{\Lambda}_1, \mathbf{\Lambda}_2, \ldots, \mathbf{\Lambda}_n; \boldsymbol{\mu})}{\partial \boldsymbol{\mu}} = \sum_{i=1}^{n} \mathbf{\Lambda}_i - \mathbf{I}.$$

The proof follows by making all derivatives equal to zero, and in the first n equations dividing both sides by 2 and adding $\mathbf{Cov}(\boldsymbol{x}_0 - \boldsymbol{x}_i)$. □

Surprisingly, this system of normal equations has the same form as the normal equations for ordinary kriging in Theorem 4.4, except that not only the covariance matrices are matrices, but all the elements in the matrices are also matrices.

Minimum Mean Square Error

The minimum mean square error is the value of the objective function for the optimal weights.

Lemma 13.5

Let $\mathbf{\Lambda}_i$ be the weights that are the solution to the system of equations in Theorem 13.3 and let $\mathbf{Cov}(\cdot)$ be the matrix in Assumption 13.2. Then

$$\sum_{i=1}^{n}\sum_{j=1}^{n}\mathbf{\Lambda}_i\mathbf{Cov}(\boldsymbol{x}_j-\boldsymbol{x}_i)\mathbf{\Lambda}_j' = \sum_{i=1}^{n}\mathbf{Cov}(\boldsymbol{x}_i-\boldsymbol{x}_0)\,\mathbf{\Lambda}_i' - \boldsymbol{\mu}\,.$$

Proof

From Theorem 13.3

$$\sum_{j=1}^{n}\mathbf{\Lambda}_j\mathbf{Cov}(\boldsymbol{x}_i-\boldsymbol{x}_j) = \mathbf{Cov}(\boldsymbol{x}_i-\boldsymbol{x}_0) - \boldsymbol{\mu},\ \text{for } i=1,\,2,\,\ldots,\,n\,.$$

Multiplying the ith equation by $\mathbf{\Lambda}_i'$ and adding them all together:

$$\sum_{i=1}^{n}\sum_{j=1}^{n}\mathbf{\Lambda}_i\mathbf{Cov}(\boldsymbol{x}_j-\boldsymbol{x}_i)\mathbf{\Lambda}_j' = \sum_{i=1}^{n}\mathbf{Cov}(\boldsymbol{x}_i-\boldsymbol{x}_0)\,\mathbf{\Lambda}_i' - \boldsymbol{\mu}\sum_{i=1}^{n}\mathbf{\Lambda}_i'\,.$$

The proof follows because $\sum_{i=1}^{n}\mathbf{\Lambda}_i' = \left[\sum_{i=1}^{n}\mathbf{\Lambda}_i\right]'$ and, by the last equation in Theorem 13.3, $\left[\sum_{i=1}^{n}\mathbf{\Lambda}_i\right]' = \mathbf{I}$. □

Theorem 13.4

Let $\mathbf{\Lambda}_i$ be the weights that are the solution to the system of equations in Theorem 13.3, let $\mathbf{Cov}(\cdot)$ be the matrix in Assumption 13.2, and let $\boldsymbol{\mu}$ be the Lagrange multiplier in Definition 13.8. Then the minimum mean square error is

$$\sigma^2(\boldsymbol{x}_0) = \mathrm{Tr}\left(\mathbf{Cov}(0) - \sum_{i=1}^{n}\mathbf{\Lambda}_i\mathbf{Cov}(\boldsymbol{x}_0-\boldsymbol{x}_i) - \boldsymbol{\mu}\right).$$

Proof

The value of the error variance for any weights is given by Theorem 13.2:

$$\sigma^2(\boldsymbol{x}_0) = \mathrm{Tr}\left\{\sum_{i=1}^{n}\mathbf{\Lambda}_i\sum_{j=1}^{n}\mathbf{\Lambda}_j'\mathbf{Cov}(0) - \sum_{i=1}^{n}\mathbf{\Lambda}_i\mathbf{Cov}(\boldsymbol{x}_0-\boldsymbol{x}_i)\sum_{j=1}^{n}\mathbf{\Lambda}_j' \right.$$
$$\left. - \sum_{i=1}^{n}\mathbf{\Lambda}_i\sum_{j=1}^{n}\mathbf{Cov}(\boldsymbol{x}_0-\boldsymbol{x}_j)\,\mathbf{\Lambda}_j' + \sum_{i=1}^{n}\sum_{j=1}^{n}\mathbf{\Lambda}_i\mathbf{Cov}(\boldsymbol{x}_j-\boldsymbol{x}_i)\,\mathbf{\Lambda}_j'\right\}',$$

but by Lemma 13.5,

$$\sigma^2(\boldsymbol{x}_0) = \mathrm{Tr}\left\{\mathbf{Cov}(0) - \sum_{i=1}^{n} \boldsymbol{\Lambda}_i \mathbf{Cov}(\boldsymbol{x}_0 - \boldsymbol{x}_i) \right.$$
$$\left. - \sum_{i=1}^{n} \mathbf{Cov}(\boldsymbol{x}_0 - \boldsymbol{x}_i)\boldsymbol{\Lambda}_i' + \sum_{i=1}^{n} \mathbf{Cov}(\boldsymbol{x}_i - \boldsymbol{x}_0)\boldsymbol{\Lambda}_i' - \boldsymbol{\mu}\right\}.$$

By Corollary 13.1, cancellation of the third and fourth terms on the right-hand side of the equation proves the theorem. □

Algorithm

Definition 13.9

Let $\mathbf{Cov}(\cdot)$ be the matrix in Assumption 13.2, let $\mathbf{I}$ be an identity matrix of order p, and let $\mathbf{0}$ be a null matrix of order p. Then

$$\mathbf{K} = \begin{bmatrix} \mathbf{Cov}(\boldsymbol{x}_1 - \boldsymbol{x}_1) & \mathbf{Cov}(\boldsymbol{x}_1 - \boldsymbol{x}_2) & \cdots & \mathbf{Cov}(\boldsymbol{x}_1 - \boldsymbol{x}_n) & \mathbf{I} \\ \mathbf{Cov}(\boldsymbol{x}_2 - \boldsymbol{x}_1) & \mathbf{Cov}(\boldsymbol{x}_2 - \boldsymbol{x}_2) & \cdots & \mathbf{Cov}(\boldsymbol{x}_2 - \boldsymbol{x}_n) & \mathbf{I} \\ \cdots & \cdots & \cdots & \cdots & \mathbf{I} \\ \mathbf{Cov}(\boldsymbol{x}_n - \boldsymbol{x}_1) & \mathbf{Cov}(\boldsymbol{x}_n - \boldsymbol{x}_2) & \cdots & \mathbf{Cov}(\boldsymbol{x}_n - \boldsymbol{x}_n) & \mathbf{I} \\ \mathbf{I} & \mathbf{I} & \cdots & \mathbf{I} & \mathbf{0} \end{bmatrix}. \quad \square$$

Definition 13.10

Let $\mathbf{Cov}(\cdot)$ be the matrix in Assumption 13.2 and let $\mathbf{I}$ be an identity matrix of order p. Then

$$\mathbf{k} = \begin{bmatrix} \mathbf{Cov}(\boldsymbol{x}_1 - \boldsymbol{x}_0) & \mathbf{Cov}(\boldsymbol{x}_2 - \boldsymbol{x}_0) & \cdots & \mathbf{Cov}(\boldsymbol{x}_n - \boldsymbol{x}_0) & \mathbf{I} \end{bmatrix}'. \quad \square$$

Definition 13.11

Let $\boldsymbol{\Lambda}_k$ be the matrix of weights in Definition 13.2 and let $\boldsymbol{\mu}$ be the matrix of Lagrange multipliers in Definition 13.8. Then

$$\mathbf{W} = \begin{bmatrix} \boldsymbol{\Lambda}_1' & \boldsymbol{\Lambda}_2' & \cdots & \boldsymbol{\Lambda}_n' & \boldsymbol{\mu} \end{bmatrix}'. \quad \square$$

Algorithm 13.1

This is an algorithm for ordinary cokriging under Assumptions 13.1, 13.2, and 13.3.

1. Calculate matrix $\mathbf{K}$ in Definition 13.9.

2. Calculate matrix $\mathbf{k}$ in Definition 13.10.
3. Solve the system of equations

$$\mathbf{KW} = \mathbf{k},$$

where $\mathbf{W}$ is the matrix in Definition 13.11.

4. Compute the estimate

$$\hat{\mathbf{Z}}(\boldsymbol{x}_0) = \sum_{i=1}^{n} \boldsymbol{\Lambda}_i \mathbf{Z}(\boldsymbol{x}_i),$$

where $\boldsymbol{\Lambda}_k$ are the weights given by $\mathbf{W}$.

5. Calculate the estimation variance by

$$\boldsymbol{\sigma}^2(\boldsymbol{x}_0) = \text{diagonal}\left(\mathbf{Cov}(0) - \mathbf{k}'\mathbf{W}\right). \quad \square$$

Although $\mathbf{Cov}(\cdot)$ is nonsymmetric, the coefficient matrix $\mathbf{K}$ is symmetric. Also, except for the null matrix in the lower right-hand corner, all other entries can be inverted, so the coefficient matrix can be reduced to lower or upper triangular form by operating on these matrices, thus simplifying the computations (Myers, 1982).

By eliminating columns in matrix $\mathbf{k}$, cokriging can be used to estimate a selected number of regionalized variables, even as few as one. Estimation of only one attribute under these conditions is different from estimation of the same attribute by kriging, because although the final result is an estimate for a single attribute, the estimation uses information from all the attributes.

The data vector may also be modified. Clearly, changing this vector requires that corresponding modifications be made to other matrices. In the limiting case, the cokriging estimation of one regionalized variable using data only from that regionalized variable is the same as ordinary kriging.

Structural Analysis

The structural analysis required by cokriging is much more demanding than that of kriging because:

i) Besides the p covariances required by ordinary kriging, the user must model additional $p(p-1)$ cross-covariances, an increase in the number of models by a factor of $(p-1)$. And, from the optimization,
ii) $\mathbf{Cov}(\cdot)$ must be positive definite. Individual selection of permissible models for all covariances and cross-covariances is not sufficient to produce a positive definite $\mathbf{Cov}(\cdot)$.

A common way to conduct structural analysis today is through the use of the so-called linear coregionalization model (Goovaerts, 1997). The modeling is customarily done in terms of semivariograms and cross-semivariograms and is then converted to covariances and cross-covariances which assure a positive definite $\mathbf{Cov}(\cdot)$. Less traveled paths not covered here include spectral methods based on Bochner's theorem (Christakos, 1992, p. 64–69, 184; Wackernagel, 1995, Chapter 21; Yao and Journel, 1998).

The kernel of linear coregionalization modeling is the testing of the positive semidefinite condition of smaller coefficient matrices $\mathbf{B}_l$ of order p.

Definition 13.12

Let $\mathbf{v}$ be any vector. Then matrix $\mathbf{B}_l$ is positive semidefinite (Golub and Van Loan, 1989, p. 146) if

$$\mathbf{v}'\mathbf{B}_l\mathbf{v} \geq 0 . \quad \square$$

If a matrix is positive semidefinite, its determinant, all of its principal minor determinants, and all of its eigenvalues are nonnegative.

Definition 13.13

Let $Z_i(\boldsymbol{x})$ and $Z_j(\boldsymbol{x})$ be two random functions in a second order stationary coregionalization. Then the cross-semivariogram is

$$\gamma_{ij}(\boldsymbol{h}) = \frac{1}{2}\mathrm{E}\Big[\{Z_i(\boldsymbol{x}) - Z_i(\boldsymbol{x}+\boldsymbol{h})\}\{Z_j(\boldsymbol{x}) - Z_j(\boldsymbol{x}+\boldsymbol{h})\}\Big] . \quad \square$$

Definition 13.14

Let the $\mathbf{B}_l$'s be positive semidefinite coefficient matrices of order p with nonnegative elements and let the $\gamma_l(\boldsymbol{h})$'s be permissible semivariogram models—such as those in Definitions 5.2 to 5.9—with unit coefficients a_i and α_i. A linear coregionalization model is the matrix

$$\boldsymbol{\Gamma}(\boldsymbol{h}) = \sum_{l=1}^{L} \mathbf{B}_l \gamma_l(\boldsymbol{h})$$

whose diagonal elements $\gamma_{ii}(\boldsymbol{h})$ are the semivariogram models; the remaining terms $\gamma_{ij}(\boldsymbol{h})$ are the cross-semivariogram models. $\square$

An example linear coregionalization model for the case of two attributes is

$$\boldsymbol{\Gamma}(\boldsymbol{h}) = \begin{bmatrix} 2 & 1 \\ 1 & 3 \end{bmatrix} + \begin{bmatrix} 4 & 2 \\ 2 & 3 \end{bmatrix}\left(1 - e^{-\frac{3h}{200}}\right) + \begin{bmatrix} 1 & 1 \\ 1 & 2 \end{bmatrix}\left(\frac{3}{2}\frac{h}{150} - \frac{1}{2}\left(\frac{h}{150}\right)^3\right).$$

Practical application of the method necessarily introduces further approximations and restrictions in addition to those made by the individually permissible models. Most significantly, a free selection of the type of permissible models and their parameters becomes difficult. The second semivariogram model parameter, the sill in the case of the transitive models, goes into the factor $\gamma_l(\boldsymbol{h})$.

Commonly a different type of model or nested models fit different experimental semivariograms, as well as different ranges. It is hard, if not impossible, to perform a linear coregionalization without minimizing diversity in terms of models and sills, and the reduction is based entirely on the user's judgment.

The first semivariogram model parameter, the sill in the case of the transitive models, goes into the coefficient matrices $\mathbf{B}_l$. Collectively, all sills in matrix $\mathbf{B}_l$ may not generate a positive semidefinite matrix, which leads to a further compromise that should both simultaneously satisfy the requirements of linear coregionalization modeling and minimize deviations from the initial individual fitting of the semivariogram models. The following theorem plays a key role in preparing positive semidefinite coefficient matrices.

Theorem 13.5

Let $\mathbf{B}_l$ be a symmetric positive semidefinite matrix. Then

$$\left|b_{ij}^l\right| \leq \sqrt{b_{ii}^l b_{jj}^l}, \text{ for } i \neq j\,.$$

Proof

By Definition 13.12, if $\mathbf{B}_l$ is a positive semidefinite matrix, then $\mathbf{v}'\mathbf{B}_l\mathbf{v} \geq 0$ for any vector $\mathbf{v}$. Suppose without losing generality that $i = 1$ and $j = 2$ and consider the inequality

$$0 \leq \begin{bmatrix} v \\ 1 \end{bmatrix}' \begin{bmatrix} b_{11}^l & b_{12}^l \\ b_{12}^l & b_{22}^l \end{bmatrix} \begin{bmatrix} v \\ 1 \end{bmatrix} = b_{11}^l v^2 + b_{12}^l v + b_{22}^l\,,$$

which is true because

$$\begin{bmatrix} b_{11}^l & b_{12}^l \\ b_{12}^l & b_{22}^l \end{bmatrix}$$

is also positive semidefinite (Golub and Van Loan,1989, p.146). In this quadratic equation in v, the determinant $4\left(b_{12}^l\right)^2 - 4b_{11}^l b_{22}^l$ must be negative for the inequality to hold. □

Corollary 13.2

Let $\mathbf{B}_l$ be a real, symmetric matrix of order 2. If $\left|b_{12}^l\right| \leq \sqrt{b_{11}^l b_{22}^l}$, then the matrix is positive semidefinite.

Proof

The submatrix in the proof of Theorem 13.5 is now equal to the entire matrix. Because $\left|b_{12}^l\right| \leq \sqrt{b_{11}^l b_{22}^l}$, $4\left(b_{12}^l\right)^2 - 4b_{11}^l b_{22}^l$ is negative and the inequality holds. □

Except for the special case of symmetric positive semidefinite matrices of order 2, the converse of Theorem 13.5 is not true. If $\left|b_{ij}^l\right| \leq \sqrt{b_{ii}^l b_{jj}^l}$, for $i \neq j$ holds, it does not imply that the matrix is positive semidefinite.

Theorem 13.6

If $\left|b_{ij}^l\right| \leq \sqrt{b_{ii}^l b_{jj}^l}$, for $i \neq j$, then the matrix is not necessarily positive semidefinite.

Proof

Consider the matrix

$$\begin{bmatrix} 124 & & \\ 195 & 310 & \\ 158 & 212 & 229 \end{bmatrix}.$$

All principal minor determinants of order 2 are positive:

$$b_{11}^l b_{22}^l - \left(b_{12}^l\right)^2 = 415 \qquad b_{11}^l b_{33}^l - \left(b_{13}^l\right)^2 = 3432 \qquad b_{22}^l b_{33}^l - \left(b_{23}^l\right)^2 = 26046\,,$$

yet the eigenvalues are −4.6465, 53.7504, and 613.8993. □

In general, however, if all principal minor determinants of order 2 are nonnegative, the matrix is indeed positive semidefinite. Most of the exceptions occur when at least one of the determinants is small and at least one of the other determinants is several orders of magnitude larger, such as in the counter example above. As inconclusive as it might be, Theorem 13.5 is the best guidance available for the selection of nested models and the diagnosis of incompatible coefficients in the linear coregionalization model. Hints that follow from Theorem 13.5 are:

Hint 13.1

If $b_{ij}^l > 0$, then $b_{ii}^l > 0$ and $b_{jj}^l > 0$. □

A structure making a contribution to the nested model of cross-semivariogram $\gamma_{ij}(\boldsymbol{h})$ must contribute also to the modeling of both semivariograms $\gamma_{ii}(\boldsymbol{h})$ and $\gamma_{jj}(\boldsymbol{h})$. The converse of this first hint is not true.

Hint 13.2

$b_{ii}^l > 0$ and $b_{jj}^l > 0$ do not have any bearing on b_{ij}^l. □

If a structure makes a contribution to semivariograms $\gamma_{ii}(\boldsymbol{h})$ and $\gamma_{jj}(\boldsymbol{h})$, such a structure may or may not contribute to the modeling of $\gamma_{ij}(\boldsymbol{h})$.

Hint 13.3

If $b_{ii}^l = 0$, then $b_{ij}^l = 0$, for $j = 1, 2, \ldots, p$. □

If a structure does not contribute to the modeling of semivariogram $\gamma_{ii}(\boldsymbol{h})$, then it cannot contribute to any of the cross-semivariograms $\gamma_{ij}(\boldsymbol{h})$ involving the ith attribute.

The final step in the modeling is to check the positive semidefiniteness of all coefficient matrices $\mathbf{B}_l$. After making $\left|b_{ij}^l\right| \leq \sqrt{b_{ii}^l b_{jj}^l}$, for all l and $i \neq j$, all the coefficient matrices will automatically be positive semidefinite, or close, to comply with the requirements. Minor adjustments of the coefficients related to the smallest determinants are all that is required. In this case, when fitting semivariogram models, it works to the user's advantage that different suites of parameters provide comparable results. One can try the adjustments by hand or with the help of a computer program such as that developed by Goulard and Voltz (1992).

Algorithm 13.1, however, requires covariances and cross-covariances instead of semivariograms of cross-semivariograms. The following results allow the execution of all required transformations.

Theorem 13.7

Let $Z_i(\boldsymbol{x})$ and $Z_j(\boldsymbol{x})$ be two random functions in a second order stationary coregionalization. Then

$$\gamma_{ij}(\boldsymbol{h}) = \mathrm{Cov}_{ij}(0) - \frac{1}{2}\mathrm{Cov}_{ij}(\boldsymbol{h}) - \frac{1}{2}\mathrm{Cov}_{ji}(\boldsymbol{h}) \,.$$

Proof

From Definition 13.13,

$$\gamma_{ij}(\boldsymbol{h}) = \frac{1}{2}\mathrm{E}\big[\{Z_i(\boldsymbol{x}) - Z_i(\boldsymbol{x}+\boldsymbol{h})\}\{Z_j(\boldsymbol{x}) - Z_j(\boldsymbol{x}+\boldsymbol{h})\}\big]$$

and expanding,

$$\begin{aligned}\gamma_{ij}(\boldsymbol{h}) = {} & \frac{1}{2}\mathrm{E}\big[\{Z_i(\boldsymbol{x})\,Z_j(\boldsymbol{x})\}\big] - \frac{1}{2}\mathrm{E}\big[\{Z_i(\boldsymbol{x}+\boldsymbol{h})\,Z_j(\boldsymbol{x})\}\big] \\ & - \frac{1}{2}\mathrm{E}\big[\{Z_i(\boldsymbol{x})\,Z_j(\boldsymbol{x}+\boldsymbol{h})\}\big] + \frac{1}{2}\mathrm{E}\big[\{Z_i(\boldsymbol{x}+\boldsymbol{h})\,Z_j(\boldsymbol{x}+\boldsymbol{h})\}\big] .\end{aligned}$$

The expression does not change by adding and subtracting the product of the means:

$$\begin{aligned}\gamma_{ij}(\boldsymbol{h}) = {} & \frac{1}{2}\mathrm{E}\big[\{Z_i(\boldsymbol{x})\,Z_j(\boldsymbol{x})\}\big] - m_i m_j - \frac{1}{2}\mathrm{E}\big[\{Z_i(\boldsymbol{x}+\boldsymbol{h})\,Z_j(\boldsymbol{x})\}\big] + m_i m_j \\ & - \frac{1}{2}\mathrm{E}\big[\{Z_i(\boldsymbol{x})\,Z_j(\boldsymbol{x}+\boldsymbol{h})\}\big] + m_i m_j \\ & \qquad + \frac{1}{2}\mathrm{E}\big[\{Z_i(\boldsymbol{x}+\boldsymbol{h})\,Z_j(\boldsymbol{x}+\boldsymbol{h})\}\big] - m_i m_j ,\end{aligned}$$

which according to Assumption 13.2 is

$$\begin{aligned}\gamma_{ij}(\boldsymbol{h}) = {} & \frac{1}{2}\mathrm{Cov}\big(Z_i(\boldsymbol{x}), Z_j(\boldsymbol{x})\big) - \frac{1}{2}\mathrm{Cov}\big(Z_j(\boldsymbol{x}+\boldsymbol{h}), Z_i(\boldsymbol{x})\big) \\ & - \frac{1}{2}\mathrm{Cov}\big(Z_i(\boldsymbol{x}), Z_j(\boldsymbol{x}+\boldsymbol{h})\big) + \frac{1}{2}\mathrm{Cov}\big(Z_i(\boldsymbol{x}+\boldsymbol{h}), Z_j(\boldsymbol{x}+\boldsymbol{h})\big) ,\end{aligned}$$

and the invariance under translation proves the theorem. □

Corollary 13.3

Let $\gamma_{ij}(\boldsymbol{h})$ be a cross-semivariogram with sill C and let $\mathrm{Cov}_{ij}(\boldsymbol{h})$ be a symmetric cross-covariance. Then

$$\gamma_{ij}(\boldsymbol{h}) = C - \mathrm{Cov}_{ij}(\boldsymbol{h}) \; .$$

Proof

The proof follows directly from Theorem 13.7 by considering that because the semivariogram is transitive, $\mathrm{Cov}_{ij}(0) = C$ and because of the symmetry of the cross-covariance, $\mathrm{Cov}_{ij}(\boldsymbol{h}) = \mathrm{Cov}_{ji}(\boldsymbol{h})$. □

Algorithm 13.2

This is a procedure for the linear coregionalization modeling of covariances and cross-covariances of p attributes.

1. Follow the steps in Algorithms 5.1 or 5.2 to model each semivariogram and cross-semivariogram individually.

2. Following Hints 13.1–13.3, look for tendencies in models and parameters to try to minimize the number of structures to consider in the modeling and still have proper fitting.

3. Check all principal minor determinants of order 2 to make sure that they are all nonnegative.

4. Verify that all coefficient matrices are positive semidefinite. If some are not, make changes to increase the smallest determinants or reduce the largest ones until the condition is achieved. For matrices resistant to change, the quickest solution in the long run may be to go back to Step 2.

5. Use Corollary 13.3 to covert all of the semivariograms and cross–semivariograms to covariances and cross-covariances. □

EXERCISE 13.1

Use the cumulative sandstone thickness coregionalization in Appendix D to:

1. Model all semivariograms and cross-semivariograms, employing the linear coregionalization model.

2. Use the data from the closest four wells and the results from the linear coregionalization model to predict the outcome of drilling a well at (1588000, 498000) according to cokriging and ordinary kriging.

3. Remove the measurement for the upper "J" sandstone at well 253 and repeat the calculations.

Solution

1. Figures 13.1 and 13.2 and Table 13.1 give the results for the modeling of all semivariograms and cross-semivariograms after completing the first step of Algorithm 13.2.

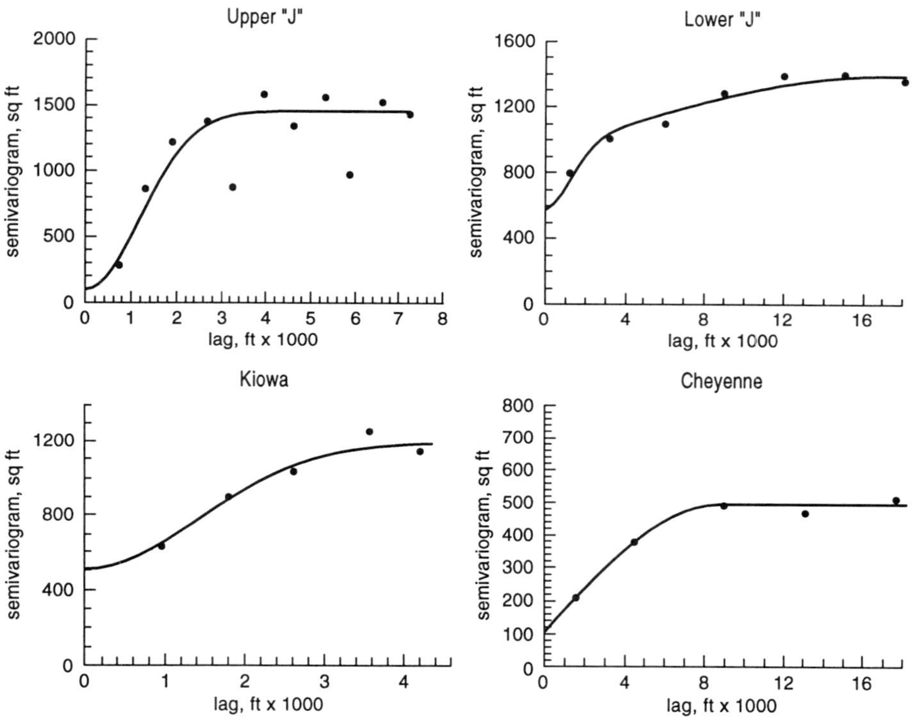

Figure 13.1. Experimental semivariograms and first pass semivariogram models for the variables in Exercise 13.1.

Table 13.1. Semivariograms and cross-semivariograms of a first pass modeling.

Attributes	Nugget	Model 1	Model 2
Upper "J"	105.1	Gau (1350, 2924)	
Lower "J"	580.0	Gau (345, 3200)	Sph (468.2, 17252)
Kiowa	510.4	Gau (680.9, 3476)	
Cheyenne	107.1	Sph (387, 4425)	
Upper "J", lower "J"	122.8	Gau (680, 3050)	
Upper "J", Kiowa	239.4	Gau (593.9, 4574)	
Upper "J", Cheyenne	43.6	Gau (244.2, 3231)	
Lower "J", Kiowa	242.1	Gau (347.1, 3646)	
Lower "J", Cheyenne	45.9	Gau (118, 4101)	
Kiowa, Cheyenne	80.0	Exp (172, 4528)	

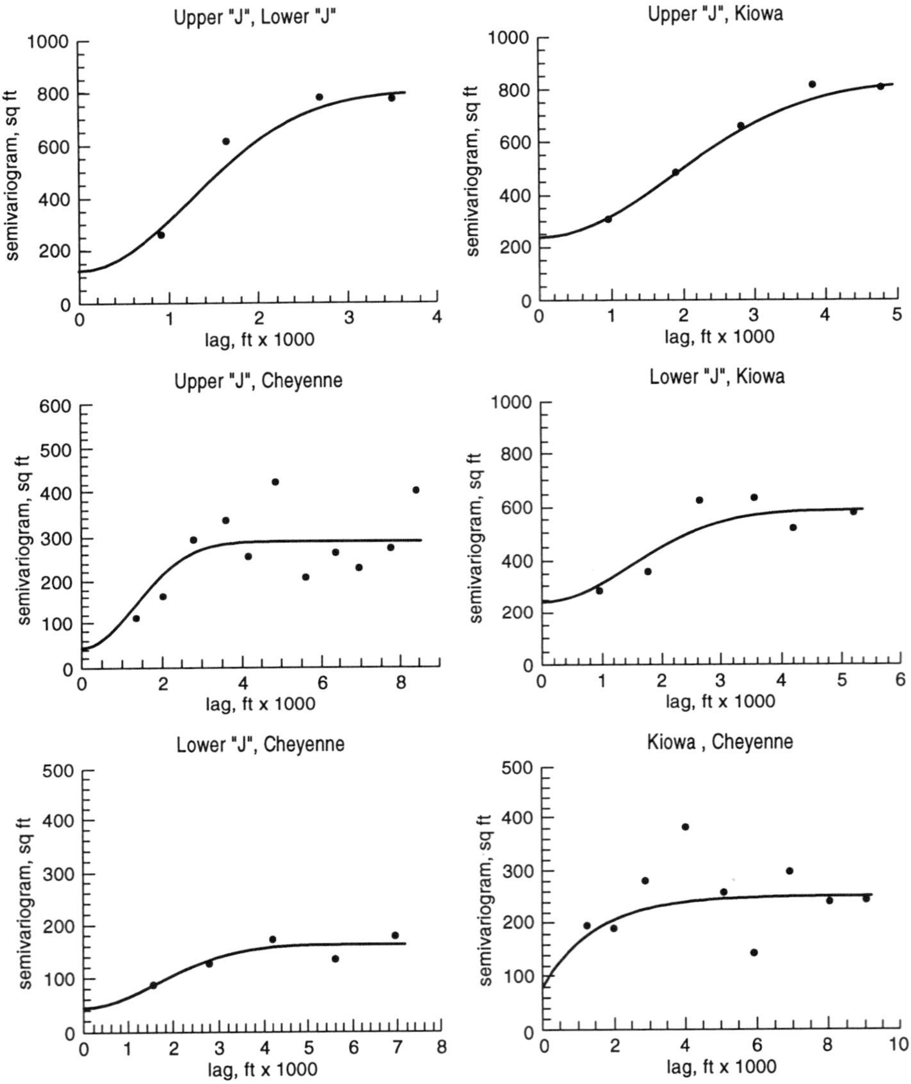

Figure 13.2. Experimental cross-semivariograms and first pass cross–semivariogram models for the variables in Exercise 13.1.

Inspecting the results, it is clear that the dominant structure is of Gaussian type; if minimizing the number of structures is desirable, a range of 3,200 ft seems like a reasonable compromise. The linear coregionalization model that results after retaining the Sph (468.2, 17252) structure and forcing all other structures to be Gaussian with a range of 3,200 ft is

$$\boldsymbol{\Gamma}(\boldsymbol{h}) = \begin{bmatrix} 105.1 & & & \\ 98.2 & 580.0 & & \\ 156.5 & 186.1 & 452.0 & \\ 0.0 & 9.3 & 177.6 & 172.1 \end{bmatrix}$$

$$+ \begin{bmatrix} 1256.0 & & & \\ 789.7 & 345.0 & & \\ 607.2 & 395.1 & 730.0 & \\ 290.0 & 150.5 & 66.0 & 324.1 \end{bmatrix} \text{Gau}(3200)$$

$$+ \begin{bmatrix} 0 & & & \\ 0 & 468.2 & & \\ 0 & 0 & 0 & \\ 0 & 0 & 0 & 0 \end{bmatrix} \text{Sph}(17252) .$$

The third step in Algorithm 13.2 reveals that the only principal minor determinant of order 2 that is negative is $b_{11}^2 b_{22}^2 - \left(b_{12}^2\right)^2$, which is equal to -190306. Going back to the model fitting of $\gamma_{11}(\boldsymbol{h})$, $\gamma_{12}(\boldsymbol{h})$, and $\gamma_{22}(\boldsymbol{h})$, adjustments compatible both with the fitting and the positive semidefinite condition produce the following final model,

$$\boldsymbol{\Gamma}(\boldsymbol{h}) = \begin{bmatrix} 105.1 & & & \\ 146.3 & 580.0 & & \\ 156.5 & 186.1 & 452.0 & \\ 0.0 & 9.3 & 177.6 & 172.1 \end{bmatrix}$$

$$+ \begin{bmatrix} 1400.0 & & & \\ 650.0 & 345.0 & & \\ 607.2 & 395.1 & 730.0 & \\ 290.0 & 150.5 & 66.0 & 324.1 \end{bmatrix} \text{Gau}(3200)$$

$$+ \begin{bmatrix} 0 & & & \\ 0 & 468.2 & & \\ 0 & 0 & 0 & \\ 0 & 0 & 0 & 0 \end{bmatrix} \text{Sph}(17252) ,$$

whose coefficient matrices are all positive semidefinite according to Table 13.2.

Table 13.2. Eigenvalues of coefficient matrices in Exercise 13.1.

Matrix	Eigenvalues			
$\mathbf{B}_1$	6.1	110.4	393.7	799.0
$\mathbf{B}_2$	10.2	219.2	423.5	2146.2
$\mathbf{B}_3$	0.0	0.0	0.0	0.0

Finally, by Step 5, the modeling in terms of covariances and cross–covariances is:

$$C(\boldsymbol{h}) = \begin{bmatrix} 1505.1 & & & \\ 796.3 & 1393.2 & & \\ 763.7 & 581.2 & 1182.0 & \\ 290.0 & 159.8 & 243.6 & 496.2 \end{bmatrix}$$

$$- \left\{ \begin{bmatrix} 105.1 & & & \\ 146.3 & 580.0 & & \\ 156.5 & 186.1 & 452.0 & \\ 0.0 & 9.3 & 177.6 & 172.1 \end{bmatrix}, \text{ if } |\boldsymbol{h}| > 0 \right\}$$

$$- \begin{bmatrix} 1400.0 & & & \\ 650.0 & 345.0 & & \\ 607.2 & 395.1 & 730.0 & \\ 290.0 & 150.5 & 66.0 & 324.1 \end{bmatrix} \left(1 - e^{-3\left(\frac{h}{3200}\right)^2}\right)$$

$$- \left\{ \frac{1}{2} \begin{bmatrix} 0 & & & \\ 0 & 468.2 & & \\ 0 & 0 & 0 & \\ 0 & 0 & 0 & 0 \end{bmatrix} \left(\left(\frac{3h}{17252}\right) - \left(\frac{h}{17252}\right)^3 \right), \text{ if } |\boldsymbol{h}| < 17252 \right\}.$$

2. The four wells closest to (1588000, 49800) are wells 247, 251, 252, and 253. Using the well label as the subindex instead of the sequential number, the matrix of distances is

$$\begin{bmatrix} |\boldsymbol{x}_0 - \boldsymbol{x}_0| & & & & \\ |\boldsymbol{x}_0 - \boldsymbol{x}_{247}| & |\boldsymbol{x}_{247} - \boldsymbol{x}_{247}| & & & \\ |\boldsymbol{x}_0 - \boldsymbol{x}_{252}| & |\boldsymbol{x}_{247} - \boldsymbol{x}_{252}| & |\boldsymbol{x}_{252} - \boldsymbol{x}_{252}| & & \\ |\boldsymbol{x}_0 - \boldsymbol{x}_{251}| & |\boldsymbol{x}_{247} - \boldsymbol{x}_{251}| & |\boldsymbol{x}_{252} - \boldsymbol{x}_{251}| & |\boldsymbol{x}_{251} - \boldsymbol{x}_{251}| & \\ |\boldsymbol{x}_0 - \boldsymbol{x}_{253}| & |\boldsymbol{x}_{247} - \boldsymbol{x}_{253}| & |\boldsymbol{x}_{252} - \boldsymbol{x}_{253}| & |\boldsymbol{x}_{251} - \boldsymbol{x}_{253}| & |\boldsymbol{x}_{253} - \boldsymbol{x}_{253}| \end{bmatrix}$$

$$= \begin{bmatrix} 0 & & & & \\ 2153 & 0 & & & \\ 2006 & 4055 & 0 & & \\ 2325 & 4303 & 2109 & 0 & \\ 1273 & 2388 & 1978 & 3266 & 0 \end{bmatrix}.$$

From Algorithms 13.1 and 13.2,

$$\begin{aligned} \mathbf{Cov}\,(\boldsymbol{x}_{247} - \boldsymbol{x}_{247}) &= \mathbf{Cov}\,(\boldsymbol{x}_{251} - \boldsymbol{x}_{251}) \\ &= \mathbf{Cov}\,(\boldsymbol{x}_{252} - \boldsymbol{x}_{252}) \\ &= \mathbf{Cov}\,(\boldsymbol{x}_{253} - \boldsymbol{x}_{253}) \\ &= \begin{bmatrix} 1505.1 & & & \\ 796.3 & 1393.2 & & \\ 763.7 & 581.2 & 1182.0 & \\ 290.0 & 159.8 & 243.6 & 496.2 \end{bmatrix} \end{aligned}$$

$$\mathbf{Cov}\left(\boldsymbol{x}_{247}-\boldsymbol{x}_{252}\right)=\mathbf{Cov}\left(\boldsymbol{x}_{252}-\boldsymbol{x}_{247}\right)=\begin{bmatrix} 11.32 & & & \\ 5.2556 & 308.95 & & \\ 4.9097 & 3.1938 & 5.9026 & \\ 2.3448 & 1.2169 & 0.5336 & 2.6206 \end{bmatrix}$$

$$\mathbf{Cov}\left(\boldsymbol{x}_{247}-\boldsymbol{x}_{251}\right)=\mathbf{Cov}\left(\boldsymbol{x}_{251}-\boldsymbol{x}_{247}\right)=\begin{bmatrix} 6.1733 & & & \\ 2.8663 & 298.19 & & \\ 2.6774 & 1.7421 & 3.2190 & \\ 1.2787 & 0.6637 & 0.2909 & 1.4291 \end{bmatrix}$$

$$\mathbf{Cov}\left(\boldsymbol{x}_{247}-\boldsymbol{x}_{253}\right)=\mathbf{Cov}\left(\boldsymbol{x}_{253}-\boldsymbol{x}_{247}\right)=\begin{bmatrix} 263.31 & & & \\ 122.25 & 436.49 & & \\ 114.20 & 74.309 & 137.30 & \\ 54.542 & 28.306 & 12.413 & 60.956 \end{bmatrix}$$

$$\mathbf{Cov}\left(\boldsymbol{x}_{252}-\boldsymbol{x}_{251}\right)=\mathbf{Cov}\left(\boldsymbol{x}_{251}-\boldsymbol{x}_{252}\right)=\begin{bmatrix} 380.30 & & & \\ 176.57 & 476.49 & & \\ 164.94 & 107.33 & 198.30 & \\ 78.777 & 40.882 & 17.929 & 88.040 \end{bmatrix}$$

$$\mathbf{Cov}\left(\boldsymbol{x}_{252}-\boldsymbol{x}_{253}\right)=\mathbf{Cov}\left(\boldsymbol{x}_{253}-\boldsymbol{x}_{252}\right)=\begin{bmatrix} 444.99 & & & \\ 206.60 & 497.69 & & \\ 193.00 & 125.58 & 232.03 & \\ 92.176 & 47.836 & 20.978 & 103.01 \end{bmatrix}$$

$$\mathbf{Cov}\left(\boldsymbol{x}_{251}-\boldsymbol{x}_{253}\right)=\mathbf{Cov}\left(\boldsymbol{x}_{253}-\boldsymbol{x}_{251}\right)=\begin{bmatrix} 61.568 & & & \\ 28.585 & 352.03 & & \\ 26.703 & 17.375 & 32.104 & \\ 12.753 & 6.6187 & 2.9024 & 14.253 \end{bmatrix}$$

$$\mathbf{Cov}\left(\boldsymbol{x}_{247}-\boldsymbol{x}_{0}\right)=\begin{bmatrix} 360.19 & & & \\ 167.23 & 469.79 & & \\ 156.22 & 101.65 & 187.81 & \\ 74.611 & 38.721 & 16.980 & 83.384 \end{bmatrix}$$

$$\mathbf{Cov}\left(\boldsymbol{x}_{252}-\boldsymbol{x}_{0}\right)=\begin{bmatrix} 430.91 & & & \\ 200.07 & 493.12 & & \\ 186.89 & 121.61 & 224.690 & \\ 89.260 & 46.323 & 20.314 & 99.756 \end{bmatrix}$$

$$\mathbf{Cov}\left(\boldsymbol{x}_{251}-\boldsymbol{x}_{0}\right)=\begin{bmatrix} 287.31 & & & \\ 133.39 & 444.93 & & \\ 124.61 & 81.083 & 149.81 & \\ 59.514 & 30.886 & 13.545 & 66.512 \end{bmatrix}$$

$$\mathbf{Cov}\left(\boldsymbol{x}_{253}-\boldsymbol{x}_{0}\right)=\begin{bmatrix} 870.93 & & & \\ 404.36 & 631.10 & & \\ 377.73 & 245.79 & 454.13 & \\ 180.41 & 93.625 & 41.058 & 201.62 \end{bmatrix}.$$

Hence, from $\mathbf{K}^{-1}\mathbf{k}$, which in this case is equal to

$$\begin{bmatrix} \mathbf{Cov}(\boldsymbol{x}_{247},\boldsymbol{x}_{247}) & \mathbf{Cov}(\boldsymbol{x}_{347},\boldsymbol{x}_{252}) & \mathbf{Cov}(\boldsymbol{x}_{247},\boldsymbol{x}_{251}) & \mathbf{Cov}(\boldsymbol{x}_{247},\boldsymbol{x}_{253}) & \mathbf{I} \\ \mathbf{Cov}(\boldsymbol{x}_{252},\boldsymbol{x}_{247}) & \mathbf{Cov}(\boldsymbol{x}_{252},\boldsymbol{x}_{252}) & \mathbf{Cov}(\boldsymbol{x}_{252},\boldsymbol{x}_{251}) & \mathbf{Cov}(\boldsymbol{x}_{252},\boldsymbol{x}_{253}) & \mathbf{I} \\ \mathbf{Cov}(\boldsymbol{x}_{251},\boldsymbol{x}_{247}) & \mathbf{Cov}(\boldsymbol{x}_{251},\boldsymbol{x}_{252}) & \mathbf{Cov}(\boldsymbol{x}_{251},\boldsymbol{x}_{251}) & \mathbf{Cov}(\boldsymbol{x}_{251},\boldsymbol{x}_{253}) & \mathbf{I} \\ \mathbf{Cov}(\boldsymbol{x}_{253},\boldsymbol{x}_{247}) & \mathbf{Cov}(\boldsymbol{x}_{253},\boldsymbol{x}_{252}) & \mathbf{Cov}(\boldsymbol{x}_{253},\boldsymbol{x}_{251}) & \mathbf{Cov}(\boldsymbol{x}_{253},\boldsymbol{x}_{253}) & \mathbf{I} \\ \mathbf{I} & \mathbf{I} & \mathbf{I} & \mathbf{I} & \mathbf{0} \end{bmatrix}^{-1}$$

$$\times \begin{bmatrix} \mathbf{Cov}(\boldsymbol{x}_{247},\boldsymbol{x}_0) \\ \mathbf{Cov}(\boldsymbol{x}_{252},\boldsymbol{x}_0) \\ \mathbf{Cov}(\boldsymbol{x}_{251},\boldsymbol{x}_0) \\ \mathbf{Cov}(\boldsymbol{x}_{253},\boldsymbol{x}_0) \\ \mathbf{I} \end{bmatrix}, \text{ one obtains}$$

$$\boldsymbol{\Lambda}'_{247} = \begin{bmatrix} 0.1625 & -0.0388 & -0.0189 & -0.0161 \\ 0.0117 & 0.2637 & 0.0050 & -0.0031 \\ 0.0147 & -0.0041 & 0.2145 & 0.0191 \\ -0.0109 & -0.0036 & 0.0215 & 0.2045 \end{bmatrix}$$

$$\boldsymbol{\Lambda}'_{252} = \begin{bmatrix} 0.0922 & -0.0537 & -0.0328 & -0.0287 \\ 0.0187 & 0.2354 & 0.0079 & -0.0048 \\ 0.0264 & 0.0005 & 0.1833 & 0.0334 \\ -0.0195 & -0.0081 & 0.0377 & 0.1655 \end{bmatrix}$$

$$\boldsymbol{\Lambda}'_{251} = \begin{bmatrix} 0.1664 & -0.0253 & -0.0119 & -0.0077 \\ 0.0078 & 0.2373 & 0.0021 & -0.041 \\ 0.0044 & -0.0082 & 0.1929 & 0.0130 \\ -0.0014 & -0.0029 & 0.0188 & 0.1892 \end{bmatrix}$$

$$\boldsymbol{\Lambda}'_{253} = \begin{bmatrix} 0.5788 & 0.1179 & 0.0636 & 0.0526 \\ -0.0382 & 0.2636 & -0.0150 & 0.0120 \\ -0.0455 & 0.0119 & 0.4093 & -0.0655 \\ 0.0318 & 0.0146 & -0.0780 & 0.4407 \end{bmatrix}$$

$$\boldsymbol{\mu} = \begin{bmatrix} -48.362 & & & \\ -46.045 & -137.61 & & \\ -47.945 & -51.896 & -122.40 & \\ -4.8390 & -4.5547 & -44.855 & -47.485 \end{bmatrix}.$$

In this instance,

$$Z_{247} = \begin{bmatrix} 11.5 \\ 0.0 \\ 0.0 \\ 44.0 \end{bmatrix}, \quad Z_{251} = \begin{bmatrix} 11.0 \\ 32.5 \\ 39.5 \\ 124.5 \end{bmatrix}, \quad Z_{252} = \begin{bmatrix} 9.0 \\ 7.0 \\ 31.5 \\ 130.5 \end{bmatrix}, \quad Z_{253} = \begin{bmatrix} 53.0 \\ 26.0 \\ 4.5 \\ 133.5 \end{bmatrix}.$$

Thus,

$$\begin{aligned} \hat{\mathbf{Z}}(\boldsymbol{x}_0) &= \boldsymbol{\Lambda}_{247} Z(\boldsymbol{x}_{247}) + \boldsymbol{\Lambda}_{251} Z(\boldsymbol{x}_{251}) + \boldsymbol{\Lambda}_{252} Z(\boldsymbol{x}_{252}) + \boldsymbol{\Lambda}_{253} Z(\boldsymbol{x}_{253}) \\ &= \begin{bmatrix} 36.45 & 21.37 & 15.49 & 116.7 \end{bmatrix}', \end{aligned}$$

with estimation variance

$$\begin{aligned}\boldsymbol{\sigma}^2(\boldsymbol{x}_0) &= \text{diagonal}\big(\mathbf{Cov}(0) - \boldsymbol{\Lambda}'_{247}\mathbf{Cov}(\boldsymbol{x}_{247}-\boldsymbol{x}_0) - \boldsymbol{\Lambda}'_{251}\mathbf{Cov}(\boldsymbol{x}_{251}-\boldsymbol{x}_0)\\ &\qquad -\boldsymbol{\Lambda}'_{252}\mathbf{Cov}(\boldsymbol{x}_{252}-\boldsymbol{x}_0) - \boldsymbol{\Lambda}'_{253}\mathbf{Cov}(\boldsymbol{x}_{253}-\boldsymbol{x}_0) - \boldsymbol{\mu}\big)\\ &= \big[\,30.3^2 \quad 31.45^2 \quad 31.60^2 \quad 20.11^2\,\big]'.\end{aligned}$$

According to Algorithm 4.2, the cumulative sandstone thicknesses and their estimation standard deviations are:

$$\begin{aligned}\hat{\mathbf{Z}}_{OK}(\boldsymbol{x}_0) &= [\,33.50 \quad 17.28 \quad 14.93 \quad 112.7\,]'\\ \boldsymbol{\sigma}_{OK}(\boldsymbol{x}_0) &= [\,30.37 \quad 31.69 \quad 31.69 \quad 20.22\,]'\end{aligned}$$

In cases such as the one in the exercise in which all sites are fully sampled, the cokriging and the ordinary kriging estimations are alike. Additionally, the cokriging estimation variances are always smaller than those for ordinary kriging. However, the improvements are so minuscule that they do not compensate for the additional effort of running the cokriging.

3. Well 253 is the well closest to the estimation location. The measurement of 53 ft at that well carries a weight of 0.5788 in the estimation of the cumulative thickness of the upper "J" sandstone. According to the conventions in the example, the thickness of the upper "J" sandstone at well 253 is the first measurement for the fourth site. Censoring this value is equivalent to ignoring $Z_1(\boldsymbol{x}_{253})$ and dropping its weights $\left(\lambda_{11}^{253}, \lambda_{21}^{253}, \lambda_{31}^{253}, \text{ and } \lambda_{41}^{253}\right)$ from the cokriging system of equations. The new cokriging estimate changes to 20.08 ft with a standard deviation of 33.13 ft, while the ordinary kriging estimate goes to 10.45 ft, with a standard deviation of 36.89 ft.

The lesson to be learned from this illustrative case is that ordinary kriging is more sensitive to the absence of a single measurement in a coregionalization. In particular, the cokriging estimation variance is significantly smaller than the ordinary kriging variance, a conclusion that holds for more severe cases of undersampling.

Cokriging, through its spatial dependence, is able to partially compensate for the lack of a few measurements. The improvements are significant, are worth the effort required by structural analysis, and explain the natural niche of cokriging in geostatistics—the estimation of undersampled coregionalizations (Journel and Huijbregts, 1978, p. 324; Atkinson *et al.*, 1992). There is one rare exception though: *intrinsic coregionalization*, which is the name given to the case where all semivariograms and cross-semivariograms are proportional to each other, a situation that arises if:

i) The linear coregionalization model comprises a single coefficient matrix $\mathbf{B}_1$.
ii) There are at least two coefficient matrices, but $\mathbf{B}_i = \alpha\mathbf{B}_j$, for all i and j.

If an intrinsic coregionalization, also referred to as an *isotopic* coregionalization, is fully sampled, the cokriging estimates and the cokriging variances are identical to those given by ordinary kriging (Journel and Huijbregts, 1978,

p. 174-175; Wackernagel, 1995, p. 149). Occurrences of intrinsic coregionalizations are rare; limiting cases are more common. The more closely a fully sampled coregionalization approaches an intrinsic coregionalization, the less significant are the improvements of ordinary cokriging over ordinary kriging. □

Regionalized Compositions

Attributes of a nonnegative coregionalization that sum to a constant are called a *regionalized composition.* This is, for example, the case in closed systems such as percentages of water, oil, and rock in an oil reservoir or the petrographic analysis of all the minerals present in a rock in which the sum of all components adds to 100%. Direct cokriging of such systems is not possible because the collinearity of the variables violates Assumption 13.3, whose purpose is to prevent the attempt to solve systems of equations when the coefficient matrix in Definition 13.9 is singular. In addition, spurious correlations derived from the closure of the system will result in covariances and cross-covariances that distort structural analyses (Aitchison, 1986, p. 48–49).

Pawlowsky (1989; Pawlowsky-Glahn and Olea, *n.d.*), integrating geostatistics and results from random sampling of compositional data (Aitchison, 1986), has proposed the most satisfactory solution to the problem by the use of the additive log-ratio transformation, followed by the additive generalized logistic transformation to bring the results back to the attribute space.

Definition 13.15

Let $Z_1(\boldsymbol{x})$, $Z_2(\boldsymbol{x})$, $\ldots$, $Z_k(\boldsymbol{x})$, $\ldots$, $Z_p(\boldsymbol{x})$ be a set of p random functions defined over the same domain. The additive log-ratio transformation is

$$Y_k(\boldsymbol{x}) = \ln\left(\frac{Z_k(\boldsymbol{x})}{Z_l(\boldsymbol{x})}\right), \quad \text{for } k \neq l\,,$$

where $Z_l(\boldsymbol{x})$ is any of the random variables modeling the coregionalization. □

Notice that the transformation reduces the number of variables by 1, and requires full sampling at every site.

Definition 13.16

Let $Z_1(\boldsymbol{x})$, $Z_2(\boldsymbol{x})$, $\ldots$, $Z_k(\boldsymbol{x})$, $\ldots$, $Z_p(\boldsymbol{x})$ be a set of random functions defined over the same domain and let $\hat{Y}_1(\boldsymbol{x})$, $\hat{Y}_2(\boldsymbol{x})$, $\ldots$, $\hat{Y}_k(\boldsymbol{x})$, $\ldots$, $\hat{Y}_{p-1}(\boldsymbol{x})$ be estimates for their transformation in Definition 13.15. The additive generalized logistic back transformation is the ratio

$$\hat{Z}_k(\boldsymbol{x}) = \frac{e^{\hat{Y}_k(\boldsymbol{x})}}{1 + \sum\limits_{j \neq l} e^{\hat{Y}_j(\boldsymbol{x})}}, \quad \text{for } k \neq l$$

$$\hat{Z}_l(\boldsymbol{x}) = \frac{1}{1 + \sum\limits_{j \neq l} e^{\hat{Y}_j(\boldsymbol{x})}},$$

where the lth variable is the same as that in the denominator of Definition 13.15. □

Aitchison (1986, p. 93–94) has proved that the results of the back transformation are the same, regardless of the selection of variable $Z_l(\boldsymbol{x})$ going into the denominator of the additive log-ratio transformation in Definition 13.15. Other appeals of the approach are:

1. The additive log-ratio transformation destroys the collinearity among the p original variables $Z_i(\boldsymbol{x})$, producing a new coregionalization of $(p-1)$ variables $Y_i(\boldsymbol{x})$ that fulfills all cokriging assumptions—in particular Assumption 13.3—thus opening the possibility for cokriging the coregionalization of the transformation following Algorithms 13.1 and 13.2.

2. The structural analysis of the additive log-ratio transformed variables is free of spurious correlations that may distort the analysis of the original variables.

3. The transformation is defined in the $(p-1)$-dimensional real space instead of the p-dimensional simplex of the attribute space.

4. In general, the transformation is closer to being a partial realization of a vectorial multinormal function than is the sampling.

5. Combined use of the transformations in Definitions 13.15 and 13.16 allows production of estimates in the interval $(0,\ a_0)$, where a_0 is the constant in Assumption 13.3.

6. Cokriging of the additive log-ratio transformations in Definition 13.15 combined with the additive generalized logistic back transformation in Definition 13.16 produces estimates $\hat{Z}_k(\boldsymbol{x})$ that add exactly to the constant a_0 in Assumption 13.3.

Limitations of the application of the transformation include:

1. Zeros lead to unacceptable transformations, either in terms of arguments for the logarithm in Definition 13.15, or in the denominator of the ratio.

2. Transformation 13.15 requires a full sampling at every site. In the presence of an undersampling, all sites lacking a measurement of one or more attributes have to be dropped, thus reducing the sampling size and precluding the opportunity for cokriging to operate at it best—that is, when some of the measurements are missing.

Chapter 14

REGIONALIZED CLASSIFICATION

An important task for earth scientists is to solve inverse problems by measuring regionalized attributes. The results can then be used to postulate natural processes that may have generated the phenomena we observe today. Because natural processes are fairly complex and samplings are rarely large enough, simplifying reality by the imposition of a manmade order commonly plays an important part in inverse modeling.

Classification of specimens into groups is one of these simplifying models. In our context, we will be dealing with sites instead of specimens or objects. In a regionalized classification, the sampling domain is a multidimensional geographic space, usually a portion of planet Earth with dimension three. If one surveys a coregionalization by making several measurements per site, the survey defines another multidimensional space, a space with as many dimensions as the number of attributes considered in the sampling, commonly four or larger. In such a space each site can be depicted as a point whose coordinates are the value of the attributes at that site. The assumption of regionalized classification is that the points in the attribute space are not uniformly and regularly distributed. Instead, as illustrated in Figure 14.1, there are zones with few or no points at all and other locations with high concentrations of points. Notice that in this example it is not possible to model the relationship among the measurements through regression.

The aim of regionalized classification is to explore the assumption that observations form distinct clusters in attribute space and to take advantage of the hypothesis when grounds are found to support it. In that case, regionalized classification breaks the observations into groups that are as internally homogeneous as possible and as different as possible from each other. Proximity in attribute space does not guarantee proximity in geographical space, yet in practice one finds that observations in the same attribute space cluster tend to remain close in the geographical space to various degrees. If contrasting colors

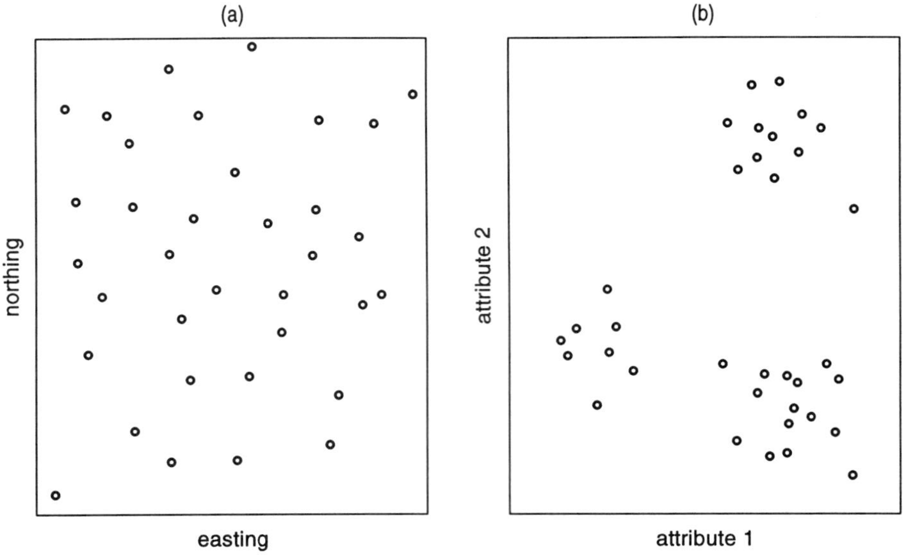

Figure 14.1. Idealized coregionalized sampling. (a) Location of sampling sites in a two-dimensional space in a survey comprising measurement of two attributes per site. (b) Mapping of the observations in the attribute space showing three clusters.

are assigned to each group, the resulting group map in the geographic space resembles a geologic map, which is a kind of regionalized classification map denoting rock age and type.

Another way to view regionalized classification is as a tool to condense maps. All attributes of a coregionalization can be mapped separately and inspected to find relationships among features on the different maps. The task is simplified by the use of a light table, but becomes progressively difficult as the number of attributes in the survey increases. Regionalized classification is a convenient replacement of the light table for producing an objective and accurate summary map, regardless of the number of attributes.

There is nothing conceptually new in regionalized classification. What is novel is the joint application of several standard techniques. Harff and Davis (1990) published the first complete formulation of the method combining geostatistical elements with some inspirational ideas of Soviet geologists (Voronin, 1967; Rodionov, 1981; Kogan, 1986).

Typification

In a nutshell, regionalized classification is the probabilistic assignment of sites to groups employing discriminant analysis. Discriminant analysis requires a training set that in regionalized classification ordinarily is prepared using cluster analysis. Preparation of maps may require the interpolation of group probabilities, which may be done by kriging.

In classical applications of discriminant analysis, a training set including direct indicators of group membership is employed. Medical science provides good examples. For instance, a group of sick patients may be carefully examined to prepare a clinical record relating internal conditions to their external symptoms and laboratory analyses, a costly and lengthy classification only possible after performing surgery or autopsy. In this context, based only on external symptoms and laboratory analyses, the task of discriminant analysis is to determine the probability that other patients suffer the internal conditions considered in the training set.

In earth sciences, what prevails is the equivalent of knowing only the external symptoms and never knowing the group assignments. Even worse, the nature and number of the groups are seldom known. Determining the groups is as important in such studies as the assignments themselves. In such a situation, the training set assignments are replaced by a typification using another method able to decide both the number and characteristics of the groups. The best alternative is cluster analysis, which aside from any prior information not contained in the data, solves the following problem:

Given a coregionalization sampling of size n comprising p attributes, is there any evidence for clustering of the sites in the attribute space into groups scattered around centroids, as against the alternative hypothesis that they are an unstructured coregionalization?

There are several ways to define and find the clusters, depending partly on the metric used to determine proximity among the sites in the p dimensional attribute space, the Euclidean distance being the most common and the only one considered here. The geographical distance is not considered in cluster analysis.

There have been many studies comparing various methods of cluster analysis using artificial data sets containing known clusters produced by Monte Carlo methods. In most of these studies the Ward's minimum variance method has been the one with the best overall performance on reproducing the known clusters (SAS Institute, 1990, p. 56).

Ward's Method

Given n sites, this heuristic method progresses by reducing the number of clusters one at a time starting from the trivial extreme case of one site per cluster and ending at the other trivial extreme case in which one cluster comprises all the sites. At each cluster reduction, the method merges the two clusters resulting in the smallest increase in the total sum of squares of the distances of each point to its cluster centroid. Sites clustered at a previous clustering step are never unmerged.

Definition 14.1

Let $\mathbf{z}_{ik}$ be a vector of attribute values at site i assigned to cluster k whose size is n_k and mean is $\bar{\mathbf{z}}_k$. Then the error sum of squares for cluster k is

$$E_k = \sum_{i=1}^{n_k} \|\mathbf{z}_{ik} - \bar{\mathbf{z}}_k\|^2 . \quad \square$$

Definition 14.2

Let E_k be the error sum of squares in Definition 14.1 and let g be the total number of clusters. The total within-group error sum of squares is

$$E = \sum_{k=1}^{g} E_k . \quad \square$$

Table 14.1 and Figure 14.2 illustrate the method for a simple case involving a single attribute. At the beginning each measurement coincides with its centroid, and both the error per cluster and, consequently, the total error are zero. In the first round of mergers, only the merging sites contribute to the total sum, so $E = E_k$.

For large numbers of clusters, instead of computing the new total sum of errors directly as the sum of all clusters in Definition 14.2, there are savings in the calculations by updating the new total sum of errors as the lowest sum from the previous round, plus the new error for the new merging clusters, minus the individual cluster errors for the merging clusters. Even larger savings arise by working directly with the square distances between centroids instead of the within-square distances E_k (Anderberg, 1973, p. 142–145; Kendall, 1986, p. 37).

Algorithm 14.1

This is an algorithm to perform cluster analysis using Ward's method.

1. Start with as many clusters as there are sites, each cluster consisting of exactly one site. At this stage the value of E in Definition 14.2 is zero.

2. Reduce the number of clusters by one by merging those two that minimize the increase of the total error E.

3. If the sites are in more than one cluster, go back to Step 2.

4. Display the results in the form of an inverted tree showing at each stage which two clusters were merged and the corresponding total error E of the merged clusters or the total number of clusters g. $\square$

Most statistical packages include implementations of several methods for cluster analysis, including Ward's (IMSL, 1987; SAS Institute, 1990). Ward's method has the following characteristics:

(a) It is biased toward finding clusters with the same number of sites in each cluster (SAS Institute, 1990, p. 56).

Table 14.1. Ward's method showing error sum of squares, total within-group error sum of squares for new partitions (underlined), and mean for best mergers.

Number of clusters	Possible partitions of measurements	E_k	E	$\bar{\mathbf{z}}_k$
6	(0.1) (1) (3) (7) (8) (10)			
5	(0.1, 1) (3) (7) (8) (10)	0.405	0.405	0.55
	(0.1, 3) (1) (7) (8) (10)	4.205	4.205	
	(0.1, 7) (1) (3) (8) (10)	23.805	23.805	
	(0.1, 8) (1) (3) (7) (10)	31.205	31.205	
	(0.1, 10) (1) (3) (7) (8)	49.005	49.005	
	(0.1) (1, 3) (7) (8) (10)	2	2	
	(0.1) (1, 7) (3) (8) (10)	1	18	
	(0.1) (1, 8) (3) (7) (10)	24.5	24.5	
	(0.1) (1, 10) (3) (7) (8)	40.5	40.5	
	(0.1) (1) (3, 7) (8) (10)	8	8	
	(0.1) (1) (3, 8) (7) (10)	12.5	12.5	
	(0.1) (1) (3, 10) (7) (8)	24.5	24.5	
	(0.1) (1) (3) (7, 8) (10)	0.5	0.5	
	(0.1) (1) (3) (7, 10) (8)	4.5	4.5	
	(0.1) (1) (3) (7) (8, 10)	2	2	
4	(0.1, 1, 3) (7) (8) (10)	4.407	4.407	
	(0.1, 1, 7) (3) (8) (10)	28.14	28.14	
	(0.1, 1, 8) (3) (7) (10)	37.407	37.407	
	(0.1, 1, 10) (3) (7) (8)	59.94	59.94	
	(0.1, 1) (3, 7) (8) (10)	8	8.405	
	(0.1, 1) (3, 8) (7) (10)	12.5	12.905	
	(0.1, 1) (3, 10) (7) (8)	24.5	24.905	
	(0.1, 1) (3) (7, 8) (10)	0.5	0.905	7.5
	(0.1, 1) (3) (7, 10) (8)	4.5	4.905	
	(0.1, 1) (3) (7) (8, 10)	2	2.405	
3	(0.1, 1, 3) (7, 8) (10)	4.407	4.907	1.367
	(0.1, 1, 7, 8) (3) (10)	49.208	49.208	
	(0.1, 1, 10) (3) (7, 8)	59.94	60.44	
	(0.1, 1) (3, 7, 8) (10)	14	14.405	
	(0.1, 1) (3, 10) (7, 8)	24.5	24.905	
	(0.1, 1) (3) (7, 8, 10)	4.667	5.077	
2	(0.1, 1, 3, 7, 8) (10)	50.048	50.048	
	(0.1, 1, 3, 10) (7, 8)	60.308	60.808	
	(0.1, 1, 3) (7, 8, 10)	4.667	9.074	8.333
1	(0.1, 1, 3, 7, 8, 10)	81.875	81.875	4.85

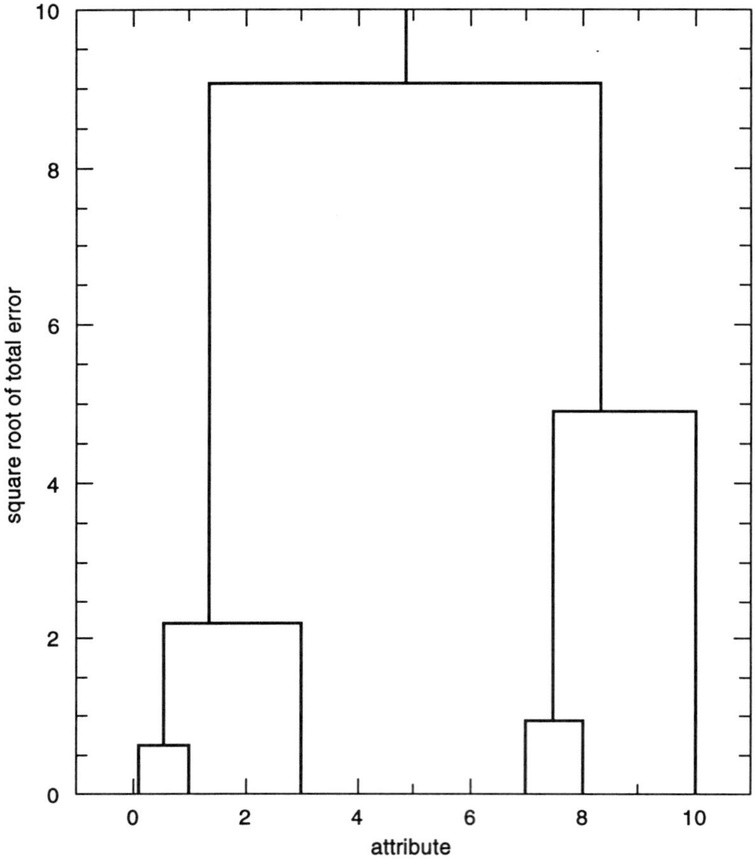

Figure 14.2. The Ward's method tree for the example in Table 14.1.

(b) Because sites clustered at a previous clustering step are never unmerged, the calculations are relatively simple, but they often result in suboptimal partitions conditioned to the previous steps. For example, at the three-cluster level Ward's method produces the clusters (1, 2), (4, 6), (9, 10) for this univariate sampling of size 6. In the next stage the method produces the clusters (1, 2, 4, 6) and (9, 10) with total error $E = 15.25$, which is larger than the smallest total error $E = 13.333$ associated with clusters (1, 2, 4) and (6, 9, 10), an unfeasible partition after selecting (1, 2), (4, 6), (9, 10) in the previous stage.

Despite its suboptimality, the method has been quite successful reproducing clusters in blind tests.

(c) If one is not interested in the initial small clusters, Ward's method cannot avoid calculations for those initial partitions which are without interest. For example for a sampling of size 1,000, if the interest is restricted to partitions with 5 to 20 clusters, it is possible to stop the calculations after reaching a partition into 5 clusters, but it is not possible to avoid the much more numerous calculations for partitioning the sampling from 999 to 21 clusters.

(d) The total sum of errors increases monotonically as the number of clusters decreases.

(e) Ward's method is not robust with respect to survey errors and outliers (Mojena, 1988).

Discriminant Analysis

In practice, the answer to the question of how to group vectorial observations $\mathbf{z}$ is never straightforward. First there is the problem of deciding how many groups. This can be settled to a large extent on the basis of external information or rules pertaining to the total sum of errors that can be obtained running cluster analysis.

A second and more intractable problem is the group assignment for observations that are not like any of the typical observations in a group and at the same time poorly resemble more than one group. Cluster analysis assignments for such problematic observations are unstable and vary depending on the measure of similarity and the clustering method.

If the reliability of group assignments is a concern, cluster analysis does not offer easy answers. That is the realm of discriminant analysis, which decides the assignment of any vector $\mathbf{z}$ based on allocation probabilities. Discriminant analysis, however, requires a training classification, which in regionalized classification is the one generally produced by cluster analysis.

The basis of discriminant analysis for assigning observations to one of a given number of groups is the minimization of the total misallocation cost. The procedure assumes that it is possible to partition the attribute space in as many mutually exclusive and exhaustive regions R_i as there are groups. A site is said to belong to group i when the vector $\mathbf{z}$ of attributes associated with the site falls in R_i.

Theorem 14.1

Let $\mathbf{Z}$ be a vectorial random function with probability density function $f(\mathbf{z})$ and let $\mathbf{z}(\boldsymbol{x})$ be a realization of $\mathbf{Z}$ at site $\boldsymbol{x}$, $\mathbf{z}$ for short. Let π_i be the proportional share of observations in the ith group whose probability density function is $f_i(\mathbf{z})$. Then the probability $p_i(\mathbf{z})$ that the site characterized by $\mathbf{z}$ belongs to the ith group is

$$p_i(\mathbf{z}) = \frac{\pi_i f_i(\mathbf{z})}{f(\mathbf{z})} \,.$$

Proof

The proof directly follows from Bayes' Theorem. If one considers that π_i can be regarded as the *a priori* probability of sampling group i, then $p_i(\mathbf{z})$ is the posterior probability that $\mathbf{z}$ belongs to group i. □

The optimality in terms of the misallocation cost is assured by assigning $\mathbf{z}$ to the group with the highest probability (McLachlan, 1992, p. 7).

The kernel of discriminant analysis is the calculation of the probabilities in Theorem 14.1. As is the case in statistics, there are nonparametric and parametric methods. Multivariate normal methods prevail by far among the latter and overall.

Definition 14.3

Let $\mathbf{z}(\boldsymbol{x})$ be a realization of a vectorial random function at site $\boldsymbol{x}$, $\mathbf{z}$ for short. Let $\bar{\mathbf{z}}_i$ be the vectorial mean and $\boldsymbol{\Sigma}_i$ the covariance matrix of group i. Then the Mahalanobis distance from $\mathbf{z}$ to the centroid of group i is the squared weighted distance

$$\delta_i^2(\mathbf{z}) = (\mathbf{z} - \bar{\mathbf{z}}_i)' \, \boldsymbol{\Sigma}_i^{-1} (\mathbf{z} - \bar{\mathbf{z}}_i) \, . \quad \square$$

Theorem 14.2

Let $\mathbf{Z}$ be a normal vectorial random function with heteroscedastic normal group distributions. If π_i is the *a priori* probability of group i and $\delta_i^2(\mathbf{z})$ is the Mahalanobis distance in Definition 14.3, then the probability $p_i(\mathbf{z})$ that the site characterized by $\mathbf{z}$ belongs to the ith group is

$$p_i(\mathbf{z}) = \frac{\pi_i \left|\boldsymbol{\Sigma}_i\right|^{-1/2} e^{-\delta_i^2(\mathbf{z})/2}}{\sum_{i=1}^{g} \pi_i \left|\boldsymbol{\Sigma}_i\right|^{-1/2} e^{-\delta_i^2(\mathbf{z})/2}} \, .$$

Proof

From Theorem 14.1,

$$p_i(\mathbf{z}) = \frac{\pi_i f_i(\mathbf{z})}{f(\mathbf{z})} \, .$$

If the probability distribution for the ith group is normal,

$$f_i(\mathbf{z}) = (2\pi)^{-\frac{p}{2}} \left|\boldsymbol{\Sigma}_i\right|^{-1/2} e^{-\delta_i^2(\mathbf{z})/2} \, ,$$

where p is the dimension of the attribute space. Considering that in addition $f(\mathbf{z}) = \sum_{i=1}^{g} \pi_i f_i(\mathbf{z})$ (McLachlan, 1992, p. 5), then

$$p_i(\mathbf{z}) = \frac{(2\pi)^{-p/2} \pi_i \left|\boldsymbol{\Sigma}_i\right|^{-\frac{1}{2}} e^{-\delta_i^2(\mathbf{z})/2}}{\sum_{i=1}^{g} (2\pi)^{-p/2} \pi_i \left|\boldsymbol{\Sigma}_i\right|^{-\frac{1}{2}} e^{-\delta_i^2(\mathbf{z})/2}} \, .$$

Cancellation of the $(2\pi)^{-p/2}$ terms proves the theorem. $\square$

Theorem 14.3

Let $\mathbf{Z}$ be a normal vectorial random function with homoscedastic normal group distributions. If π_i is the *a priori* probability of group i and $\delta_i^2(\mathbf{z})$ is the Mahalanobis distance in Definition 14.3 computed using the common covariance matrix $\boldsymbol{\Sigma}$, then the probability $p_i(\mathbf{z})$ that the site characterized by $\mathbf{z}$ belongs to the ith group is

$$p_i(\mathbf{z}) = \frac{e^{\ln(\pi_i) - 0.5\delta_i^2(\mathbf{z})}}{\sum_{i=1}^{g} e^{\ln(\pi_i) - 0.5\delta_i^2(\mathbf{z})}}.$$

Proof

From Theorem 14.2, considering that all covariance matrices are the same:

$$p_i(\mathbf{z}) = \frac{\pi_i \left|\boldsymbol{\Sigma}\right|^{-1/2} e^{-\delta_i^2(\mathbf{z})/2}}{\left|\boldsymbol{\Sigma}\right|^{-1/2} \sum_{i=1}^{g} \pi_i e^{-\delta_i^2(\mathbf{z})/2}}.$$

Proof follows by the cancellation of the covariance matrix determinant term and consolidation of exponents after the trivial transformation $\pi_i = e^{\ln(\pi_i)}$. □

Use of the model in Theorem 14.3 results in hyperplanes for the boundaries of the group regions R_i, thus the term *linear* discriminant analysis, while *quadratic* discriminant analysis deals with different covariance matrices and second order surfaces arising from the heteroscedastic model.

Whatever the model of discriminant analysis used for regionalized classification, the model requires a training set to determine the number of groups and to assign realizations to the different groups in order to have some data for estimating the centroid and the covariance matrix of the groups.

Algorithm 14.2

This is a procedure for the calculation of group probabilities for vectorial measurements. The method employs a normal model of discriminant analysis.

1. Run Algorithm 14.1 using the whole sampling of the coregionalization.
2. Based on the total sum of errors, some external criteria, or both, select a classification for the sampling.
3. Use the vectorial measurements to estimate all group centroids and covariance matrices and to make a decision about the *a priori* group probabilities. The alternatives are:

(a) If the sampling properly represents the true group sizes, the best estimate of the *a priori* probabilities are the relative proportions of the n_i sites per group,

$$\pi_i = n_i / \sum_{i=1}^{g} n_i.$$

(b) If the proportions of sites per group are regarded as not being indicative of the *a priori* group probabilities and the user believes that he or she has some better external information to assess the probabilities, the user may employ such external assessment.

(c) If the proportions of sites per group are regarded as not being indicative of the *a priori* group probabilities and the user lacks ways to assess them, they can be made equal to $1/g$, in which case they are ignored in the calculations.

4. Compare the group covariance matrices and decide whether they are sufficiently similar to assume homoscedasticity.

5. Calculate the Mahalanobis distance in Definition 14.3 for measurement $\mathbf{z}$. Use the average of all group covariances if the assumption is that the covariance is homoscedastic.

6. Compute for measurement $\mathbf{z}$ the probability $p_i(\mathbf{z})$ of belonging to each of the groups

$$p_i(\mathbf{z}) = \frac{e^{-0.5D_i^2(\mathbf{z})}}{\sum_{i=1}^{g} e^{-0.5D_i^2(\mathbf{z})}},$$

where, if

$D_i^2(\mathbf{z}) = \delta_i^2(\mathbf{z})$, the discriminant analysis is linear with unknown or equal *a priori* group probabilities;

$D_i^2(\mathbf{z}) = \delta_i^2(\mathbf{z}) - 2\ln(\pi_i)$, the discriminant analysis is linear with known and different *a priori* group probabilities;

$D_i^2(\mathbf{z}) = \delta_i^2(\mathbf{z}) + \ln|\boldsymbol{S}_i|$, the discriminant analysis is quadratic with unknown or equal *a priori* group probabilities;

$D_i^2(\mathbf{z}) = \delta_i^2(\mathbf{z}) + \ln|\boldsymbol{S}_i| - 2\ln(\pi_i)$, the discriminant analysis is quadratic with known and different *a priori* group probabilities.

The choice of generalized distance $D_i^2(\mathbf{z})$ must be consistent with the decision of homoscedasticity made in Step 4. The term $\delta_i^2(\mathbf{z})$ is the Mahalanobis distance in Definition 14.3; and $\boldsymbol{S}_i$ is the estimate of the covariance matrix for group i.

7. Go back to Step 5 until group probabilities have been assigned to all sites in the training set. □

Major statistical computer packages such as IMSL (1987) and SAS (SAS Institute, 1990) have implementations of Algorithm 14.2 along with several other procedures.

Statisticians concur that quadratic discriminant analysis indeed provides superior results if the group covariances are considerably different and the group sizes are large. Quadratic discriminant analysis, however, is more sensitive to deviations from multinormality and assignment errors in the training set (Lachenbruch, 1982).

Allocation by Extension

The use that regionalized classification makes of discriminant analysis is to a certain degree analogous to the use of crossvalidation in kriging, and different from the classical use of discriminant analysis. Classical use of discriminant analysis employs the training set for calibration and then proceeds to classify vectorial measurements without assignments.

In regionalized classification, the interest is in the calculation of the probabilities $p_i(\mathbf{z})$ for the same realizations in the training set already classified by the prior cluster analysis. After one obtains all the probabilities $p_i(\mathbf{z})$, reallocation of the sites to the group with the highest probability is a trivial endeavor. This reallocation offers an opportunity to check results and compare methods in case the user wants to consider more than one type of cluster analysis or discriminant analysis. Results from cluster and discriminant analysis should be comparable only with minor variations that should be employed to select methods and parameters yielding the most consistent results.

The final step in regionalized classification is the mapping of groups, which can be accomplished by arbitrarily assigning colors or black and white patterns to the groups. As arbitrary as the color or pattern selection may be, it always helps to select a combination of alternatives that maximizes contrast to the eye, which customarily requires some experimentation by trial and error.

Most mapping procedures require a collection of values regularly spaced at short intervals, which is rarely the case with coregionalization samplings. In addition, interpolation presumes that the variable is continuous. In such circumstances, Harff and Davis (1990) recommend mapping the group probabilities $p_i(\mathbf{z})$ by treating them as regionalized variables and then using the allocation rule to produce the discontinuous group map. Remember that $\mathbf{z}$ is a shorthand for $\mathbf{z}(\boldsymbol{x})$, where $\boldsymbol{x}$ is the location of the site. Then the probabilities are actually regionalized variables $p(\mathbf{z}(\boldsymbol{x}))$ that depend on location. One does group allocation node by node, performing a grid-to-grid operation in which each node is assigned to the group with the highest probability.

Considering that at each site the probabilities sum to 1.0, cokriging of the probabilities is not possible because of violation of Assumption 13.3. In addition, use of alr transformation in Definition 13.15 is not feasible because of the numerous zeros, both in the numerator, resulting in a null argument for the logarithm, and in the denominator, producing unacceptable ratios.

Ordinary or universal kriging, the default geostatistical options of choice, suffer from their inability to restrict the estimates to an interval—0–1 in this instance—let alone being unable to force the probabilities to sum to 1. Estimates outside the 0–1 interval, however, are rare and never far away from the interval. A common solution for forcing a vector in a coregionalization to add to 1 is to rescale the values. Because such correction does not change the ranking of the membership probabilities, the allocation is insensitive to the rescaling.

Alternative strategies involving the interpolation of the generalized distances (Bohling, 1997) or the coregionalization itself instead of the membership probabilities are deceptive. Although the probabilities end up honoring all constraints, the use of estimates instead of true values results in propagation of unaccounted errors. Allocation in regionalized classification remains open to improvements.

Algorithm 14.3

This is a procedure for the regionalized classification of fully sampled coregionalizations involving p attributes.

1. Assign the sites to one and only one of g groups either by using Algorithm 14.1 or any other cluster analysis procedure, or on the basis of external information.
2. For each site calculate the group probabilities either by Algorithm 14.2 or any other discriminant analysis procedure deemed appropriate.
3. If the mapping procedure does not require a regular grid of values, go to Step 4. Otherwise, use some form of kriging to produce grids of estimated values for every group probability.
4. For every site or node, assign the site or node to the group with the largest probability.
5. Prepare a map showing the group assignment. This is the regionalized classification of the area under study, conditional on the values and attributes in the sampling.
6. Prepare a map of the highest probability per node. This is the probability that the site in the regionalized classification indeed belongs to the group in the regionalized classification.
7. Prepare a censored map eliminating nodes or sites likely to be misclassified, which is more likely to occur when the highest probability is only marginally larger than the second highest probability. □

EXERCISE 14.1

Figure 14.3 contains maps for the cumulative sandstone thicknesses in Appendix D. Use the sampling to prepare the following:

1. a clustering of the data based on Ward's method,
2. a discriminant analysis using the result from clustering analysis as the training set,
3. maps of group membership probabilities,
4. a map with the regionalized classification of the aquifer,
5. the map for the membership probability associated to each allocation,
6. a censored map eliminating all allocations for which the difference between the two highest probabilities is less than 0.1 times the sum of the kriging standard errors.

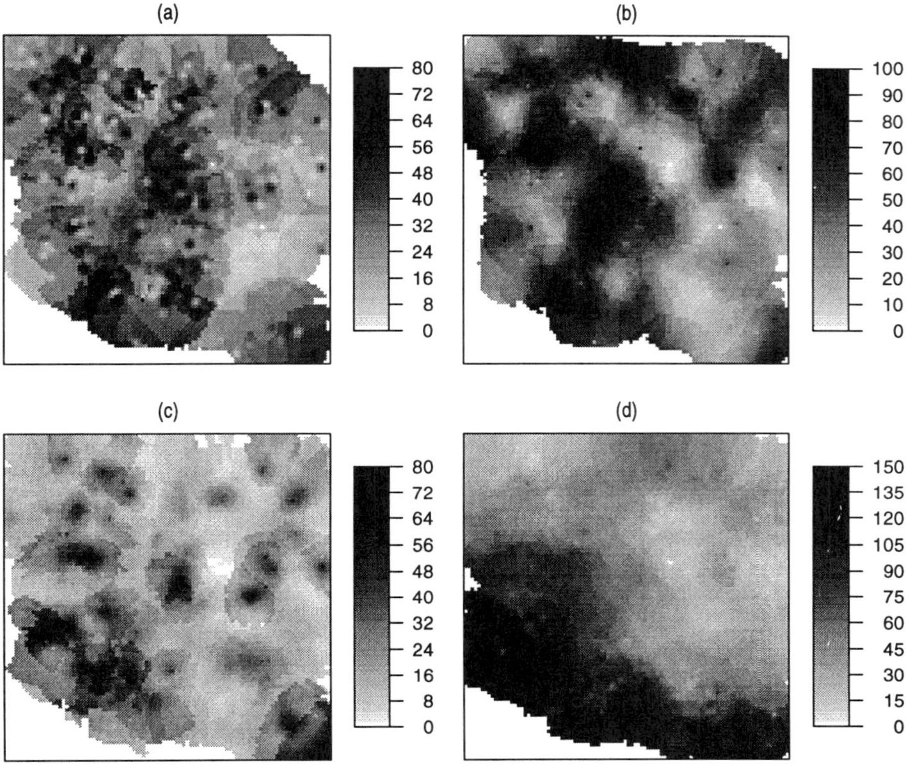

Figure 14.3. Cumulative thickness maps (in ft) for units comprising the Dakota aquifer, south-central Hodgeman County, Kansas: (a) upper "J"; (b) lower "J"; (c) Kiowa; (d) Cheyenne. A color rendition is available at Internet site http://www.kgs.ukans.edu/Mathgeo/Books/Geostat/index.html.

Solution

1. For any real multivariate sampling it is not possible to prepare as detailed a tree as the one in Figure 14.2. Figure 14.4 is a possible alternative rendition restricting the attention to the final stages of the clustering, which customarily are the ones with the partitions of interest. A bar diagram with the mean value of the attributes per cluster is one way to reduce the dimensionality of the data. In Figure 14.4, the vertical axis goes from 0–150 ft and the thickness sequence from left to right is upper "J", lower "J", Kiowa, and Cheyenne cumulative sandstone thickness. The numerical value denotes the number of wells in the cluster.

Naming clusters for easy reference in the text is another challenge in the display of results. Here I decided to take advantage of the uniqueness of cluster size and label each cluster by its size.

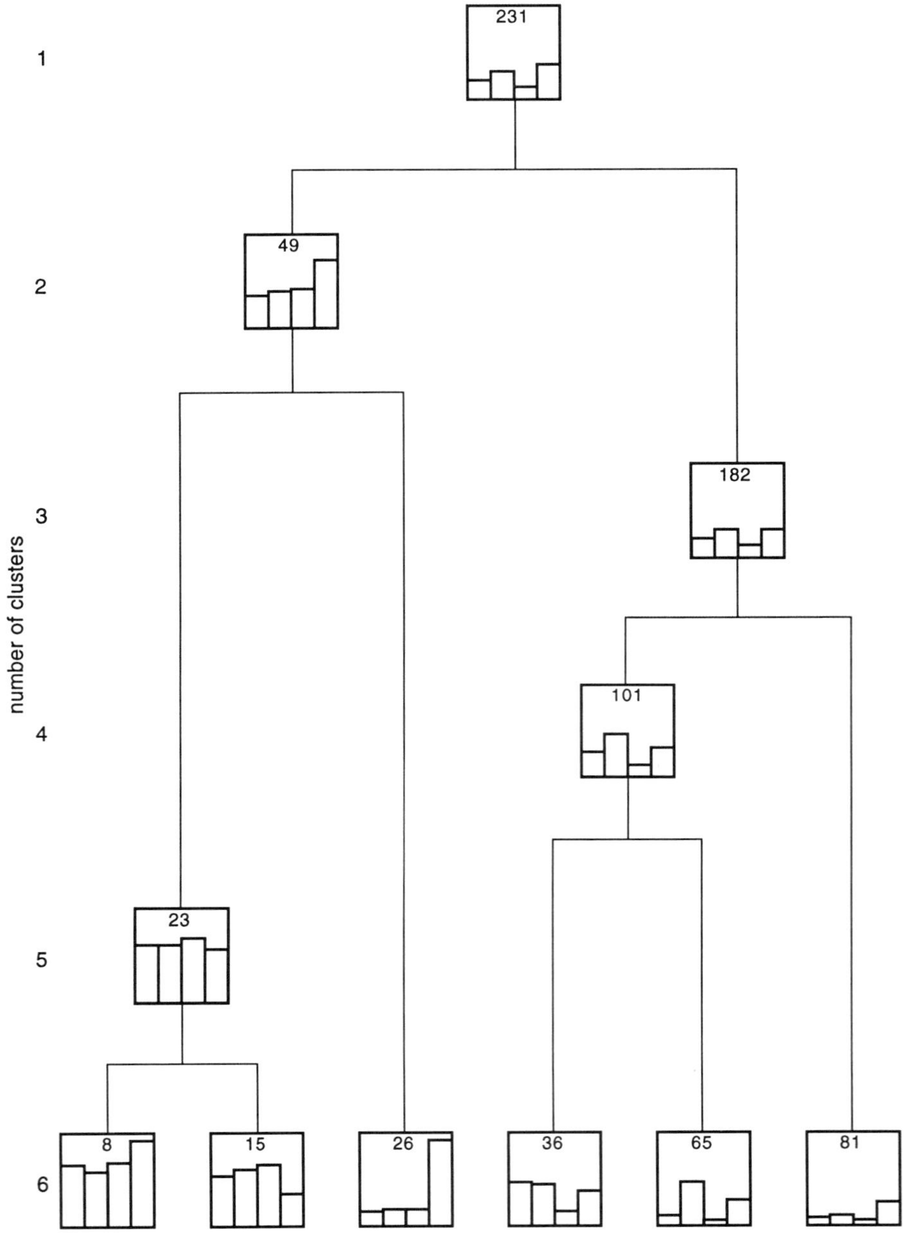

Figure 14.4. Final portion of the Ward's method tree for the training set in Appendix D. The horizontal axis does not have a scale. The boxes are bar diagrams for the coordinates of the centroids in a vertical scale of 0–150 ft.

Remember that Ward's method progresses from largest to smallest number of clusters. Traveling the tree in the opposite direction from its generation, one can observe that:

i) The topmost bar diagram provides a pictorial representation of the mean value per attribute for the whole sampling whose size is 231 in this case. The last merger is between a group of 49 wells with the rest of the 182 wells. The smaller cluster has the wells with the most sandstone, especially for the Kiowa.
ii) Cluster 49 results from the merging of clusters 23 and 26 that mostly share large amounts of Cheyenne sandstone.
iii) At the three-cluster level, Ward's method merges clusters 81 and 101 which basically share all average values except for the second attribute from the left—cumulative sandstone thickness for the lower "J".
iv) The previous four-cluster stage merges clusters 36 and 65, which differ only in the amount of upper "J" sandstone.
v) The five-cluster stage merges clusters 8 and 15, whose main discrepancy is the amount of cumulative sandstone in the Cheyenne.
vi) So far all breaking clusters relate to interesting characteristics on which to base the discriminant analysis in the next step. The six-cluster stage is no exception, but the partition contains two clusters that are too small for a regionalized classification. Thus, to avoid dealing with small clusters, it is not advisable to continue analyzing the declustering of the tree.

The training set for discriminant analysis should be the partition into five clusters, with the sizes and means given in Table 14.2

2. According to the testing performed by the SAS program (SAS Institute, 1990) used in the processing, the within-group covariance matrices in Table 14.3 are sufficiently different to make the quadratic discriminant analysis the model of choice.

Table 14.4 shows a 97% agreement between the Ward's method clusters and the groups based on the posterior probabilities of membership. The group sizes are in the bottom line. Although they are slightly different than the corresponding cluster sizes, I have kept the cluster labels to avoid cluttering the exposition.

Considering that in this case the relative cluster sizes in the sampling are a good approximation to the true relative sizes, one should use the relative proportions in Table 14.2 as estimates for the *a priori* group probabilities.

3. Figure 14.5 shows the continuous variation for all membership probabilities across the study area. Table 14.5 contains the semivariogram models and Figure 14.6 displays both the models and the experimental semivariograms. To minimize extrapolation outside the convex hull of the sampling, estimates exceeding a certain kriging variance threshold were ignored. From top to bottom and from left to right, maximum variance values for the maps in Figure 14.5 are 0.13, 0.055, 0.16, 0.235, and 0.235.

Table 14.2. Size, proportion, and means for the partition into five clusters.

Group size	Proportion	Attribute	Mean
23	0.10	Upper "J" cumulative sandstone	90.4
		Lower "J" cumulative sandstone	92.8
		Kiowa cumulative sandstone	102.9
		Cheyenne cumulative sandstone	83.4
26	0.11	Upper "J" cumulative sandstone	22.3
		Lower "J" cumulative sandstone	28.1
		Kiowa cumulative sandstone	29.0
		Cheyenne cumulative sandstone	141.6
36	0.16	Upper "J" cumulative sandstone	70.2
		Lower "J" cumulative sandstone	65.0
		Kiowa cumulative sandstone	23.6
		Cheyenne cumulative sandstone	55.0
65	0.28	Upper "J" cumulative sandstone	17.6
		Lower "J" cumulative sandstone	69.6
		Kiowa cumulative sandstone	9.4
		Cheyenne cumulative sandstone	43.0
81	0.35	Upper "J" cumulative sandstone	12.0
		Lower "J" cumulative sandstone	12.3
		Kiowa cumulative sandstone	7.5
		Cheyenne cumulative sandstone	35.6

4. Figure 14.7 displays the results of the grid allocations node by node. Remember that minimizing the misclassification error discriminant analysis allocates any vector of values to the group with the maximum group membership probability.

5. The group membership probabilities of the allocations are in Figure 14.8. They represent the highest group membership probability per node.

6. Considering that the allocation is done for the most part based on estimated probabilities instead of probabilities based on measurements and calculated by Algorithm 14.2, close membership probabilities have potential for misclassification. The map in Figure 14.9 repeats the one in Figure 14.8 by eliminating those nodes whose difference between the highest and second highest probabilities is less than 0.1 times the sum of their kriging standard deviation. □

Evaluation of the results of the regionalized classification requires borrowing from the basic geology of the Dakota aquifer and referring to either Figure 14.4 or Table 14.2 to remember the predominant characteristics of each group.

Table 14.3. Within-group covariance matrices for cumulative sandstone thickness.

Group		UJ	LJ	K	Ch
23	UJ	1532.1			
	LJ	344.6	615.9		
	K	−224.3	186.3	692.7	
	Ch	606.2	146.5	59.6	2312.4
26	UJ	317.1			
	LJ	53.2	490.8		
	K	127.3	140.5	965.0	
	Ch	−204.6	−17.9	−444.3	1543.1
36	UJ	436.8			
	LJ	317.6	1336.4		
	K	−73.1	214.0	330.7	
	Ch	−91.2	−176.3	21.3	511.1
65	UJ	254.9			
	LJ	57.3	329.8		
	K	−30.2	42.4	96.7	
	Ch	44.1	115.4	79.6	765.4
81	UJ	148.5			
	LJ	29.6	155.8		
	K	5.7	45.5	138.8	
	Ch	−39.7	29.8	97.1	616.5

Table 14.4. Resubstitution summary.

	Group				
Cluster	23	26	36	65	81
23	23	0	0	0	0
26	0	25	0	1	0
36	0	0	33	1	2
65	0	0	0	65	0
81	0	2	0	1	78
Total	23	27	33	68	80

In the area comprising the sampling, the aquifer is really two aquifers, the Cheyenne sandstone and the Dakota formation, which are separated by the Kiowa shale. The Cheyenne is saturated with salty water coming from the dissolution of underlying Permian salt beds while the Dakota formation is the unit of economic interest yielding fresh water. Throughout the area it is then important that the Cheyenne and the Dakota not be connected. At sites in group 23, a sandy Kiowa unit might result in an unavoidable leak in

Table 14.5. Semivariogram models for group membership probabilities.

Group	Type	Nugget	Sill	Range (ft)
23	Spherical	0.002	0.103	3137
26	Spherical	0.019	0.054	20251
36	Spherical	0.065	0.127	7140
65	Spherical	0.130	0.201	13050
81	Spherical	0.116	0.200	14104

the sealing, and drilling and completions should be approached with caution at sites in group 26 where the Cheyenne is thick.

The third group to avoid is the most predominant group 81, where for some intriguing geological reason there has never been much deposition of sand. Thus the reasons to avoid drilling group 81 are economic rather than environmental.

Finally, groups 36 and 65 should be the primary targets for groundwater supply in the area. If one accepts the postulation that a meandering river deposited the Dakota Formation in flood plains, it could be suggested that group 65 developed along a drainage system flowing to the northwest. The exceptionally favorable areas associated with group 36 tend to occur on the flanks of group 65.

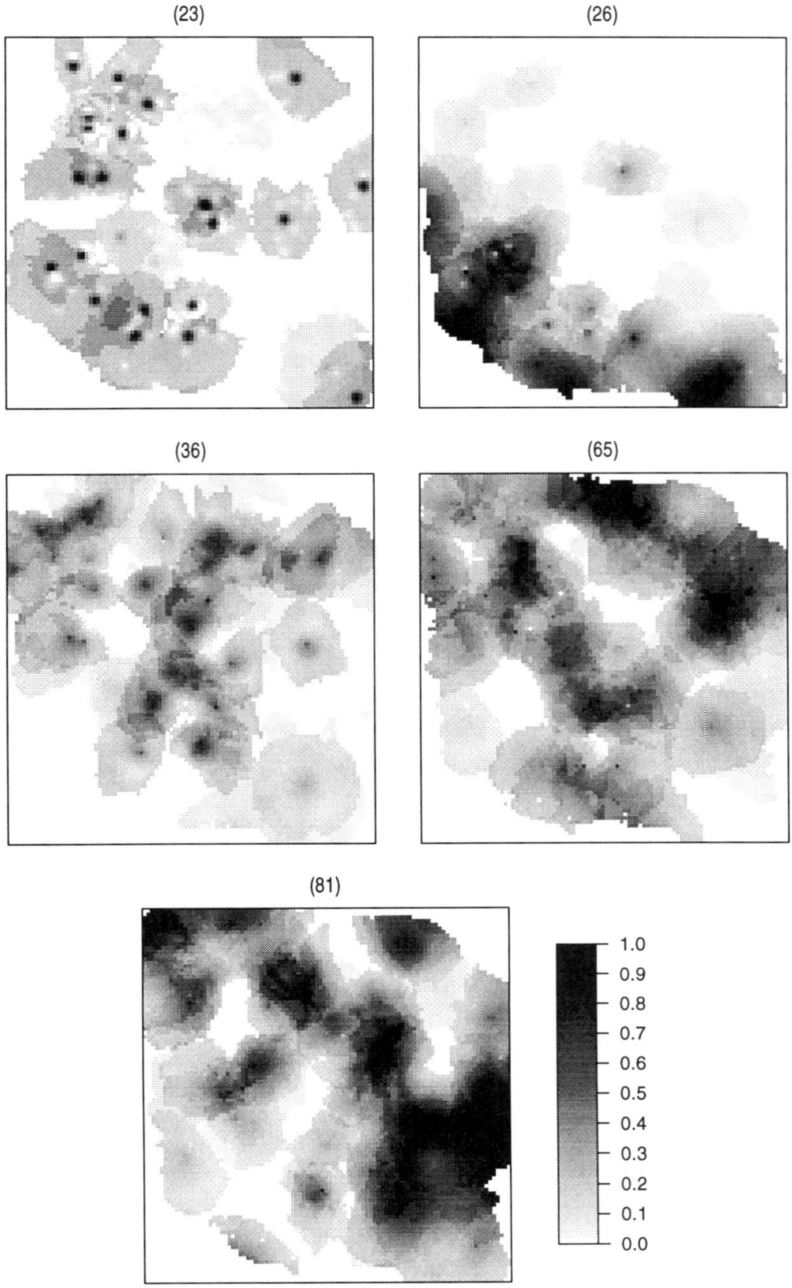

Figure 14.5. Membership probability maps for all five groups in regionalized classification of Dakota aquifer, south-central Hodgeman County, Kansas. Visit http://www.kgs.ukans.edu/Mathgeo/Books/Geostat/index.html for a color rendition available on the Internet.

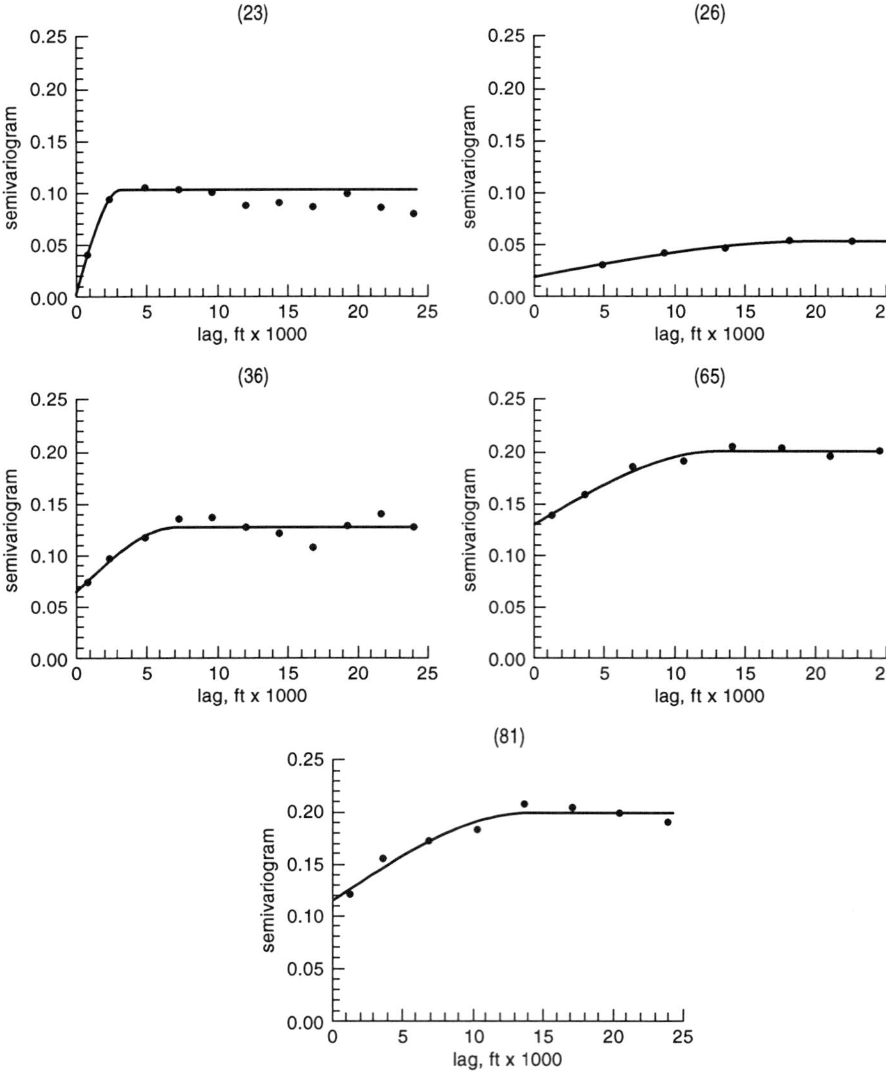

Figure 14.6. Probability semivariograms for all five groups in the regionalized classification of the Dakota aquifer, south-central Hodgeman County, Kansas.

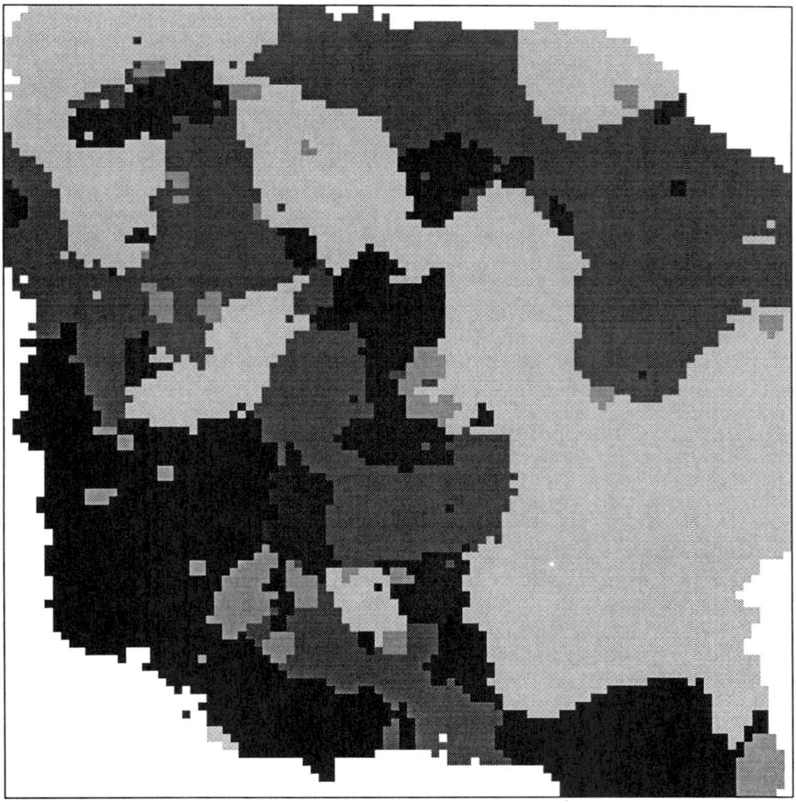

Figure 14.7. Regionalized classification of the Dakota aquifer in south-central Hodgeman County, Kansas. The lightest, most predominant shading denotes group 81, second darkest is 23, followed by 65 and 36. The darkest shading denotes group 26. Viewing the color rendition on Internet is highly recommended: http://www.kgs.ukans.edu/Mathgeo/Books/Geostat/index.html.

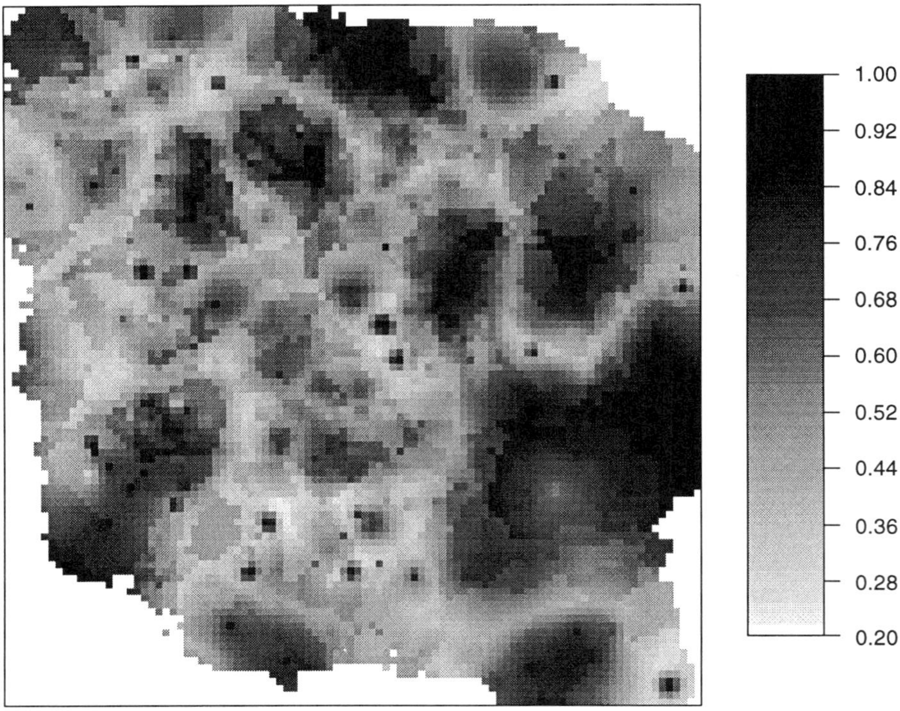

Figure 14.8. Group probabilities in the regionalized classification of the Dakota aquifer, south-central Hodgeman County, Kansas. For a color rendition, visit http://www.kgs.ukans.edu/Mathgeo/Books/Geostat/index.html.

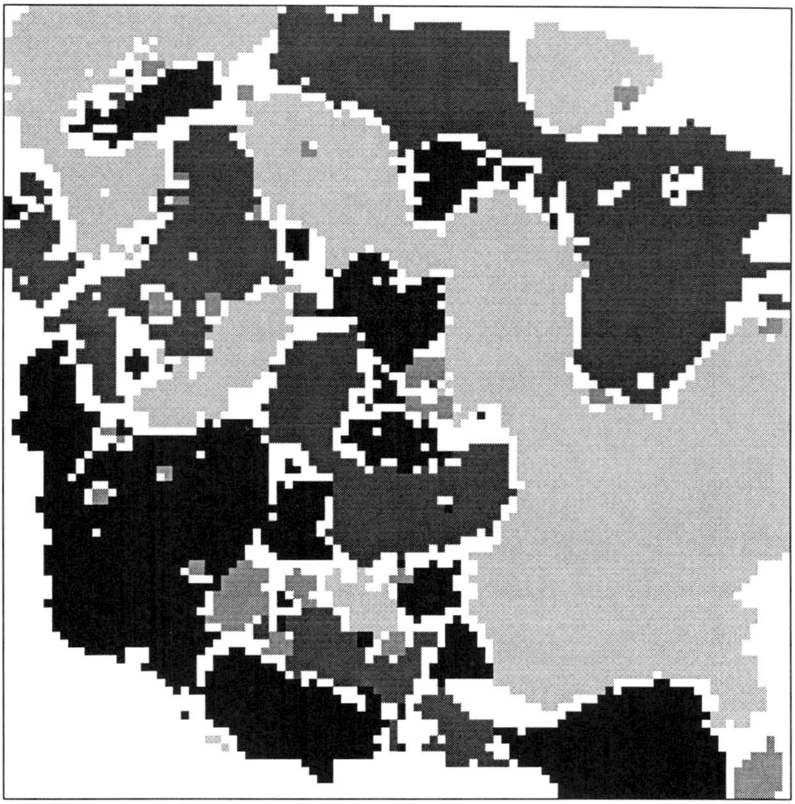

Figure 14.9. Regionalized classification of Dakota aquifer, south-central Hodgeman County, Kansas, eliminating maximum uncertainty allocations. The lightest and most predominant shading denotes group 81, second darkest is 23, followed by 65 and 36. The darkest shading denotes group 26. Visiting Internet site http://www.kgs.ukans.edu/Mathgeo/Books/Geostat/index.html for a color rendition is highly recommended.

References

Numbers in brackets denote chapter in which work is cited; letters denote appendix.

Aarts, E. and J. Korst, 1989, *Simulated Annealing and Boltzman Machines:* John Wiley & Sons, Chichester, United Kingdom, 272 pp. [9]

Aitchison, J., 1986, *The Statistical Analysis of Compositional Data:* Chapman & Hall Ltd., London, 416 pp. [13]

Anderberg, M.R., 1973, *Cluster Analysis for Applications:* Academic Press, New York, 359 pp. [14]

Atkinson, P.M., R. Webster, and P.J. Curran, 1992, Cokriging with ground-based radiometry: *Remote Sensing of Environment,* v. 41, no. 1, p. 45–60. [13]

Bard, Y., 1970, Comparison of gradient methods for the solution of nonlinear parameter estimation problems: *SIAM Journal on Numerical Analysis,* Society for Industrial and Applied Mathematics, v. 7, no. 1, p. 157–186. [5]

Barnes, R.J., 1991, The variogram sill and the sample variance: *Mathematical Geology,* v. 23, no. 4, p. 673–678. [5]

Bohling, G.C., 1997, GSLIB-style programs for discriminant analysis and regionalized classification: *Computers & Geosciences,* v. 23, no. 7, p. 739–761. [14]

Brooker, P.I., 1985, Two-dimensional simulation by turning bands: *Mathematical Geology,* v. 17, no. 1, p. 81–90. [9]

Christakos, G., 1984, On the problem of permissible covariance and variogram models: *Water Resources Research,* v. 20, no. 2, p. 251–265. [5]

Christakos, G., 1992, *Random Field Models in Earth Sciences:* Academic Press, San Diego, California, 474 pp. [9, 13]

Christensen, R., 1991, *Linear Models for Multivariate, Time Series, and Spatial Data:* Springer–Verlag, 317 pp. [1, 5]

Clark, I., 1979, *Practical Geostatistics:* Applied Science Publishers Ltd., London, England, 129 pp. [5]

Cooper, L., 1974, *Applied Nonlinear Programming for Engineers and Scientists:* Aloray Publisher, Englewood, New Jersey, 251 pp. [4, 6]

Cressie, N.A.C., 1985, Fitting variogram models by weighted least squares: *Mathematical Geology,* v. 17, no. 5, p. 563–568. [5]

Cressie, N.A.C., 1990, The origins of kriging: *Mathematical Geology,* v. 22, no. 3, p. 239–252. [4]

Cressie, N.A.C., 1991, *Statistics for Spatial Data:* John Wiley & Sons, Inc., New York, 900 pp. [1, 3, 5]

Cressie, N.A.C. and M.O. Grondona, 1992, A comparison of variogram estimation with covariogram estimation, *in* Mardia, K.V., ed., *The Art of Statistical Science:* John Wiley & Sons, Ltd., New York, p. 191–208. [5]

Damsleth, E., C.B. Tjølsen, H. Omre, and H.H. Haldorsen, 1992, A two-stage stochastic model applied to a North Sea reservoir: *Journal of Petroleum Technology,* v. 44, no. 4, p. 402–408, 486. [9]

David, M., 1977, *Geostatistical Ore Reserve Estimation:* Developments in Geomathematics 2, Elsevier Scientific Company, Amsterdam, 364 pp. [9]

Davis, G.J. and M.D. Morris, 1997, Six factors which affect the condition number of matrices associated with kriging: *Mathematical Geology,* v. 29, no. 5, p. 669–683. [4,5]

Davis, M.W., 1987, Production of conditional simulations via the LU triangular decomposition of the covariance matrix: *Mathematical Geology,* v. 19, no. 2, p. 91–98. [9]

Deutsch, C.V. and P.W. Cockerham, 1994, Practical considerations in the application of simulated annealing to stochastic simulation: *Mathematical Geology,* v. 26, no. 1, p. 67–82. [9]

Deutsch, C.V. and A.G. Journel, 1992, *GSLIB—Geostatistical Software Library and User's Guide:* Oxford University Press, New York, 340 pp., 2 diskettes. [1, 3, 5, 9]

Dubrule, O., 1994, Estimating or choosing a geostatistical model, *in* Dimitrakopoulos, R., ed., *Geostatistics for the Next Century:* Kluwer Academic Publishers, Dordrecht, The Netherlands, p. 3–14. [5]

Englund, E. and A. Sparks, 1988, *Geo-EAS (Geostatistical Environmental Assessment Software): User's Guide:* EPA Report #60018-91/008, Las Vegas, Nevada, 172 pp. [1, 5]

Galli, A., H. Beucher, G. Le Loc'h, and B. Doligez, 1994, The pros and cons of the truncated Gaussian method, *in* Armstrong, M. and P.A. Dowd, eds., *Geostatistical Simulations:* Kluwer Academic Publishers, Dordrecht, The Netherlands, 217–233. [9]

Goldberg, A.S., 1962, Best linear unbiased prediction in the generalized linear regression model: *Journal of the American Statistical Association,* v. 57, no. 298, p. 369–375. [4]

Golub, G.H. and C.F. Van Loan, 1989, *Matrix Computations, 2nd Ed.:* The Johns Hopkins University Press, 642 pp. [9, 13]

Goovaerts, P., 1997, *Geostatistics for Natural Resources Evaluation:* Oxford University Press, 483 pp. [13]

Gotway, C.A., 1991, Fitting semivariogram models by weighted least squares: *Computers & Geosciences,* v. 17, no. 1, p. 171–172. [5]

Goulard, M. and M. Voltz, 1992, Linear coregionalization model: Tools for estimation and choice of cross-variogram matrix: *Mathematical Geology,* v. 24, no. 3, p. 269–286. [13]

Haldorsen, H.H. and E. Damsleth, 1990, Stochastic modeling: *Journal of Petroleum Technology,* v. 42, no. 4, p. 404–412. [9]

Harff, J. and J.C. Davis, 1990, Regionalization in geology by multivariate classification: *Mathematical Geology,* v. 22, no. 5, p. 573–588. [14]

Helterbrand, J.D. and N. Cressie, 1994, Universal cokriging under intrinsic coregionalization: *Mathematical Geology,* v. 26, no. 2, p. 205–226. [13]

Hillier, F.S. and G.J. Lieberman, 1995, *Introduction to Operations Research:* McGraw–Hill, Inc., New York, 998 pp. [4, 6, 8, 12, 13]

IMSL, 1987, *Stat/Library—Fortran Subroutines for Statistical Analysis:* IMSL, Houston, Texas, 1231 pp. plus indices. [3]

Isaaks, E.H. and R.M. Srivastava, 1988, Spatial continuity measures for probabilistic and deterministic geostatistics: *Mathematical Geology,* v. 20, no. 4, p. 313–341. [5]

Isaaks, E.H. and R.M. Srivastava, 1989, *An Introduction to Applied Geostatistics:* Oxford University Press, New York, 561 pp. [3, 12]

Itô, K. (ed.), 1987, *Encyclopedic Dictionary of Mathematics:* English transl. from *3rd* Japanese Ed., Massachusetts Institute of Technology, 4 vols., 2148 pp. [2, 4]

Jian, X., R.A. Olea, and Y.-S. Yu, 1996, Semivariogram modeling by weighted least squares: *Computers & Geosciences,* v. 22, no. 4, p. 387–397. [5]

Jones, T.A., D.E. Hamilton, and C.R. Johnson, 1986, *Contouring Geologic Surfaces with the Computer:* Van Nostrand Reinhold Company, New York, 314 pp. [C]

Journel, A.G., 1969, *Etude sur l'Estimation d'une Variable Regionalisée—Application à la Cartographie Sous-Marine:* Service Central Hydrographique de la Marine, Paris, 86 pp. [6]

Journel, A.G., 1986, Geostatistics: Models and tools for the earth sciences: *Mathematical Geology,* v. 18, no. 1, p. 119–140. [2]

Journel, A.G. and C.J. Huijbregts, 1978, *Mining Geostatistics:* Academic Press, London, England, 600 pp. [1, 5, 12, 13]

Kendall, M., 1980, *Multivariate Analysis, 2nd Ed.:* Charles Griffin & Company Ltd., High Wycombe, England, 210 pp. [14]

Kirkpatrick, S., C.D. Gellat, Jr., and M.P. Vecchi, 1983, Optimization by simulated annealing: *Science,* v. 220, no. 4598, p. 671–680. [9]

Kitanidis, P.K., 1997, *Introduction to Geostatistics: Applications to Hydrology:* Cambridge University Press, New York, 249 pp. [5]

Kogan, R.I., 1986, *Interval'nye ocenki v Geologicheskich Issledovanijach:* Nedra Press, Moscow, 335 pp. [14]

Kolmogorov, A.N., 1939, Sur l'interpolation et extrapolation des suites stationnaires: *Comptes Rendus Académie des Sciences,* Paris, v. 208, p. 2043–2045. [2]

Kolmogorov, A.N., 1941, The distribution of energy in local isotropic turbulence: *Doklady Akademii Nauk SSSR,* Moscow, v. 32, p. 19–31. [2]

Krige, D.G., 1951, A statistical approach to some basic mine valuation problems on the Witwatersrand: *Journal of the Chemical, Metallurgical, and Mining Society of South Africa,* v. 52, p. 119–139. [2]

Laarhoven, P.J.M. van and E.H.L. Aarts, 1987, *Simulated Annealing: Theory and Applications:* D. Reidel Publishing Company, Dordrecht, The Netherlands, 186 pp. [9]

Lachenbruch, P.A., 1982, Discriminant analysis, *in* Kotz, S. and N.L. Johnson, eds., *Encyclopedia of Statistical Sciences, Vol. 2:* John Wiley & Sons, New York, p. 389–397. [14]

Levenberg, K., 1944, A method for the solution of certain non-linear problems in least squares: *Quarterly of Applied Mathematics,* v. II, no. 2, p. 164–168. [5]

Luenberger, D.G., 1969, *Optimization by Vector Space Methods:* John Wiley & Sons, New York, 326 pp. [2]

Luster, G.R., 1985, *Raw Materials for Portland Cement—Applications of conditional simulation of coregionalization:* Doctoral dissertation, Stanford University, 531 pp. [9]

Mantoglou, A. and J.L. Wilson, 1982, The turning bands method for simulation of random fields using line generation by a spectral method: *Water Resources Research,* v. 18, no. 15, p. 1379–1394. [9]

Marquardt, D.E., 1963, An algorithm for least squares estimation of nonlinear parameters: *Journal of the Society for Industrial and Applied Mathematics,* v. 11, no. 2, p. 431–441. [5]

Matheron, G., 1962, *Traité de Géotatistique Appliquée, Tome I: Mémoires du Bureau de Recherches Géologiques et Minières:* Editions Technip, Vol. 14, Paris, 333 pp. [1, 2]

Matheron, G., 1963, *Traité de Géotatistique Appliquée, Tome II: Mémoires du Bureau de Recherches Géologiques et Minières:* Editions Technip, Vol. 24, Paris, 171 pp. [2, 12]

Matheron, G., 1973, Intrinsic random functions and their applications: *Advances in Applied Probability,* No. 5, p. 439–468. [9]

McBratney, A.B. and R. Webster, 1986, Choosing functions for semi-variograms of soil properties and fitting them to sampling estimates: *Journal of Soil Science,* v. 37, no. 4, p. 617–639. [5]

McLachlan, G.J., 1992, *Discriminant Analysis and Statistical Pattern Recognition:* John Wiley & Sons, Inc., New York, 526 pp. [14]

Metropolis, N., A.W. Rosenbluth, M.N. Rosenbluth, A.H. Teller, and E. Teller, 1953, Equation of state calculations by fast computing machines: *The Journal of Chemical Physics,* v. 21, no. 6, p. 1087–1092. [9]

Mojena, R., 1988, Ward's clustering algorithm, *in* Kotz, S. and N.L. Johnson, eds., *Encyclopedia of Statistical Sciences, Vol. 9:* John Wiley & Sons, New York, p. 529–532. [14]

Myers, D.E., 1982, Matrix formulation of cokriging: *Mathematical Geology,* v. 14, no. 3, p. 249–257. [13]

Myers, D.E., 1991, Interpolation and estimation with spatially located data: *Chemometrics and Intelligent Laboratory Systems,* v. 11, no. 3, p. 209–228. [4]

Olea, R.A., 1984, Sampling design optimization for spatial functions: *Mathematical Geology,* v. 16, no. 4, p. 369–392. [2]

Olea, R.A. (ed.), 1991, *Geostatistical Glossary and Multilingual Dictionary* (auspices of International Association for Mathematical Geology, 1984–1989 Committee on Geostatistics—G. Christakos, I. Clark, M. David, A.G. Journel, D.G. Krige, R.A. Olea, Chm.): Oxford University Press, Studies in Mathematical Geology No. 3, 177 pp. [1, 5, 8]

Olea, R.A., 1996, XVAN: A computer program for the analysis of spatial estimation errors: *Computers & Geosciences,* v. 22, no. 4, p. 445–448. [7]

Pawlowsky, V., 1989, Cokriging of regionalized compositions: *Mathematical Geology,* v. 21, no. 5, p. 513–521. [13]

Pawlowsky-Glahn, V. and R.A. Olea, *n.d.,* Spatial Correlation and Characterization of Compositional Data in the Geosciences, *in prep.* [13]

Press, W.H., S.A. Teukolsky, W.T. Vetterling, and B.P. Flannery, 1992, *Numerical Recipes in Fortran, 2nd Ed.:* Cambridge University Press, New York, 963 pp. [3, 5]

Radon, J., 1917, Über die Bestimmung von Funktionen durch ihre Integralwerte langs gewisser Mannigfaltigkeiten: *Berichte Sachs. Akad. der Wissenschaften,* Leipzig, Math. Phys. Kl., v. 69, p. 262–267. [9]

Rivoirard, J., 1990, A review of lognormal estimators for *in situ* reserves: *Mathematical Geology,* v. 22, no. 2, p. 213–221. [3]

Rodionov, D.A., 1981, *Statisticheskie Rezhenija v Geologii:* Nedra Press, Moscow, 231 pp. [14]

Rogers, G.S., 1980, *Matrix Derivatives: Lecture Notes on Statistics, Vol. 2:* Marcel Dekker, Inc., New York, 209 pp. [13]

SAS Institute, 1990, SAS/STAT *User's Guide, Version 6, Vol. 1:* SAS Institute Inc., Cary, North Carolina, 943 pp. [14]

Tompson, A.F.B., R. Ababou, and L.W. Gelhar, 1989, Implementation of the three-dimensional turning bands random field generator: *Water Resources Research,* v. 25, no. 10, p. 2227–2243. [9]

Voronin, J.A., 1967, *Geologija i Matematika:* Nauka Press, Novosibirsk, 253 pp. [14]

Wackernagel, H., 1995, *Multivariate Geostatistics:* Springer–Verlag, Berlin, 256 pp. [13]

Wiener, N., 1949, *The Extrapolation, Interpolation, and Smoothing of Stationary Time Series with Engineering Applications:* Published jointly by the Technology Press of the Massachusetts Institute of Technology, Cambridge, Massachusetts, John Wiley & Sons, Inc., New York, and Chapman & Hall, Ltd., London, 163 pp. [2]

Wold, H., 1938 [1954], *A Study in the Analysis of Stationary Time Series:* Almqvist & Wiksell, Stockholm, Sweden, 211 p.; *2nd* Edition, 1954, with Appendix by Peter Whittle, 236 pp. [2]

Yao, T. and A. Journel, 1998, Automatic modeling of (cross) covariance tables using fast Fourier transform: *Mathematical Geology,* v. 30, no. 6, p. 589–615. [13]

Appendix A

West Lyons Field Sampling

West Lyons field is in west-central Kansas. The reservoir occurs in Mississippian (Lower Carboniferous) sediments deposited in the shallow epicontinental seas that covered much of North America in the Late Paleozoic. The data, collected as part of an engineering study by Kewanee Oil Company, are on the Internet at http://www.kgs.ukans.edu/Mathgeo/Books/Geostat/index.html and may be downloaded from that site.

Table A.1. West Lyons field sampling.

Well	Easting (miles)	Northing (miles)	Porosity (%)	Well	Easting (miles)	Northing (miles)	Porosity (%)
3	2.097	4.611	12.7	5	1.863	4.370	14.1
6	2.112	4.371	11.6	7	2.175	4.317	11.2
10	1.634	4.136	15.2	11	1.880	4.126	14.7
12	2.096	4.129	13.3	14	1.613	3.885	17.2
15	1.841	3.883	14.7	16	2.113	3.881	12.4
18	1.610	3.628	13.4	19	1.861	3.627	13.7
20	2.088	3.627	13.9	21	2.358	3.627	14.5
23	1.365	3.381	13.1	24	1.614	3.382	14.5
25	1.863	3.380	13.2	26	2.088	3.380	11.9
29	1.342	3.270	13.4	30	1.177	3.131	12.1
31	1.363	3.131	13.8	32	1.614	3.130	13.6
33	1.862	3.132	11.5	34	2.117	3.129	12.5
36	1.118	2.883	12.4	37	1.363	2.880	14.7
38	1.614	2.881	11.7	39	1.861	2.882	11.5
40	1.924	2.943	11.5	41	2.113	2.878	11.4
44	1.151	2.628	12.8	45	1.362	2.628	12.1
46	1.615	2.631	10.6	47	1.862	2.629	11.5

(*cont.*)

Table A.1. West Lyons field sampling (*concluded*).

Well	Easting (miles)	Northing (miles)	Porosity (%)	Well	Easting (miles)	Northing (miles)	Porosity (%)
48	2.088	2.626	10.8	49	1.140	2.381	11.4
50	1.373	2.380	11.4	51	1.615	2.378	11.7
52	1.862	2.379	11.3	53	2.127	2.388	10.5
54	2.363	2.377	10.3	57	1.142	2.127	10.7
58	1.364	2.129	11.1	59	1.616	2.127	10.5
60	1.864	2.127	13.9	61	2.092	2.124	10.6
62	2.363	2.125	9.9	63	2.612	2.124	10.5
65	1.121	1.878	10.9	66	1.369	1.880	11.7
67	1.676	1.937	11.2	68	1.866	1.880	11.7
69	2.116	1.881	10.4	70	2.339	1.875	10.4
73	1.123	1.626	11.5	74	1.369	1.627	11.6
75	1.614	1.628	11.6	76	1.867	1.627	11.1
77	2.117	1.626	10.3	80	0.887	1.396	18.2
81	1.157	1.371	11.3	82	1.372	1.373	11.5
83	1.616	1.374	10.9	84	1.868	1.386	11.3
86	1.137	1.133	10.8	87	1.373	1.122	11.0
88	1.616	1.123	10.5	89	1.867	1.121	11.9
91	1.121	0.872	8.0	92	1.400	0.874	11.1
93	1.592	0.899	10.5	94	1.849	0.872	10.4
95	1.120	0.616	9.5	96	1.397	0.619	9.7
97	1.623	0.639	10.2	99	1.374	0.371	9.7

Appendix B

High Plains Aquifer Sampling

This sampling is a subset of a larger annual measurement of water levels in the Kansas portion of the High Plains aquifer. The survey was conducted by the State Board of Agriculture of Kansas and the United States Geological Survey during the period December 1980 to March 1981. The High Plains aquifer is late Tertiary to Quaternary and consists of alluvial, dune-sand, and valley fill deposits from streams flowing eastward from the Rocky Mountains. In the following table, L.S.E. stands for land surface elevation, W.D. for water depth, and W.T.E. for water table elevation. The data are on the Internet at http://www.kgs.ukans.edu/Mathgeo/Books/Geostat/index.html and may be downloaded from that site.

Table B.1. Kansas High Plains aquifer sampling.

ID	County	Legal description of location	Easting (miles)	Northing (miles)	L.S.E. (ft)	W.D. (ft)	W.T.E. (ft)
15	Cheyenne	2S 40W 28DBA 1	33.02509	211.16249	3452.0	113.5	3338.5
16	Cheyenne	2S 41W 27BBD 1	27.60393	211.75963	3620.0	201.0	3419.0
17	Cheyenne	2S 42W 14DDD 1	23.47554	213.19019	3693.0	232.5	3460.5
18	Cheyenne	3S 37W 19BBC 1	48.10212	206.00024	3468.0	228.8	3239.2
19	Cheyenne	3S 37W 21DDD 1	50.91596	205.15459	3422.0	218.2	3203.8
20	Cheyenne	3S 37W 36ADB 1	53.70786	203.69003	3381.0	201.4	3179.6
21	Cheyenne	3S 39W 20DAC 1	37.90396	205.87445	3450.0	141.8	3308.2
22	Cheyenne	3S 39W 24DDD 1	41.98035	205.47758	3505.0	220.0	3285.0
23	Cheyenne	3S 39W 32BDB 1	37.34115	204.27779	3490.0	154.2	3335.8
24	Cheyenne	3S 40W 9BAA 2	32.67124	208.68553	3358.0	20.5	3337.5
27	Cheyenne	3S 40W 35AAC 1	34.86293	204.49368	3445.0	101.2	3343.9
28	Cheyenne	3S 41W 16AAC 1	27.06479	207.79530	3570.0	134.3	3435.7
29	Cheyenne	3S 41W 33ABB 1	26.69602	204.94556	3594.0	173.5	3420.5
30	Cheyenne	3S 42W 4AAA 1	21.35551	210.17715	3727.0	231.1	3495.9

(*cont.*)

Table B.1. High Plains aquifer sampling (*continued*).

ID	County	Legal description of location		Easting (miles)	Northing (miles)	L.S.E. (ft)	W.D. (ft)	W.T.E. (ft)
31	Cheyenne	4S 37W 14BAB	1	52.11569	201.01073	3405.0	200.9	3204.1
32	Cheyenne	4S 37W 25DCA	1	53.37287	198.23169	3374.0	151.1	3222.9
33	Cheyenne	4S 38W 4BAC	1	44.25118	203.15585	3509.0	215.4	3293.6
34	Cheyenne	4S 38W 20CCC	1	42.85724	199.47672	3485.0	155.6	3329.4
35	Cheyenne	4S 38W 21ADC	1	44.61671	199.91013	3491.0	185.9	3305.1
36	Cheyenne	4S 39W 2DBC	1	40.50639	202.79491	3520.0	200.8	3319.2
37	Cheyenne	4S 39W 15CCA	1	39.05595	200.73232	3492.0	150.4	3341.6
38	Cheyenne	4S 39W 18CAB	1	36.21449	201.08405	3530.0	156.9	3373.1
39	Cheyenne	4S 39W 27CCA	1	38.97412	198.74550	3502.0	135.4	3366.6
40	Cheyenne	4S 40W 22BCB	1	32.94634	200.45740	3520.0	125.2	3394.9
41	Cheyenne	4S 41W 16DAA	1	26.91493	201.44768	3403.0	15.8	3387.2
42	Cheyenne	4S 41W 25BCB	1	28.93127	199.61497	3571.0	142.1	3428.9
43	Cheyenne	4S 41W 31ACA	1	24.58318	198.80919	3552.0	98.7	3453.3
44	Cheyenne	4S 42W 26BDB	1	22.27446	199.90916	3487.8	18.6	3469.2
45	Cheyenne	5S 37W 15DBB	1	51.11561	194.58136	3397.0	152.1	3244.9
46	Cheyenne	5S 38W 13BAD	1	47.03086	195.10007	3390.0	77.5	3312.5
47	Cheyenne	5S 38W 22ACB	1	45.12255	194.04909	3437.0	97.6	3339.4
48	Cheyenne	5S 38W 26CCA	1	45.68245	192.53780	3460.0	117.7	3342.3
49	Cheyenne	5S 39W 11CBC	1	39.72585	195.85860	3530.0	152.6	3377.4
50	Cheyenne	5S 39W 18CCC	1	35.69878	194.75821	3630.0	216.1	3413.9
51	Cheyenne	5S 39W 25CDA	1	40.96551	192.70605	3533.0	132.1	3400.9
52	Cheyenne	5S 40W 4CBD	1	31.93300	197.13242	3653.0	207.9	3445.1
53	Cheyenne	5S 40W 15ACB	1	33.23453	195.46733	3650.0	226.2	3423.8
54	Cheyenne	5S 40W 27BBA	1	32.78741	193.74037	3658.0	204.2	3453.8
55	Cheyenne	5S 41W 12ADC	1	29.53741	196.47363	3679.0	211.9	3467.1
56	Cheyenne	5S 41W 20DAA	1	25.63076	194.52692	3742.0	225.4	3516.6
57	Cheyenne	5S 42W 4AAB	1	20.70889	198.23430	3515.0	17.9	3497.1
58	Cheyenne	5S 42W 14CBC	1	21.83602	195.56645	3687.0	153.6	3533.4
59	Cheyenne	5S 42W 36CCB	1	22.69358	192.41327	3800.0	212.8	3587.2
61	Decatur	1S 26W 18DDB	1	114.35582	216.42046	2413.0	31.5	2381.5
62	Decatur	1S 29W 3DDB	1	99.57481	218.75893	2539.0	31.6	2507.4
63	Decatur	1S 29W 19BDD	1	96.20944	216.23335	2572.0	24.0	2547.9
64	Decatur	1S 30W 34DDD	1	93.66951	213.81834	2610.0	35.1	2574.9
65	Decatur	2S 26W 11BBA	1	117.58112	212.08537	2509.0	85.2	2423.8
66	Decatur	2S 28W 13ABA	1	107.20194	211.35425	2487.0	29.9	2457.1
67	Decatur	2S 29W 24BCC	1	100.57094	210.15475	2682.2	94.8	2587.4
68	Decatur	2S 30W 26DCC	1	94.14078	208.82616	2835.0	137.2	2697.8
69	Decatur	3S 26W 30CBB	2	113.27920	202.75702	2610.0	126.0	2484.0
70	Decatur	3S 28W 6DCB	1	101.96583	206.76294	2571.0	44.8	2526.1
71	Decatur	3S 28W 32BCA	1	102.47420	202.27351	2749.0	142.7	2606.3

(*cont.*)

Table B.1. High Plains aquifer sampling (*continued*).

ID	County	Legal description of location	Easting (miles)	Northing (miles)	L.S.E. (ft)	W.D. (ft)	W.T.E. (ft)
72	Decatur	3S 29W 12BBA 1	100.60022	206.54837	2556.1	31.8	2524.3
73	Decatur	3S 29W 17DCB 1	96.98643	204.89520	2587.0	24.6	2562.4
74	Decatur	3S 29W 31DCC 1	95.91154	201.81171	2633.0	26.9	2606.1
75	Decatur	3S 30W 3CBA 1	92.73382	207.24669	2807.0	101.1	2705.9
76	Decatur	4S 26W 8DDD 1	115.04419	199.34091	2455.7	31.9	2423.8
77	Decatur	4S 26W 19DCA 1	113.76634	197.51254	2464.0	17.2	2446.8
78	Decatur	4S 27W 17DAC 1	108.97080	198.74921	2648.0	105.2	2542.8
79	Decatur	4S 27W 33BBB 1	109.15633	196.38132	2528.0	19.5	2508.6
80	Decatur	4S 28W 15AAA 1	105.14063	199.46748	2700.0	92.2	2607.8
81	Decatur	4S 28W 30DDD 1	102.08223	196.68979	2726.0	91.2	2634.8
82	Decatur	4S 30W 7BBB 1	89.45439	200.85693	2697.0	15.3	2681.7
83	Decatur	5S 26W 5ADD 1	114.93300	194.86301	2607.0	127.6	2479.4
84	Decatur	5S 26W 26DDA 1	117.77794	190.42242	2437.0	23.6	2413.4
85	Decatur	5S 26W 33DCC 1	115.41256	189.37129	2475.0	18.9	2456.1
86	Decatur	5S 27W 21CCA 1	109.15672	191.65317	2675.0	98.7	2576.3
87	Decatur	5S 28W 7BBC 1	101.15601	194.47743	2644.0	20.0	2624.0
88	Decatur	5S 28W 10BBB 1	104.13999	194.51997	2600.0	10.9	2589.1
89	Decatur	5S 28W 14ADD 1	105.96646	193.10306	2723.0	134.5	2588.5
90	Decatur	5S 28W 17DAC 1	102.85614	192.93822	2734.0	101.7	2632.4
91	Decatur	5S 29W 11BAA 1	99.54774	194.63818	2670.0	13.1	2656.9
92	Decatur	5S 29W 22CBB 1	98.11600	192.18419	2686.0	15.8	2670.2
372	Gove	11S 26W 4CDC 1	116.29624	158.43758	2583.0	63.1	2519.9
373	Gove	11S 27W 16AAA 1	110.98627	157.46828	2713.0	110.3	2602.7
374	Gove	11S 27W 36BCC 1	112.96054	154.06345	2676.0	77.2	2598.8
375	Gove	11S 29W 4DAD 1	99.28117	159.16090	2844.0	113.5	2730.5
376	Gove	11S 30W 27ABB 1	93.82134	155.98224	2922.0	130.3	2791.7
377	Graham	6S 21W 19CDC 1	143.88550	184.84438	2305.0	102.2	2202.8
378	Graham	6S 22W 19CCC 1	137.67519	184.90807	2400.0	108.5	2291.5
379	Graham	6S 22W 28ACA 1	140.27769	184.50719	2360.0	114.3	2245.7
380	Graham	6S 23W 13BBB 1	136.72041	186.79459	2350.0	57.9	2292.1
381	Graham	6S 24W 28BAB 1	127.98153	184.98848	2479.0	99.2	2379.8
382	Graham	6S 24W 29BAB 1	126.98682	185.00719	2510.0	123.1	2386.9
383	Graham	6S 24W 35DDD 1	130.55019	183.07536	2492.0	146.6	2345.4
384	Graham	6S 25W 28CBC 1	121.73193	184.47917	2540.0	107.3	2432.7
385	Graham	6S 25W 35CCA 1	123.82757	183.31992	2512.0	98.8	2413.2
386	Graham	7S 22W 10BBC 1	140.59123	181.64490	2217.0	8.8	2208.2
387	Graham	7S 22W 19BBB 1	137.56969	179.81061	2295.0	38.9	2256.1
388	Graham	7S 24W 8CBA 1	126.77754	181.52431	2519.0	126.5	2392.5
389	Graham	7S 25W 24BBB 1	124.62090	180.06818	2495.0	86.2	2408.8
390	Graham	9S 22W 19BBB 1	137.32347	167.87659	2416.0	95.7	2320.4

(*cont.*)

Table B.1. High Plains aquifer sampling (*continued*).

ID	County	Legal description of location	Easting (miles)	Northing (miles)	L.S.E. (ft)	W.D. (ft)	W.T.E. (ft)
391	Graham	9S 24W 22BAA 1	128.70850	168.04491	2491.0	93.9	2397.1
392	Graham	9S 25W 14DDD 1	124.20194	168.24323	2534.0	91.4	2442.6
833	Logan	11S 32W 4ACD 1	81.24866	159.91974	3059.0	111.7	2947.3
834	Logan	11S 32W 19AAB 1	79.29759	157.38274	3073.0	102.9	2970.1
835	Logan	11S 36W 6ADD 2	55.74755	160.73698	3380.0	162.5	3217.5
836	Logan	11S 36W 6DBB 2	55.36806	160.62888	3386.0	172.6	3213.4
968	Rawlins	1S 33W 29CCC 1	73.20178	215.37129	2992.0	114.0	2878.0
969	Rawlins	2S 31W 3CAD 1	87.23461	213.25276	2665.0	20.3	2644.7
970	Rawlins	2S 32W 14DCA 1	82.50986	211.26065	2723.0	23.2	2699.8
971	Rawlins	2S 32W 20DCD 1	79.54680	210.22064	2735.0	15.4	2719.6
972	Rawlins	2S 35W 13ABB 1	65.67006	212.50679	3178.0	152.6	3025.4
973	Rawlins	2S 35W 34CAA 1	63.44318	209.10126	3064.0	31.0	3033.0
974	Rawlins	2S 36W 13DDD 1	60.06818	211.83514	3286.0	190.6	3095.4
975	Rawlins	2S 36W 15CDD 1	57.60354	211.90451	3334.0	213.4	3120.6
976	Rawlins	2S 36W 18CCB 1	54.28070	212.12453	3380.0	230.0	3150.0
977	Rawlins	2S 36W 36BAA 1	59.49897	209.73721	3263.0	176.7	3086.3
978	Rawlins	3S 33W 3DCC 1	75.39347	207.35251	2823.0	30.5	2792.5
979	Rawlins	3S 33W 8CDC 1	73.13605	206.42188	2855.0	27.9	2827.1
980	Rawlins	3S 34W 3ABB 1	69.49242	208.39978	2882.0	15.9	2866.1
981	Rawlins	3S 34W 26BAC 1	70.09028	204.27596	2900.0	14.9	2885.1
982	Rawlins	3S 34W 33BCC 1	67.82110	203.10432	2945.0	18.5	2926.5
983	Rawlins	3S 35W 24CBB 1	64.91588	205.06747	3001.0	28.5	2972.5
984	Rawlins	3S 36W 14CBB 1	58.01420	206.29720	3332.0	221.4	3110.6
985	Rawlins	3S 36W 17CCC 1	55.03520	206.01144	3375.0	210.9	3164.1
986	Rawlins	4S 31W 16ABD 1	86.09967	199.83231	2761.0	13.0	2748.0
987	Rawlins	4S 31W 25DDD 1	89.21709	196.99953	2755.0	18.7	2736.3
988	Rawlins	4S 33W 18DDA 1	72.52644	199.60220	3068.0	85.6	2982.4
989	Rawlins	4S 33W 28DCA 1	74.19326	197.56693	3125.0	155.9	2969.1
990	Rawlins	4S 34W 33CBC 1	67.60480	196.89149	3160.0	119.7	3040.3
991	Rawlins	4S 35W 6DCD 1	60.45597	201.86806	3252.0	154.8	3097.2
992	Rawlins	4S 35W 13DAD 1	65.60748	199.94058	3002.0	16.4	2985.6
993	Rawlins	4S 35W 29DDD 1	61.55627	197.84683	3219.0	154.0	3065.0
994	Rawlins	4S 36W 6BBB 1	53.92708	202.93695	3370.0	190.0	3180.0
995	Rawlins	5S 31W 10DDA 1	87.16438	194.19531	2820.0	45.7	2774.3
996	Rawlins	5S 31W 20CCA 1	84.39804	192.28889	2865.0	37.4	2827.6
997	Rawlins	5S 31W 23DDD 1	88.08633	192.04869	2950.0	122.1	2827.9
998	Rawlins	5S 32W 14CDD 1	81.71686	193.24323	3020.0	130.8	2889.2
999	Rawlins	5S 33W 29BDA 1	72.76231	192.13731	3042.0	20.0	3022.0
1000	Rawlins	5S 34W 1BBB 1	70.55563	196.42740	3137.0	116.0	3021.0
1001	Rawlins	5S 34W 28ADC 1	68.18063	192.14781	3207.0	135.1	3071.9

(*cont.*)

Table B.1. High Plains aquifer sampling (*continued*).

ID	County	Legal description of location	Easting (miles)	Northing (miles)	L.S.E. (ft)	W.D. (ft)	W.T.E. (ft)
1002	Rawlins	5S 35W 10CDD 1	62.94145	194.81029	3267.0	167.6	3099.4
1003	Rawlins	5S 36W 21BCD 1	55.68056	193.55966	3220.0	20.0	3200.0
1234	Sheridan	6S 27W 3DCD 1	111.60567	188.48280	2560.0	34.9	2525.1
1235	Sheridan	6S 27W 8DCA 1	109.60543	187.65971	2588.0	20.8	2567.2
1236	Sheridan	6S 27W 19DAB 1	108.69437	185.93782	2610.0	30.8	2579.2
1237	Sheridan	6S 27W 27BCC 1	110.89686	185.00931	2716.0	159.4	2556.6
1238	Sheridan	6S 28W 28ADD 1	104.83452	185.16824	2682.0	46.8	2635.2
1239	Sheridan	6S 29W 10DBC 1	99.60456	188.04247	2823.0	131.1	2691.9
1240	Sheridan	6S 29W 24ABB 1	101.53125	186.62839	2781.0	102.3	2678.7
1241	Sheridan	6S 29W 33CDA 1	98.39173	183.96124	2828.0	104.5	2723.5
1242	Sheridan	6S 30W 2BCA 1	94.34391	189.52928	2780.0	22.8	2757.2
1243	Sheridan	6S 30W 13BAA 1	95.54009	187.76160	2875.0	126.7	2748.3
1244	Sheridan	6S 30W 14CCD 1	94.27328	186.91888	2884.0	109.2	2774.8
1245	Sheridan	6S 30W 24DDC 2	95.86261	185.88857	2898.0	138.2	2759.8
1246	Sheridan	7S 26W 6AAB 1	114.55729	183.29419	2634.0	129.8	2504.2
1247	Sheridan	7S 26W 12BAC 1	118.93845	182.04451	2559.0	98.4	2460.6
1248	Sheridan	7S 26W 15BDD 1	117.07079	180.85069	2622.0	142.2	2479.8
1249	Sheridan	7S 26W 19BBC 1	113.74282	180.20013	2625.0	123.3	2501.7
1250	Sheridan	7S 26W 28CAB 1	115.91517	178.76468	2634.0	154.4	2479.6
1251	Sheridan	7S 27W 24BBD 1	112.87737	180.22231	2639.0	128.9	2510.1
1252	Sheridan	7S 28W 8BDC 1	103.14765	182.22333	2808.0	160.6	2647.4
1253	Sheridan	7S 28W 19BBA 1	101.99171	180.63542	2800.0	147.1	2652.9
1254	Sheridan	7S 28W 21ABB 1	104.34249	180.57063	2774.0	155.9	2618.1
1255	Sheridan	7S 28W 36ABA 1	107.38273	178.49336	2725.0	139.3	2585.7
1256	Sheridan	7S 29W 2DBD 1	100.57860	183.03732	2838.0	151.8	2686.2
1257	Sheridan	7S 29W 21ABB 1	98.42858	180.72293	2860.0	149.8	2710.2
1258	Sheridan	7S 29W 27CCC 1	98.87034	178.84390	2869.0	177.1	2691.9
1259	Sheridan	7S 29W 30ABA 1	96.55903	179.77374	2886.6	151.9	2734.7
1260	Sheridan	8S 26W 14DAA 1	118.39418	174.70265	2398.0	16.4	2381.6
1261	Sheridan	8S 26W 16CDD 1	115.92953	174.39923	2447.0	36.0	2411.0
1262	Sheridan	8S 27W 11DCD 1	112.26002	175.49921	2504.0	10.4	2493.6
1263	Sheridan	8S 27W 33BBD 1	109.69800	172.32497	2554.0	19.9	2534.1
1264	Sheridan	8S 28W 9ABC 1	104.23516	176.46024	2766.0	136.0	2630.0
1265	Sheridan	8S 29W 1DCB 1	101.27975	176.91501	2823.0	145.8	2677.2
1266	Sheridan	8S 29W 29BAA 1	97.13605	173.78354	2844.0	98.4	2745.6
1267	Sheridan	8S 30W 11CBC 1	93.85859	176.22948	2941.0	177.0	2764.0
1268	Sheridan	8S 30W 13DAA 1	95.69942	175.31519	2891.0	138.9	2752.1
1269	Sheridan	8S 30W 30ABC 1	90.31005	173.82915	2962.0	127.8	2834.2
1270	Sheridan	9S 27W 12CCC 1	112.47791	169.50900	2678.0	110.3	2567.7
1271	Sheridan	9S 27W 19DDD 1	108.33349	167.62460	2750.0	128.5	2621.5

(*cont.*)

Table B.1. High Plains aquifer sampling (*continued*).

ID	County	Legal description of location	Easting (miles)	Northing (miles)	L.S.E. (ft)	W.D. (ft)	W.T.E. (ft)
1272	Sheridan	9S 28W 4BCC 1	103.60425	171.24298	2677.0	27.0	2650.0
1273	Sheridan	9S 28W 6CCB 1	101.61727	170.92456	2691.0	8.3	2682.7
1274	Sheridan	9S 29W 17BAB 1	96.90593	169.80618	2854.0	99.4	2754.6
1275	Sheridan	9S 29W 26BAA 1	99.93206	167.73027	2863.0	136.4	2726.6
1276	Sheridan	9S 30W 3AAB 2	93.49164	171.88503	2933.0	143.0	2790.0
1277	Sheridan	9S 30W 16CDA 1	92.04995	169.18459	2882.0	60.5	2821.5
1278	Sheridan	10S 27W 22DBA 1	110.91667	161.94460	2568.0	19.8	2548.2
1279	Sheridan	10S 28W 29DAA 1	103.21741	161.15854	2691.0	28.9	2662.1
1280	Sheridan	10S 29W 20CCC 2	96.44839	161.97681	2776.0	13.7	2762.3
1281	Sheridan	10S 30W 8DDD 1	91.41414	164.10204	2930.0	104.8	2825.1
1282	Sheridan	10S 30W 17DAD 1	91.39418	163.35661	2895.5	68.8	2826.7
1283	Sherman	6S 37W 3BCC 1	51.73390	190.70708	3424.0	163.8	3260.2
1284	Sherman	6S 37W 7BBA 1	48.89315	190.18529	3304.0	9.5	3294.5
1285	Sherman	6S 37W 16CDD 1	51.02580	188.24353	3460.0	172.9	3287.1
1286	Sherman	6S 37W 19ABB 1	49.18134	188.18285	3476.0	160.9	3315.1
1287	Sherman	6S 38W 9ABD 1	45.44776	190.18095	3510.0	164.2	3345.8
1288	Sherman	6S 38W 20ACC 1	44.25284	187.98090	3407.0	14.8	3392.2
1289	Sherman	6S 39W 9DDD 1	39.73153	189.63351	3585.0	147.5	3437.5
1290	Sherman	6S 40W 10AAC 1	34.71583	190.55556	3641.0	160.4	3480.6
1291	Sherman	6S 40W 30DCC 1	31.38486	186.93774	3718.0	164.8	3553.2
1292	Sherman	6S 40W 35BCC 1	34.77565	186.31969	3668.0	154.6	3513.4
1293	Sherman	6S 41W 1ABB 1	30.62508	191.82243	3675.0	157.9	3517.1
1294	Sherman	6S 41W 9BCB 1	27.08215	190.73161	3759.0	190.0	3569.0
1295	Sherman	6S 41W 19DBD 1	25.60803	188.42812	3792.0	180.3	3611.7
1296	Sherman	6S 41W 27DBD 1	28.54822	187.30367	3741.0	162.1	3578.9
1297	Sherman	6S 42W 2AAA 1	24.01697	192.11134	3777.0	197.6	3579.4
1298	Sherman	6S 42W 8CBB 1	20.06495	190.78448	3841.0	208.9	3632.1
1299	Sherman	6S 42W 22DCC 1	22.46836	188.31265	3837.0	192.8	3644.2
1300	Sherman	6S 42W 30ADA 1	19.81929	188.05193	3871.0	199.7	3671.3
1301	Sherman	7S 37W 4BBC 1	50.53291	185.02501	3455.0	137.8	3317.2
1302	Sherman	7S 37W 5CCB 1	49.52494	184.43678	3472.0	138.2	3333.8
1303	Sherman	7S 37W 11ACB 1	52.94997	183.82552	3445.0	156.2	3288.8
1304	Sherman	7S 37W 17BDA 1	49.83546	182.93269	3472.0	136.5	3335.5
1305	Sherman	7S 39W 9BBB 1	38.65215	184.56400	3589.0	115.0	3474.0
1306	Sherman	7S 39W 24BAA 1	41.93498	182.45557	3587.0	146.4	3440.6
1307	Sherman	7S 39W 30CCB 1	36.50347	180.90379	3643.0	140.0	3503.0
1308	Sherman	7S 40W 6ADB 1	31.56581	185.56163	3722.0	166.1	3555.9
1309	Sherman	7S 40W 29BBA 1	31.76271	181.82031	3708.0	138.3	3569.7
1310	Sherman	7S 40W 35BBB 1	34.53346	180.72491	3650.0	125.7	3524.3
1311	Sherman	7S 40W 36BAB 1	35.75742	180.68134	3643.0	132.5	3510.5

(*cont.*)

Table B.1. High Plains aquifer sampling (*continued*).

ID	County	Legal description of location				Easting (miles)	Northing (miles)	L.S.E. (ft)	W.D. (ft)	W.T.E. (ft)
1312	Sherman	7S	41W	7BCB	1	24.82813	184.84738	3840.0	194.7	3645.3
1313	Sherman	7S	41W	10BBA	1	27.94381	184.96346	3770.0	160.0	3610.0
1314	Sherman	7S	41W	16ADC	1	27.50986	183.61134	3775.0	156.9	3618.1
1315	Sherman	7S	41W	28DBB	1	27.16651	181.50797	3774.0	126.3	3647.7
1316	Sherman	7S	42W	7DAA	1	19.67677	184.81651	3903.0	199.6	3703.4
1317	Sherman	7S	42W	17CCC	1	19.74179	183.44263	3864.0	137.1	3726.9
1318	Sherman	7S	42W	27AAB	1	22.45147	182.20541	3862.0	162.7	3699.3
1319	Sherman	8S	37W	3ADB	1	52.02509	178.88210	3476.0	157.6	3318.4
1320	Sherman	8S	37W	6ADB	1	49.06211	178.97443	3506.0	140.7	3365.3
1321	Sherman	8S	37W	21CCC	1	50.15672	175.33160	3496.0	140.7	3355.3
1322	Sherman	8S	37W	28ABC	1	50.64236	175.06825	3482.0	130.4	3351.6
1323	Sherman	8S	37W	32ABB	1	49.61774	174.22791	3468.0	97.1	3370.9
1324	Sherman	8S	38W	17CDD	1	43.67566	176.53969	3603.0	162.2	3440.8
1325	Sherman	8S	38W	24AAB	1	47.96551	176.27036	3513.0	123.2	3389.8
1326	Sherman	8S	38W	28ACC	1	44.72049	175.00900	3587.0	145.3	3441.7
1327	Sherman	8S	39W	17DCD	1	37.95541	176.74100	3672.0	183.4	3488.6
1328	Sherman	8S	39W	27AAB	1	40.03354	175.54578	3647.0	179.3	3467.7
1329	Sherman	8S	39W	28CAB	1	38.51342	175.10149	3660.0	169.6	3490.4
1330	Sherman	8S	40W	12DBA	1	36.02020	178.18118	3670.0	169.1	3500.9
1331	Sherman	8S	40W	17CDB	1	31.67053	177.09312	3727.0	131.6	3595.4
1332	Sherman	8S	40W	20CBC	3	31.38518	176.23216	3719.0	120.1	3598.9
1333	Sherman	8S	40W	20CCC	1	31.37397	175.98398	3716.0	114.6	3601.4
1334	Sherman	8S	40W	20DAA	1	32.25087	176.32513	3668.0	72.8	3595.2
1335	Sherman	8S	40W	24AAA	1	36.20305	176.67961	3692.0	185.4	3506.6
1336	Sherman	8S	40W	25AAC	1	36.03314	175.56479	3701.0	182.5	3518.5
1337	Sherman	8S	40W	35CCB	1	34.24227	174.01089	3732.0	173.6	3558.4
1338	Sherman	8S	41W	17CBA	1	25.62610	177.58594	3843.0	143.1	3699.9
1339	Sherman	8S	41W	25BBC	1	29.39015	175.81178	3754.0	115.1	3638.9
1340	Sherman	8S	42W	4BBC	1	20.60259	180.16501	3912.0	181.3	3730.7
1341	Sherman	8S	42W	15DDB	1	22.24408	177.47783	3859.0	122.1	3736.9
1342	Sherman	8S	42W	19ABB	1	19.30151	177.36766	3929.0	171.2	3757.8
1343	Sherman	8S	42W	29ACB	1	19.92053	176.08041	3904.0	124.3	3779.7
1344	Sherman	8S	42W	31DCD	1	19.20080	174.50024	3872.0	76.7	3795.3
1345	Sherman	8S	42W	34DCB	1	21.86174	174.50229	3914.0	148.7	3765.3
1346	Sherman	9S	39W	2BAB	1	40.45297	173.53732	3646.0	168.2	3477.8
1347	Sherman	9S	39W	17BBA	1	37.24487	171.65956	3697.0	152.4	3544.6
1348	Sherman	9S	40W	8CCB	1	31.19815	172.13147	3773.0	155.7	3617.3
1349	Sherman	9S	40W	8CDB	1	31.44484	172.12192	3766.0	167.0	3599.0
1350	Sherman	9S	40W	13CDC	1	35.34099	170.85701	3722.0	161.0	3561.0
1351	Sherman	9S	40W	29BBB	1	31.09635	169.89441	3782.0	156.4	3625.6

(*cont.*)

Table B.1. High Plains aquifer sampling (*continued*).

ID	County	Legal description of location	Easting (miles)	Northing (miles)	L.S.E. (ft)	W.D. (ft)	W.T.E. (ft)
1352	Sherman	9S 41W 5DCC 1	25.80398	173.21733	3860.0	166.1	3693.9
1353	Sherman	9S 41W 14BBC 1	28.21891	171.87453	3835.0	172.0	3663.0
1354	Sherman	9S 41W 28AAA 1	27.02320	170.05399	3854.0	182.7	3671.3
1355	Sherman	9S 41W 34BAB 1	27.34864	169.04491	3841.0	152.5	3688.5
1356	Sherman	9S 42W 8AAA 1	20.17598	173.32742	3943.0	155.4	3787.6
1357	Sherman	9S 42W 11CCC 1	22.27091	172.36679	3923.0	166.5	3756.5
1358	Sherman	9S 42W 14AAA 1	23.14339	172.20526	3901.0	166.5	3734.5
1359	Sherman	9S 42W 16CDD 2	20.59580	171.43996	3943.0	144.2	3798.8
1360	Sherman	9S 42W 35ABB 1	22.63613	169.23540	3916.0	139.5	3776.5
1361	Sherman	10S 37W 23ABB 1	52.22285	164.20740	3421.0	192.8	3228.2
1362	Sherman	10S 40W 10ADC 1	33.66611	166.43591	3624.0	16.6	3607.4
1363	Sherman	10S 41W 15CAD 1	27.30863	165.43434	3762.0	20.9	3741.1
1364	Sherman	10S 42W 21BBB 1	19.95131	165.35930	3963.0	106.3	3856.7
1365	Sherman	10S 42W 24BAB 1	23.21457	165.22356	3903.0	98.3	3804.6
1492	Thomas	6S 31W 3ADB 1	88.03283	189.68837	2957.0	120.1	2836.9
1493	Thomas	6S 31W 33CCD 1	86.27667	184.14677	2916.0	30.9	2885.1
1494	Thomas	6S 32W 12CBC 1	83.33239	188.46915	3020.0	120.0	2900.0
1495	Thomas	6S 32W 29CDC 1	79.54514	185.35164	3077.0	124.4	2952.6
1496	Thomas	6S 32W 34CBC 1	81.24179	184.55247	3032.0	93.8	2938.2
1497	Thomas	6S 33W 7BBB 1	72.46165	189.40816	3177.0	137.4	3039.6
1498	Thomas	6S 33W 23DDD 1	77.22980	186.41675	2997.0	14.3	2982.7
1499	Thomas	6S 33W 31CAB 1	72.56495	184.93230	3065.0	14.9	3050.1
1500	Thomas	6S 34W 11CDD 1	70.83594	188.58569	3218.0	163.2	3054.8
1501	Thomas	6S 34W 17CBC 1	67.49432	187.93924	3261.0	157.8	3103.2
1502	Thomas	6S 35W 2CDD 1	64.96323	189.76530	3245.0	130.1	3114.9
1503	Thomas	6S 35W 26ACB 1	64.96938	186.40483	3300.0	154.4	3145.6
1504	Thomas	6S 36W 6BCD 1	54.79759	190.60039	3408.0	190.9	3217.1
1505	Thomas	6S 36W 11ACC 1	59.11987	189.47089	3360.0	170.3	3189.7
1506	Thomas	6S 36W 30DCB 1	55.01444	186.24393	3417.0	154.5	3262.5
1507	Thomas	6S 36W 34DDB 1	58.21449	185.14299	3334.0	101.6	3232.4
1508	Thomas	7S 32W 7ACA 1	78.84848	183.00844	3056.0	78.5	2977.5
1509	Thomas	7S 32W 13AAA 1	83.99976	182.10432	3037.0	120.0	2917.0
1510	Thomas	7S 33W 7BDA 1	72.63266	183.19026	3203.0	165.7	3037.3
1511	Thomas	7S 33W 21DBC 1	74.66895	180.76814	3141.0	116.7	3024.3
1512	Thomas	7S 33W 35ADD 1	76.97072	178.95763	3145.0	150.3	2994.7
1513	Thomas	7S 34W 8BBB 1	67.34478	183.59052	3224.0	99.1	3124.9
1514	Thomas	7S 34W 25AAA 1	72.04672	180.47380	3167.0	114.8	3052.2
1515	Thomas	7S 34W 26DBD 1	70.79427	179.88715	3177.0	109.6	3067.4
1516	Thomas	7S 35W 9CCC 1	62.36182	182.88557	3315.0	129.6	3185.4
1517	Thomas	7S 36W 17CCC 1	55.36419	182.12965	3417.0	145.2	3271.8

(*cont.*)

Table B.1. High Plains aquifer sampling (*concluded*).

ID	County	Legal description of location	Easting (miles)	Northing (miles)	L.S.E. (ft)	W.D. (ft)	W.T.E. (ft)
1518	Thomas	7S 36W 35CBB 1	58.25497	179.41635	3341.0	88.9	3252.1
1519	Thomas	8S 31W 3CDD 1	87.30879	177.15088	3003.0	130.3	2872.7
1520	Thomas	8S 31W 20CDD 1	85.24763	174.22964	3026.0	114.7	2911.3
1521	Thomas	8S 32W 7BAA 1	78.40262	177.29623	3102.0	119.1	2982.9
1522	Thomas	8S 32W 12DBC 1	83.46031	176.52344	3057.0	113.7	2943.4
1523	Thomas	8S 32W 27DAB 1	81.63478	173.71512	3078.0	115.6	2962.4
1524	Thomas	8S 33W 2CDA 1	76.42622	177.60487	3141.0	145.8	2995.2
1525	Thomas	8S 33W 7AAB 1	72.81794	177.46820	3194.0	155.3	3038.7
1526	Thomas	8S 33W 34BBC 1	74.91335	173.29839	3168.0	149.5	3018.5
1527	Thomas	8S 34W 1BAC 1	71.36143	178.38045	3179.0	123.7	3055.3
1528	Thomas	8S 34W 6CBC 1	66.16714	178.02809	3266.0	137.8	3128.2
1529	Thomas	8S 34W 23CBD 1	70.13652	174.93497	3232.0	174.3	3057.7
1530	Thomas	8S 36W 18ABA 2	54.80698	177.05264	3428.0	128.3	3299.7
1531	Thomas	9S 31W 10BBB 1	86.76484	171.07378	2999.0	88.2	2910.8
1532	Thomas	9S 31W 22ABD 1	87.32386	168.94270	2990.0	84.8	2905.2
1533	Thomas	9S 32W 9BDA 1	80.17787	171.02068	3043.0	76.8	2966.2
1534	Thomas	9S 32W 27BCD 1	80.82378	167.89110	3076.0	117.0	2959.0
1535	Thomas	9S 33W 15ACC 1	75.29806	170.05145	3135.0	129.2	3005.8
1536	Thomas	9S 33W 26DAD 1	76.58128	167.76878	3145.0	150.4	2994.6
1537	Thomas	9S 33W 30CAA 1	72.13621	168.04466	3216.0	201.3	3014.7
1538	Thomas	9S 33W 35AAD 1	76.56408	167.27162	3145.0	152.8	2992.2
1539	Thomas	9S 34W 12ADA 1	71.74692	171.28859	3199.0	155.4	3043.6
1540	Thomas	9S 35W 32DAA 1	61.67787	167.34541	3361.0	187.0	3174.0
1541	Thomas	10S 31W 26AAA 1	88.37382	162.07260	2891.0	10.7	2880.3
1542	Thomas	10S 32W 4CAB 1	79.87989	165.80264	3091.0	120.3	2970.6
1543	Thomas	10S 32W 29DCB 1	78.98564	161.59683	3064.0	93.9	2970.1
1544	Thomas	10S 33W 3DBC 1	75.15380	165.82703	3145.0	148.5	2996.5
1545	Thomas	10S 33W 19CBD 1	71.72309	162.96425	3161.0	104.8	3056.2
1546	Thomas	10S 33W 27BBB 1	74.54624	162.48935	3119.0	104.0	3015.0
1547	Thomas	10S 34W 12BCD 1	70.80587	165.22482	3220.0	180.5	3039.5
1548	Thomas	10S 36W 36ACC 1	59.09651	161.57820	3359.0	171.8	3187.2
1767	Wallace	11S 38W 35CCC 2	46.76602	155.53662	3372.0	140.1	3231.9
1550	Wallace	11S 42W 8DDC 1	20.95668	160.45091	3953.0	107.4	3845.6

Appendix C

UNCF Sampling

This sampling relates to elevation of an unconformity (Jones *et al.*, 1986, Table B.4). The data are the property of Exxon Production Company, which reserved the right to withdraw some information. According to the senior author's recollection (Jones, 1995, personal communication), the coordinates are in meters and the depth in feet. The data are available on Internet for downloading at http://www.kgs.ukans.edu/Mathgeo/Books/Geostat/index.html.

Table C.1. UNCF sampling.

Easting (m)	Northing (m)	ID	Depth (ft)	Easting (m)	Northing (m)	ID	Depth (ft)
32000	87015	1	−7888	38600	87030	2	−8020
44000	86400	3	−7943	48200	87060	4	−8003
30650	82200	5	−7918	35030	80901	6	−7905
39476	80970	7	−7900	44198	80910	8	−7998
33530	76230	9	−7879	38024	76170	10	−7817
42440	76650	11	−8018	46955	76620	12	−7890
30350	71700	13	−7903	35195	71745	14	−7815
39650	71760	15	−7966	44180	71766	16	−7913
48800	70215	17	−7818	29240	81960	1610	−7920
32270	83760	1614	−7923	33770	84720	1616	−7918
35240	85590	1618	−7927	36740	86520	1620	−7969
39680	88350	1624	−8043	41240	89250	1626	−8059
29840	67560	2068	−7932	31280	66600	2070	−7883
32720	65610	2072	−7842	34190	64650	2074	−7842
35720	63720	2076	−7874	37100	62700	2078	−7871
31730	63000	3078	−7740	33500	63720	3080	−7799
36680	65160	3084	−7902	38300	65790	3086	−7943

(*cont.*)

Table C.1. UNCF sampling (*concluded*).

Easting (m)	Northing (m)	ID	Depth (ft)	Easting (m)	Northing (m)	ID	Depth (ft)
39830	66540	3088	−7962	41420	67200	3090	−7967
43010	67920	3092	−7931	44690	68580	3094	−7876
46250	69300	3096	−7854	30200	76500	4546	−7849
31640	75420	4548	−7880	33080	74280	4550	−7869
34520	73260	4552	−7820	37280	71070	4556	−7892
38660	70020	4558	−7965	40040	69000	4560	−8008
41450	67890	4562	−7986	42800	66900	4564	−7922
44240	65730	4566	−7852	45500	64710	4568	−7773
33800	89220	5003	−7896	34880	87720	5005	−7915
35840	86280	5007	−7943	37040	84720	5010	−7948
37940	83160	5012	−7928	40784	79260	5018	−7923
42740	75120	5022	−8004	43730	74610	5024	−8001
44660	73200	5026	−7927	45680	71700	5028	−7877
46760	70200	5030	−7857	47600	68790	5032	−7824
48500	67380	5034	−7768	49520	65940	5036	−7696
42320	89100	6192	−8032	43700	88080	6194	−7971
45080	87000	6196	−7946	46520	86010	6198	−7973
47900	85050	6200	−8003	49400	84000	6202	−8032

Appendix D

Dakota Aquifer Sampling

The Dakota aquifer sampling was prepared at the Kansas Geological Survey from interpretations of gamma ray logs. The measurements are part of a larger effort to characterize the aquifer throughout Kansas. The subset in Appendix D extends over a square area of three by three townships (9.65 km × 9.65 km) in south-central Hodgeman county, close to outcrops. The Dakota aquifer comprises—from oldest to most recent—the Cheyenne Sandstone, the Kiowa Formation, and the Dakota Formation, all Lower Cretaceous in age. The Dakota Formation is subdivided into the top "D" and the thicker "J" sandstones. Most of the "D" sandstone is missing due to fluvial erosion in the eastern side of the study area, so it was discarded from the subdivision of cumulative sandstone thickness to eliminate the influence of such modern erosion. Based on depositional trends, the "J" sandstone was further subdivided into the basal 120 ft and the remainder of the deposits below the "D" sandstone. Except for the last 16 wells, discarding of shales in all cumulative sandstone thickness calculations was done automatically by computer from digitized logs. Data for the Dakota aquifer sampling are on the Internet and may be downloaded from http://www.kgs.ukans.edu/Mathgeo/Books/Geostat/index.html.

Table D.1. Dakota aquifer sampling.

				Cumulative sandstone thickness			
ID	Easting (ft)	Northing (ft)	Formation thickness (ft)	Upper "J" (ft)	Lower "J" (ft)	Kiowa (ft)	Cheyenne (ft)
5	1631797	546606	286.0	0.0	27.0	0.0	54.4
9	1635275	545570	280.0	13.0	9.0	2.0	58.3
12	1641860	545503	279.0	95.0	112.0	83.3	68.1
31	1630041	536362	340.0	33.0	96.0	56.3	49.1
32	1631291	535016	341.0	28.5	80.0	20.5	14.5
38	1643084	532306	273.0	17.9	51.0	8.9	30.5

(*cont.*)

Table D.1. Dakota aquifer sampling (*continued*).

ID	Easting (ft)	Northing (ft)	Formation thickness (ft)	Cumulative sandstone thickness Upper "J" (ft)	Lower "J" (ft)	Kiowa (ft)	Cheyenne (ft)
40	1643137	534940		2.0	42.0	2.0	23.6
42	1648014	533313		73.0	70.0	32.6	60.1
48	1658548	530399	257.5	53.7	75.0	8.9	47.2
49	1660208	529439	230.5	6.0	34.0	8.9	52.1
50	1657520	526895		22.0	6.0	0.0	19.7
52	1658854	526844	230.0				
57	1652223	524384	257.0	2.1	58.0	25.8	21.6
59	1619712	547175	298.5	0.0	69.5	0.5	38.5
60	1607855	548835	316.5	16.0	94.0	0.0	33.0
61	1598543	549051	296.5	0.5	96.0	0.0	21.0
64	1611762	544744		38.3	84.0	1.5	70.5
68	1618670	544228	275.5	8.5	90.0	39.0	69.5
69	1617687	544248	288.5	1.5	63.0	0.0	33.0
70	1622289	544483	277.5	26.5	89.0	9.5	68.0
72	1626236	543077	283.5	46.5	73.0	0.0	50.0
75	1627484	539064	317.0	19.0	74.0	23.5	31.0
76	1620905	539390	315.0	95.5	55.5	11.0	37.0
79	1617557	536348	318.5	91.0	91.5	47.5	43.0
80	1610319	538152	256.5	0.5	0.0	0.0	50.0
81	1604730	539296	314.0	0.0	0.5	0.0	43.0
82	1606034	539265	290.5	56.5	11.5	16.0	37.0
83	1598477	541111	316.0	16.0	16.5	12.5	52.5
84	1602404	538352	334.5	166.5	120.0	102.5	76.0
85	1602364	535691	317.5	31.0	19.0	23.0	49.5
87	1598759	534150	302.0	1.0	14.5	11.0	51.0
88	1599379	532145	292.0	25.5	25.5	8.0	40.5
89	1604990	534296	259.5	3.0	11.0	0.0	32.0
91	1616880	535363	328.5	5.5	45.5	10.5	44.5
92	1618118	531339	308.5	62.5	51.0	36.5	55.0
93	1623388	531203	325.5	3.0	69.0	0.0	4.0
94	1626054	532477	303.5	5.0	0.0	5.0	7.5
95	1627411	535098	336.0	48.0	74.5	53.0	44.0
97	1627694	531443	277.0	18.5	7.0	0.0	0.0
101	1615428	527431	298.5	36.0	0.0	0.0	32.5
104	1598727	530844	302.0	38.5	58.0	0.0	35.5
105	1600941	526476	305.5	73.0	92.0	27.5	80.5
106	1616657	523435	290.0	34.5	2.5	3.0	24.5
107	1617968	522083	296.5	105.0	78.5	21.0	23.0
108	1619267	520414		3.5	6.5	58.5	53.0

(*cont.*)

Table D.1. Dakota aquifer sampling (*continued*).

ID	Easting (ft)	Northing (ft)	Formation thickness (ft)	Cumulative sandstone thickness Upper "J" (ft)	Lower "J" (ft)	Kiowa (ft)	Cheyenne (ft)
109	1623278	524586	273.0	40.5	14.0	18.5	37.5
112	1593222	549092	294.5	24.0	38.5	0.0	30.5
113	1595911	551702	295.0	20.5	31.0	0.0	8.0
114	1589279	551731	287.5	40.0	22.0	22.0	23.0
115	1587945	551746	279.5	2.5	4.5	5.5	9.0
116	1589238	549114	340.0	117.0	116.0	20.0	32.0
117	1586593	549133	289.0	0.0	41.5	0.0	0.0
118	1582610	549150	282.5	63.0	68.0	95.0	30.5
119	1581312	551771	279.0	15.0	83.0	3.5	0.0
120	1577424	549170	293.5	16.5	37.5	30.0	33.0
121	1573371	547856	285.5	1.0	54.0	9.5	22.5
122	1574698	550486	296.5	38.5	9.0	0.0	15.5
123	1572083	550477	267.0	19.0	22.0	22.5	34.5
124	1574674	547858	261.0	19.0	57.0	12.5	18.5
127	1573367	546534	308.0	25.0	20.5	19.0	32.0
128	1574670	546535	318.5	23.5	13.5	19.5	21.5
129	1572038	546533	283.5	15.5	35.0	0.0	21.5
130	1579926	542551	352.5	94.0	109.0	10.0	51.0
131	1578934	544212	325.5	40.0	66.0	0.0	11.0
132	1585232	546514	322.0	80.0	108.5	0.0	0.0
133	1594535	545127	272.0	17.5	73.5	90.0	26.5
134	1594533	543799	308.5	0.0	44.0	46.5	35.0
135	1590518	537219	327.5	13.0	67.5	0.0	13.5
136	1583908	541206	360.5	37.5	31.0	0.0	67.5
138	1581226	537274	308.5	8.5	5.0	0.0	47.0
139	1573283	541245	352.5	81.5	111.0	10.0	21.5
140	1571907	537271	307.5	9.0	4.5	0.0	47.0
141	1567911	533340	340.0	31.0	75.5	0.0	27.5
142	1579869	533288	281.0	17.0	9.0	6.5	49.5
143	1577213	533286	300.0	64.5	4.0	48.5	53.5
144	1586525	535925	319.5	40.5	66.5	19.0	41.5
145	1586523	534597	275.5	62.0	69.0	88.5	37.5
146	1586502	533263	312.0	68.0	4.5	9.5	52.0
147	1586499	531941	320.5	154.0	117.5	116.5	46.5
148	1591460	532243	324.5	22.0	71.5	12.0	29.0
149	1588483	531932	309.0	7.5	108.5	34.0	31.0
150	1597100	531552	313.5	42.5	65.0	1.5	33.5
151	1592775	530917	310.5	0.0	104.5	17.0	35.0
152	1596115	530561	284.0	46.0	86.5	96.5	40.5

(*cont.*)

Table D.1. Dakota aquifer sampling (*continued*).

ID	Easting (ft)	Northing (ft)	Formation thickness (ft)	Cumulative sandstone thickness Upper "J" (ft)	Lower "J" (ft)	Kiowa (ft)	Cheyenne (ft)
153	1597431	530224	297.0	3.0	3.0	0.0	35.0
154	1591802	529276	332.0	16.0	84.0	0.5	30.5
155	1586496	530624	321.5	38.0	53.0	10.5	13.5
156	1576865	530969	300.0	0.0	1.0	0.0	24.0
158	1566564	530705	346.5	52.5	108.5	50.0	59.0
159	1567861	528039	332.5	19.5	116.0	6.0	65.0
162	1587693	521375	293.0	3.0	89.5	41.5	74.5
163	1590455	525323	316.0	41.5	94.5	1.5	27.0
164	1587754	525342	322.0	111.5	99.0	20.5	71.5
165	1595746	525270	299.5	0.5	66.0	12.5	52.0
166	1660074	518849	252.5	0.0	37.0	0.0	1.0
167	1660023	516188	234.0	53.0	92.5	104.5	22.5
170	1645547	518976	263.0	0.0	51.0	0.0	0.0
172	1632313	513836	268.0	0.0	0.0	0.0	0.0
173	1637517	511137		46.0	55.5	1.0	31.0
174	1644409	510045		61.2	31.0	47.0	48.0
175	1645310	513673		10.0	70.5	17.5	21.5
176	1645442	512348		0.5	78.0	25.5	35.5
180	1657332	513561	228.0	1.5	5.0	1.0	2.0
181	1657231	505627		18.5	0.0	0.0	9.5
182	1655933	508283		0.0	0.0	0.0	0.0
183	1658211	503959		23.0	0.0	0.0	8.5
186	1641382	504456		0.0	0.0	0.0	8.5
187	1638777	507136		67.5	38.5	125.0	26.0
188	1629429	500645	281.5	11.0	6.0	0.0	32.5
190	1637376	501970	282.5	17.5	0.0	1.5	15.0
194	1645180	492553		0.0	0.0	0.0	20.0
196	1659722	491076		40.5	12.0	0.0	14.5
200	1612504	515610		59.5	82.5	30.5	42.5
201	1598092	510693	305.0	9.5	43.5	6.0	66.0
202	1607242	511741	306.0	10.0	80.0	0.0	53.0
203	1617728	510145	314.0	97.0	113.5	117.5	60.0
204	1613780	514247	303.5	75.0	84.5	12.5	59.0
206	1620423	512752	269.0	28.0	0.0	6.5	41.0
209	1624242	504707	313.0	85.5	108.0	43.0	42.0
210	1620180	506103		79.0	94.5	95.5	53.5
211	1621621	507067		24.0	15.0	0.0	17.5
212	1621629	505088		8.5	25.5	0.5	28.0
213	1617709	508811	313.0	36.0	49.5	4.0	39.5

(*cont.*)

Table D.1. Dakota aquifer sampling (*continued*).

ID	Easting (ft)	Northing (ft)	Formation thickness (ft)	Cumulative sandstone thickness Upper "J" (ft)	Lower "J" (ft)	Kiowa (ft)	Cheyenne (ft)
214	1603136	503894	303.5	40.5	73.5	7.5	71.0
215	1608410	502405	335.5	79.0	112.0	22.5	72.5
217	1626793	500687	312.0	24.5	94.0	0.0	15.5
219	1623072	496783	276.0	13.0	61.5	5.5	36.0
220	1613493	494335	314.5	12.0	75.5	16.0	62.0
221	1605694	498516	325.5	15.5	68.5	18.5	61.0
224	1602927	493317	325.0	50.5	88.0	37.0	70.0
225	1608186	491845	321.5	24.5	58.0	4.5	68.5
226	1609537	493126	303.5	17.5	71.0	20.5	71.0
227	1614707	490372	312.0	26.5	53.5	1.5	39.0
228	1618663	490272	311.5	20.0	61.0	16.0	30.5
229	1620032	492875	301.5	72.5	53.0	23.5	27.0
230	1590299	518562	283.0	120.5	116.5	107.5	111.0
231	1583971	518421	308.0	92.0	102.5	87.0	82.5
233	1581198	511851		56.3	38.5	29.0	87.5
234	1576550	513236		2.0	61.3	4.0	110.5
235	1584833	510794	293.5	58.0	61.0	25.0	98.5
236	1585197	514429	309.0	0.0	49.0	14.0	79.5
237	1583875	512471		4.0	11.0	8.0	71.5
239	1587483	513410	314.0	3.0	68.0	8.5	92.5
240	1589811	513091	337.5	3.5	51.5	16.0	74.5
241	1587027	511761	313.5	11.5	0.0	9.0	87.0
242	1593182	513751	317.5	12.5	22.5	18.5	91.0
243	1592497	513761	286.5	4.5	12.0	31.5	89.5
244	1584688	506796	280.0	0.0	3.0	0.0	61.5
245	1579420	506900	328.0	40.5	37.0	0.0	49.0
246	1584514	500119	308.0	34.0	27.0	49.5	126.5
247	1588487	500097	300.0	11.5	0.0	0.0	44.0
248	1594927	502433		39.0	60.5	86.5	101.0
249	1592456	497486	279.0	29.0	0.0	3.0	93.0
250	1590160	499115	303.0	42.5	19.5	16.5	115.5
251	1588760	495803	304.0	11.0	32.5	39.5	124.5
252	1586747	496434	307.5	9.0	7.0	31.5	130.5
253	1586795	498411	305.0	53.0	26.0	4.5	133.5
254	1585465	499099	304.0	3.0	25.0	42.5	118.5
257	1579042	490924	273.0	11.5	30.0	5.0	126.5
258	1576774	493989	255.5	34.0	80.5	135.5	177.5
259	1577070	492973	261.5	0.5	44.0	35.0	164.5
260	1576349	489633	336.0	0.0	31.5	0.0	41.0

(*cont.*)

Table D.1. Dakota aquifer sampling (*continued*).

ID	Easting (ft)	Northing (ft)	Formation thickness (ft)	Cumulative sandstone thickness Upper "J" (ft)	Lower "J" (ft)	Kiowa (ft)	Cheyenne (ft)
262	1637835	486794		2.5	10.5	43.0	39.5
272	1638333	476736	259.0	26.5	0.0	0.0	0.0
273	1628959	475530	283.5	0.0	0.0	0.0	15.5
275	1644813	467420	312.0	46.0	42.5	21.0	124.0
277	1650124	467367	265.0	62.0	16.0	117.5	98.5
279	1652711	466027	281.5	7.5	1.0	23.0	92.0
280	1647401	463451	295.5	25.0	0.0	2.0	147.5
281	1638137	462197	285.5	22.0	8.5	26.0	112.8
284	1634123	459611	276.0	2.0	0.0	0.0	114.5
285	1657895	459399	327.0	91.5	114.5	161.5	165.5
288	1625201	486167	275.5	0.0	3.5	0.0	29.0
289	1615874	483699	328.5	58.5	11.0	0.0	52.5
290	1614542	483740	317.0	103.5	65.5	117.5	113.5
292	1598661	478505	337.0	16.0	58.0	76.5	119.0
293	1599690	481129	335.5	61.0	18.5	3.5	80.0
294	1601363	482732	333.5	63.0	84.5	101.5	124.0
295	1610566	482514	270.5	9.0	9.0	12.5	118.0
296	1611879	481161	277.0	5.5	1.5	11.5	73.0
297	1610174	478558	275.5	0.0	18.0	0.0	83.5
300	1621988	475327		6.5	52.0	7.5	151.0
301	1613139	475836	323.5	151.5	36.0	34.5	149.5
302	1617103	477085	317.5	38.0	90.5	15.0	114.5
304	1608176	476958	313.5	3.5	53.5	14.5	91.5
306	1609481	476600	328.0	11.0	16.0	87.0	136.0
308	1602571	476085	363.0	18.0	82.0	15.5	63.5
309	1598626	476177	370.0	145.0	113.5	55.5	154.0
310	1597327	477547	363.5	7.0	88.0	10.0	89.0
311	1612109	471534	335.0	16.5	87.5	26.5	135.0
312	1588233	485531	303.0	113.5	112.5	122.0	130.0
313	1581968	487883	284.0	28.0	13.5	6.5	176.0
315	1578574	482924	275.0	1.5	25.0	7.0	222.0
319	1637587	546548	279.5	8.0	25.0	24.5	53.0
320	1637907	545549	326.5	29.0	28.5	1.5	34.5
321	1639175	538904	282.5	27.5	46.0	5.0	46.0
325	1641780	533625		84.0	87.5	6.5	36.5
326	1652918	530001		10.9	49.0	2.5	12.5
329	1605370	538951	308.0	0.0	0.0	7.0	37.0
330	1606022	537936	257.0	1.0	24.0	3.0	44.0
331	1599714	531806	291.0	16.0	31.5	1.0	33.5

(*cont.*)

Table D.1. Dakota aquifer sampling (*concluded*).

			Formation	Cumulative sandstone thickness			
ID	Easting (ft)	Northing (ft)	thickness (ft)	Upper "J" (ft)	Lower "J" (ft)	Kiowa (ft)	Cheyenne (ft)
332	1582626	550461	300.0	30.5	34.5	22.5	9.0
333	1586626	551754	278.5	29.5	31.5	2.5	17.5
334	1569436	551821	274.5	4.0	1.5	3.5	29.0
340	1641496	512402		8.0	74.5	2.0	27.0
343	1638644	497928		0.0	0.0	0.0	9.0
346	1617747	511484	315.0	78.0	109.5	112.0	72.5
348	1604307	495921		58.0	99.0	61.5	81.5
349	1613456	493020	309.0	0.0	0.0	0.0	1.0
350	1583973	518425	309.0	77.5	102.5	86.0	82.5
351	1584794	497789	309.0	108.5	111.5	131.0	68.0
357	1654281	481877	256.5	9.5	25.5	0.0	28.5
359	1609180	478592	331.0	7.5	29.5	5.0	33.0
360	1614522	482411	308.0	75.5	32.5	9.0	101.5
361	1615845	482374	324.5	104.0	47.5	0.5	92.0
362	1589619	488188	303.5	20.5	22.0	14.5	127.5
501	1624500	460500		59.0	105.0	12.0	98.0
502	1591000	461500		0.0	10.0	5.0	56.0
503	1596000	468000		55.0	66.0	9.0	157.0
504	1656000	501000		12.0	6.0	10.0	25.0
505	1638500	499000		0.0	0.0	0.0	10.0
506	1642000	487500		0.0	48.0	18.0	35.0
507	1604000	463000		19.0	22.0	7.0	208.0
508	1628500	519000		0.0	15.0	0.0	18.0
509	1594000	514000		5.0	15.0	15.0	80.0
510	1566000	507000		25.0	0.0	2.0	125.0
511	1573500	473500		10.0	20.0	20.0	270.0
512	1643500	472000		42.0	10.0	4.0	60.0
513	1627000	517000		0.0	0.0	0.0	0.0
514	1596000	513500		5.0	10.0	12.0	92.0
515	1654000	495500		7.0	35.0	18.0	45.0
516	1614500	501500		61.0	25.0	18.0	42.0

Author Index

Subject Index